TECH-GEMONY

BRAD GLOSSERMAN
DAVID S. LEE

Tech-Gemony

Understanding the New National Security Economy

OXFORD
UNIVERSITY PRESS

OXFORD
UNIVERSITY PRESS

Oxford University Press is a department of the University of Oxford. It furthers the University's objective of excellence in research, scholarship, and education by publishing worldwide.

Oxford New York
Auckland Cape Town Dar es Salaam Hong Kong Karachi
Kuala Lumpur Madrid Melbourne Mexico City Nairobi
New Delhi Shanghai Taipei Toronto

With offices in
Argentina Austria Brazil Chile Czech Republic France Greece
Guatemala Hungary Italy Japan Poland Portugal Singapore
South Korea Switzerland Thailand Turkey Ukraine Vietnam

Published in the United States of America by
Oxford University Press
546 Fifth Avenue, New York, NY 10036, United States of America

Library of Congress Cataloging-in-Publication Data is available
Brad Glosserman and David S. Lee.
Tech-Gemony: Understanding the New National Security Economy.
ISBN: 9780197875681

Printed in the United Kingdom on acid-free paper
by Bell and Bain Ltd, Glasgow

Brad
To Fan and Reed, as always

David
For Amy, Claire, and Noah, who give my journey meaning

CONTENTS

Acknowledgments ix

Introduction 1

PART I
FRAMING THE NEW NATIONAL SECURITY ECONOMY

1. Rethinking National Security in a New Type of Competition 15
2. A World of Insecurity 39

PART II
GOVERNMENT IN THE NEW NATIONAL SECURITY ECONOMY

3. China Reimagines Capitalism 67
4. Governments Respond: Playing Defense 91
5. Governments Respond: Playing Offense 119

PART III
BUSINESS IN THE NEW NATIONAL SECURITY ECONOMY

6. The New Normal for Business 155

7. Framing the Business Response 181
8. Business Purpose in the New National Security Economy 201

PART IV
A PERILOUS MOMENT

9. A Perilous Moment 217

Notes 241
Index 315

ACKNOWLEDGMENTS

Publication of this book has been a long and frustrating journey. Such are the pains of being visionary. The notion of a "new national security economy" emerged during the first Trump administration, but its contours were elusive and the intensity of the competition that shaped it was hard to envision. The COVID-19 pandemic and everything that followed convinced us—confirmed?—that we were on to something.

Putting the pieces together wasn't too difficult, but the resulting product defies easy categorization. It covers a lot of ground and doesn't fit neatly into any particular section of the catalog or bookstore. That confusion stumped several publishers, but not Michael Dwyer of Hurst. He gave us the green light and guided us through the publication process. Thanks to him and all the staff at Hurst, especially Olivia Ralphs and Letty Allen for their ace proofreading, along with the three anonymous reviewers who liked our manuscript. Thanks to Tobias Harris, author of the definitive biography of Japan's former Prime Minister Abe Shinzo, for introducing us to Michael.

Brad wants to thank his colleagues in Japan, Kokubun Toshifumi, Nagino Shotaro, and Igata Akira, for their help and patience. I could not have asked for better friends and mentors. They were instrumental in helping me navigate the political, bureaucratic, and business thickets that this book explores. Few have their insight, their knowledge, and their access. They were instrumental in bringing these ideas to light. In Honolulu, where I now live and work, David Santoro, president of Pacific Forum, has been a great source of support.

Together, we're trying to give the ideas in *Tech-Gemony* more substance. Carl Baker too deserves thanks for his wise and typically frustrating counsel. Carl has been an intellectual guide, helping me to organize and better present my thoughts. He's always frustrating and almost always right. My work is better for his questions.

Brad also must thank David for his patience, hard work, invariable good cheer, and level head.

Brad

Echoing Brad's sentiment, Michael Dwyer and the wonderful Hurst team have been helpful and gracious throughout this process. We are deeply thankful for our partnership with them.

As a self-described accidental academic, I am fortunate to call the University of Hong Kong Business School home. I have received consistent support and opportunities that have made all the difference. I am especially grateful to the students, executives, and organizations I have the privilege of engaging with. They have certainly taught me more than I might have taught them. The insights I have gained from them infuse much of this work.

Many of the seeds that grew into my contribution to *Tech-Gemony* were planted by impactful teachers like Alice Amsden, Carter Eckert, Eric Hyer, and Daniel Nielson. Each showed me what compelling scholarship, consequential teaching, and genuine care for students looked like. In so many ways, I still strive to emulate their respective examples.

We are deeply thankful to Amy Gallo and the editors at *Harvard Business Review* for the opportunity to write an article about the new national security economy. We were relatively early in our research for what would become *Tech-Gemony*, but we felt we were onto something. The response to our article confirmed this intuition and renewed our motivation.

Shortly after I completed graduate school, Brad kindly included me in a series of workshops and conferences that he organized for Pacific Forum. Possessing little knowledge but considerable curiosity at the time, these were formative experiences for me. Little did we know that our relationship would extend nearly two decades, ultimately resulting in this work. From the beginning, I was grateful for my relationship with Brad; that appreciation has only deepened over the years.

David

INTRODUCTION

This is a big book. That's not a reference to its length or weight. Rather, it includes big ideas and bold claims. We believe that technological advancement is propelling the world into a new geopolitical era and the growing pile of commentary and analysis that explores this terrain fails to grasp the enormity of the challenges and the consequences of the competition that defines this moment in history. The stakes could not be higher. And given these stakes, this book matters for business, government, and policy professionals alike. Most simply, the winner will get to write the rules of global order and will lead the world in the decades to come. That is what we mean by "Tech-Gemony."

Our analysis rests on three pillars. First, this world is profoundly interconnected, and those connections provide both opportunity and danger. The pursuit of efficiency and the desire to spread capitalism and the prosperity it produces have promoted globalization and the creation of global value chains that link nations into increasingly extensive and complex supply networks. This process is facilitated by digital technologies. These same technologies penetrate deeper into daily life, offering ever richer experiences. Indeed, there is an impetus to extend them ever further, as their value and utility are proportional to the size of the networks within which they operate. Seduced by these new opportunities, there is a tendency to ignore who or what controls the network and its constituent parts; that is a huge mistake. Control confers power and status.

Moreover, every node in a network is also a vulnerability, affording access and potential influence to any actor who can crack the code. Robert Mueller, then director of the FBI, explained in a 2007 speech that:

> In the days of the Roman empire, roads radiated from the capital city, spanning more than 52,000 miles. The Romans built these roads to access the vast areas they had conquered. But, in the end, these same roads led to Rome's downfall, for they allowed the invaders to march right up to the city gates.[1]

Susan Gordon, former US principal deputy director of National Intelligence, more succinctly captured the essence of this new environment when she warned of "threats to and through information." In this world, national security assumes new meaning, and governments, businesses, and ordinary citizens must all be alert to the implications and play their part in protecting themselves, their countries, and a still larger set of institutions and ideologies.

Second, at the very same time, this is a geopolitical moment, often described as a US–China contest but in fact a struggle between China and advanced industrial economies, often referred to as "the West." The primary arena of competition is economic, and the focus is mastery of new and emerging technologies. The companies and countries that prevail—and mastery includes both innovating and disseminating the fruits of that innovation via mass production—will dominate the twenty-first century. Centrifugal forces are at work on "the West," however, and those countries can no longer assume that they are united by shared interests and perhaps even values. While we believe that those nations should operate as a like-minded coalition, it is becoming increasingly difficult for them to do so. Governments, businesses, and individuals need to be more attentive to national security concerns, but that need not—should not—empower nationalism or a retreat to protectionism or autarky. The scale that is needed to prevail in this competition can only be achieved through their cooperation and collaboration.

Third, adjusting to these basic facts demands a rethinking of foundational assumptions about business, economics, and foreign and security policy. Many of the principles that guide decision-making in cabinets and c-suites are in desperate need of revision. Most basically, lines that have historically separated the public and private sectors need to be redrawn. In the West, there has long been a tension between the state and the market (although it has manifested in each country in different ways at different times). One commentator describes "parallel universes," or "totalizing paradigms through which to view the world that view the realm of the other as the exception; seen through the eyes of the trade regime, security is a regrettable, but sometimes necessary evil; through the eyes of the security regime, trade is a peacetime privilege, rather than right, and a wartime tool."[2]

Words fail us

That keen turn of phrase notwithstanding, decision-makers lack the vocabulary and the conceptual framework to address these problems.[3] We need to take a hard look at and question basic assumptions about relations between states and the relationship between state and society. The current meaning and definition of "national security" looks inadequate. "Globalization" too seems to be a concept in need of a rethink. At the heart of these reassessments is the need to rethink and reincorporate economic security considerations into most business and policy calculations. As former US Treasury Secretary Janet Yellen warned, "it will be increasingly difficult to separate economic issues from broader considerations of national interest, including national security."[4]

Unfortunately, economic security is an elusive concept. You'd think that there'd be clarity given its increasing use and prominence. After all, the United States has declared that "economic security is national security,"[5] and the European Union, the EU Parliament, and the European Council issued a strategy on European Economic and Security Strategy in 2023.[6] The communique from the 2023 summit of the Group of Seven leading industrialized countries declared that "ensuring economic resilience and economic security globally remains our best protection against the weaponization of economic vulnerabilities."[7]

Dig a little deeper and imprecision, if not confusion, dominates. The contours of economic security are flexible, its contents opaque. Historically, it has referred to individual concerns, such as the protection of living standards, income, and other daily necessities, such as health, education, dwellings, and social protection. In some cases, it referred to the economic dimension of national security or the ability of nations to support or sustain the material elements of hard power, including the quantity and quality of assets a country can deploy to the battlefield, along with the quality of its soldiers.

In recent years, however, the meaning of economic security has shifted. Conclusions about exactly what it means tend to depend on who is asking. Strategists and foreign policy experts debate US–China relations; international political economy experts debate interdependence and the prospects for international regimes; international economists debate the efficiency of increasingly restrictive trade and investment regimes; and lawyers and legal theorists debate the scope and significance of "national security exceptions" in international law. All tend to reflect "a deeper confusion in our discourse whether economics and national security are rivals or partners, whether economics is a tool of national security or national security a protector of economic relations."[8]

After digging through the declarations and documents, one authoritative assessment concludes, after noting a tendency to "avoid defining economic security," that it is "the protection of the survival of a nation and its people, its sovereign independence, and its economic prosperity from threats such as disruptions in the supply of critical commodities, the outflow of critical advanced technologies, and dependence on other countries."[9]

This absence of clarity and precision has prompted writers and analysts to take refuge in the notion of geoeconomics, a term popularized by Edward Luttwak in the early 1990s to note the prioritization of commercial over military methods to secure territorial and strategic control. In fact, the concept has a pedigree that stretches over 100 years and features in discussions of international political economy in the 1920s and 1930s.[10] It has experienced a resurgence in the last decade, even though it too, upon close scrutiny, remains "shrouded in ambiguity."[11]

Enter the new national security economy

Unhappy with either term, we've come up with one of our own: "the new national security economy." This paradigm is defined by four factors: 1) deep connectivity within and between societies, which has yielded benefits like globalization while introducing risks related to interdependence; 2) geopolitical competition between the West and China, in which 3) the primary arena is advanced technologies, such as AI and biotechnology; and 4) the private sector, which is a primary actor in this fight. Semiconductors are just one of many fields in which this dynamic is evident and intensifying.

There is no model or experience to prepare for this new world. Governments have long viewed technology development and proliferation through a fairly narrow lens, with concerns focusing on potential military applications. In a 1982 speech, for example, then US Secretary of Defense Caspar Weinberger warned that the US was engaged in the unfettered technological transfer and trade with the Soviet Union, which was effectively the technological "rope to hang us" as it bolstered Soviet military capability.[12] This concern manifested in strategic trade controls that attempted to limit the spread of technologies that would strengthen militaries. Mindful of Lenin's boast, governments tried to stop companies from selling enemies that rope. That effort continues, but it has been complicated by the increasing integration of new technologies into weapons systems along with the means to deliver and direct them. For some observers, the key national security challenge is anticipating which new technologies will magnify military capability and how, and then restricting their spread. This is not their book.[13]

For us, the significance and application of technology more broadly is the issue. A country's ability to master the frontiers of new technology will determine its standing in the world. This requires governments to play both offense and defense: nurturing the development of new technologies while protecting the intellectual property that is created; figuring out how to use directly and indirectly those technologies to extend national influence, while safeguarding against the weaponization of economic interdependence enabled by its application; and insulating their societies from disinformation campaigns enabled by connectivity, the decentralization of power, and the democratization of media.

At the end of World War II, there was recognition of the vital role that science and technology played in ending that conflict, and for some especially far-sighted individuals, the imperative that the United States remain at those frontiers. Most influential was Vannevar Bush, who sought to capture insights he had gleaned during his tenure as head of the Office of Scientific Research and Development, a role that put him in charge of pretty much all US federal government science during the war, directing efforts that produced the atomic bomb and microwave radar among other innovations. In a foundational essay that led to the creation of the National Science Foundation, he observed that:

> New products, new industries, and more jobs require continuous additions to knowledge of the laws of nature, and the application of that knowledge to practical purposes. Similarly, our defense against aggression demands new knowledge so that we can develop new and improved weapons. ... without scientific progress no amount of achievement in other directions can insure our health, prosperity, and security as a nation in the modern world.[14]

The oil shocks of the 1970s were eye-opening, forcing countries to confront what we now call supply chain vulnerabilities when the cartel of oil-producing nations imposed petroleum embargoes. Those incidents produced a reconceptualization of security to include economic and political instruments, and most especially supplies of raw materials. In the US, this yielded the Carter Doctrine, a declaration that the US president would use the military to protect overseas economic interests. Similar lessons were learned during the COVID-19 outbreak, when the economic disruption generated by the pandemic hammered home this new sense of vulnerability. Governments were forced to acknowledge that economic and business orthodoxy, which squeezed out every inefficiency in the pursuit of quarterly returns, had failed to pay sufficient attention to resilience. Global supply chains configured for "just in time" delivery foundered when forced to deal with "just in case." Worse, governments appeared ready to use their control of key nodes in those supply chains to push for their preferred political and policy outcomes.

The echo of the 1970s notwithstanding, this was unprecedented territory. While the West and the Soviets fought a global ideological battle throughout the Cold War, there was never the economic integration between the two that would allow one to exercise influence over the other. Today, China is deeply embedded in global supply chains and has been ready to use that leverage when it can. And the government in Beijing, with others among "the rest," complain that the US is increasingly inclined to do the same. By one count, the US has imposed three times as many sanctions as any other country or international body, with one-third of all nations targeted with some type of financial penalty. Sanctions are "an almost reflexive weapon ... seemingly irresistible," concluded one analysis.

More remarkably, in another distinction between this moment and those preceded by it, Washington is ready to use that leverage against allies and partners. President Trump's statement that "the EU is nastier than China" sent tremors through European capitals.[15] The imposition of a 50 per cent tariff on all steel and aluminum imports, 25 per cent tariffs on other imports from Mexico and Canada (with some exemptions) and a 15 per cent tariff on most EU goods—which qualifies as "favorable" treatment after extensive negotiations—prove that those aren't empty words. For Trump, the tariff is a cudgel to reduce immigration, stop drug flows, change investment, alter trade balances, or acquire territory.

Economic issues elbow their way into this new security mindset on another level. Never before has the global economic leadership of the West (and the US more specifically)—its Tech-Gemony—been challenged. China's rise has thrown that into question as the West's economic model has faltered—Global Financial Crisis, anyone?—and Beijing stepped up to meet the needs of the developing world through the Belt and Road Initiative and other efforts. That effort is buttressed by the success of China's "hybrid capitalism," which has raised hundreds of millions of people out of poverty and proved capable of generating national wealth and matching the West in developing technologies.

Most books of this sort are devoted to government decision-making. Though we discuss this, a distinctive component of our approach is the focus on corporate decision-making as well. Businesses must innovate, but they must also be alert to the ways that their increasingly far-flung activities (enabled by technology) render them vulnerable to disruption. They must expand their definition of risk.

Previously focused on business competition, they must now recognize the ways that they are impacted by the new strategic competition between countries and the geopolitical implications of their operations. Managers and executives might have once considered their activities as stateless or above national considerations. No longer. Business operations, their successes and failures, are no longer incidental to rivalries between governments but are central to them.

For us, the primary "battlespace" of the new geopolitical competition is the private sector. And to paraphrase another revolutionary theorist, even if executives don't care about geopolitical competition, they must realize that "geopolitical competition cares about them." It is hard, if not impossible, to find a sector of the economy that is immune to these pressures. Intellectual property is an easy target, but the dangers go well beyond that. Any device or application that generates personal data is a potential security threat. And theft—espionage—is no longer the only concern. An array of tactics that include "stock market, social media, and regulatory manipulation; data harvesting; national champion building; digital bank robbery; and even the destruction of corporate assets" are being deployed to prevail in this competition.[16]

Back to the future in Ukraine?

Russia's February 2022 invasion of Ukraine looks a lot like statecraft as traditionally practiced, even though the physical invasion of a sovereign state is a shock to twenty-first-century sensibilities. Russian President Vladimir Putin dispensed with the "gray zone" tactics that he used in the 2014 annexation of Crimea; rather than deploying "little green men" to obscure responsibility for an act of aggression, he sent his army into Ukraine to "liberate" the Russian-speaking peoples there and realize his dream of recreating the Russian empire.

Nevertheless, the Ukraine conflict has confirmed the importance of the new national security economy. First, there was a vicious struggle in cyberspace. While the fighting began on February 24, the day Russian troops crossed the borders and moved into Ukraine, Brad Smith, president of Microsoft, argues that the invasion actually began a day earlier, on February 23. On that day, Microsoft saw Russia launch a sudden all-out assault on Ukraine's institutions—300 coordinated Russian cyberattacks targeting the Kyiv government and financial institutions.[17]

If there was less action than anticipated,[18] experts credit strong Ukrainian defenses, honed by years of attacks by Russian hackers. Western governments have hardened networks and called for vigilance and preparation against the anticipated offensive. Warnings by Western governments, the US in particular, that they would "respond" to cyberattacks may have encouraged Russian restraint.

Second, there was the priority afforded to the contest for domination of the information space. Russia promoted a narrative of a "special military operation" to "demilitarize and denazify" Ukraine. In this telling, Moscow is waging a war of liberation to protect human rights. The West countered with its own information offensive, one designed to puncture Russian claims and brand its military action as illegitimate, brutal, and a violation of international law. Governments are using propaganda, disinformation, and selective intelligence disclosures (in the West) in this struggle. In Russia, independent media have been shut down while Orwellian techniques of information management are being deployed to shape thinking at home and abroad.

The third, and most important, way that the Ukraine conflict has confirmed the model of the new national security economy is the West's readiness to counter Russian military aggression with an economic response (leaving the actual fighting to the Ukrainians, with Western support). The sanctions campaign has denied Russia access to at least $450 billion since the war began.[19] Then US Deputy National Security Advisor Daleep Singh explained that "financial sanctions deny foreign capital to Russia, and export controls deny critical technological inputs that Russia needs to diversify its economy."[20] President Biden went further. Announcing steps he was taking in February 2022 as the war began, he declared that "we're going to impair [Russia's] ability to compete in a high-tech 21st century economy."[21]

This cuts both ways. While Russia has a newfound appreciation of the dangers of relying on the West for strategic goods, Russia's customers have had an equally rude awakening about their business partner. As one European official explained, "We thought interdependence, connectiveness, would be conducive to stability because we had correlating interests. Now, we've seen this is not the case. … Interdependence, we've now seen, can entail severe risks, if a country is ruthless enough."[22]

Other countries have been put on alert. Yu Yongding, an economist and former advisor to the People's Bank of China, professed his "shock." "We never expected that the U.S. would freeze a country's foreign currency reserves one day. And this action has fundamentally undermined national credibility in the international monetary system."[23] Many in China are now more convinced than ever that the decision to reduce reliance on and exposure to US power was correct. Indigenous capability is a must as geopolitical competition sharpens.

There is no mistaking the growing intensity of that competition. The invasion of Ukraine has transformed Western thinking. As NATO Secretary General Jens Stoltenberg explained, "Meaningful dialogue, as we strived for before, is not an option for Russia."[24]

Nor is Russia the only adversary. For many Western nations, the critical division is that between liberal and authoritarian governments, an outlook that aligns Beijing with Moscow. Warning lights were flashing after the Russia–China 4 February 2022 Joint Statement that declared a partnership with "no limits,"[25] and China's support for Russia since the Ukraine invasion has proven that was no empty promise. As criticism of Russia's invasion mounted, Chinese Foreign Ministry spokesman Zhao Lijian pushed back, arguing that the United States was constantly adding fuel to the fire, heightening tensions, coercing others to pick sides, and creating a chilling effect of "friend or foe." According to the official Xinhua news agency, Zhao charged the US with "attempting to shift blame, stoke confrontation, profit from the situation, and seek space for its plot to simultaneously suppress China and Russia."[26] As both sides sharpen the divisions between them, geopolitical Balkanization is accelerating.

That momentum is also evident in the private sector. Hundreds of companies have stopped doing business in Russia, some because of outrage at the atrocities committed, some because of fear of being sucked into the sanctions maw, and still others because of the fear of reputational damage if they continued operations. The third factor may be as important as the second, and a considerable amount of the pressure on companies to withdraw has been mounted by individuals and nonprofits: Yale Business School professor Jeffrey Sonnenfeld was instrumental in identifying and spotlighting companies doing business in Russia.

As of this writing, the war in Ukraine continues. Despite the many uncertainties about what that will entail and how it will conclude, we can be sure that it will be bloody and dangerous. We can also be sure that the outcome will be shaped by the opportunities and vulnerabilities that are an integral part of an increasingly connected world. While the invasion of Ukraine is the most violent expression of the new geopolitics, intensifying competition and periodic conflict are constants of this new order. A growing number of politicians worry that Ukraine today is Asia tomorrow. Policy-makers must weigh a new range of considerations as they develop twenty-first-century national security strategies. Understanding connectivity—the threats now posed by the great enabler of globalization—is the key to security today.

PART I

FRAMING THE NEW NATIONAL SECURITY ECONOMY

1

RETHINKING NATIONAL SECURITY IN A NEW TYPE OF COMPETITION

Since Donald Trump first became the US president, a trade war between the United States and China, the world's two leading economies, has seemed inevitable. When asked days before jetting off to meet China's supreme leader, Xi Jinping, in Busan, South Korea, ahead of the 2025 APEC summit, Trump blithely said that "we're in one now,"[1] and the world should expect it to continue. As Trump explained, "We have a 100 percent tariff [on Chinese products]. If we didn't have tariffs, we would be exposed as being a nothing."[2]

Since taking office both times, Trump has threatened tariffs on trading partners who, he was convinced, were "exploiting" American largesse and naivete. In some cases, he went ahead and imposed eye-watering levies; in others, he backed off to facilitate negotiations over trade deals. Several governments struck agreements (although it could be difficult to tell given the paltry and often confusing information that was released). In several cases, however, target governments refused to yield and retaliated with tariffs of their own.

For Trump, one of the biggest offenders is China, whose persistent trade surplus is proof (to him) of Beijing's unfair trade practices. Yet, even with the imposition of tariffs that reached as high as 145 per cent, China has not bent to Trump's will and instead fought back, countering with restrictions of its own, most notably on the export of rare earth elements, seventeen minerals with tongue-twisting names like neodymium, yttrium, and europium, that are essential to the manufacture of high-tech products: "Everything you can switch on or off likely runs on rare earths."[3] Throughout his second term, US pressure on China has been met with tit-for-tat retaliation, and the steady ratcheting up of pressure yielded howls of protest from US officials. Secretary of the Treasury Scott Bessent denounced Chinese export controls on rare earths, imposed to counter US sanctions, as "unacceptable," calling them "a bazooka pointed at the supply chains and the industrial base of the entire free world" and pitting "China versus the world."[4] US Trade Representative Jamieson Greer echoed that line, calling the new regulations "a global supply chain power grab. ... an act of economic coercion directed at every country in the world."[5] Chinese government spokespersons responded by charging the US with "rolling out 20 measures in 20 days that suppress China and seriously damage its interests." A Commerce Ministry spokesperson complained that the US position "seriously distorts and exaggerates China's measures, deliberately creating unnecessary misunderstanding and panic."[6] Trump then threatened imposition of an additional 100 per cent tariff on Chinese goods exported to the US because "they made me do it."

The pressure was relieved when Trump met Xi on the sidelines of the APEC summit in South Korea in late October. There, the two men agreed to a deal in which the US would drop its tariffs on Chinese imports to "just" 47 per cent (a 10-percentage point decrease) and suspend for one year both an investigation into China's shipbuilding sector and the application of a rule that would expand the imposition of export controls to Chinese companies' subsidiaries. In return, China would defer the export controls on rare earths announced just weeks earlier, resume purchases of soybeans (which had been suspended to pressure the US administration to negotiate), and expand cooperation on stopping the flow of fentanyl precursor chemicals into the US—which Trump has accused China of facilitating. It is by all accounts a truce, though the deep competition that shapes relations between the two countries continues.

The slide into a trade war is easily explained by Trump's twin beliefs that "tariffs is the most beautiful word to me," and that "trade wars are good and easy to win." But there is considerably more to this situation than the US president's odd aesthetic and misguided economics. China's ability to go toe to toe with the US reflects structural shifts in the international economy and the failure of US policymakers to recognize those changes.

The globalization conundrum

While there is a tendency to equate the current geopolitical moment with the Cold War, there is a fundamental difference between the two. Then, the Soviet empire and the West were two separate economic spheres, with almost no integration between them. That reflected a far darker view of the nature of the competition—a profound ideological clash with immediate and unmistakable implications for the world—as well as the fact that there was little the Soviets had to offer the West. There was neither the technological infrastructure to facilitate the sort of interaction that is today ordinary and assumed, nor the legal and institutional infrastructure to make it possible. That absence reflected the lack of will to create them; in fact, most of the West's regulatory framework regarding such matters was intended to block commerce, not promote it.[7]

Deep integration—globalization—followed the end of the Cold War. That reflected a strategy to promote development in those hitherto unconnected parts of the world in hopes that growth and prosperity would promote democratization and stability. No country benefited more from that policy than China. Its reform and opening program, coupled with access to developed world markets, yielded spectacular economic growth. By 2010, it had overtaken Japan to become the second-largest economy in the world. As part of the evolution, China has become the number one economy in terms of absolute volume of exported goods and services, surpassing even the United States and Germany.[8]

China burrowed its way into global supply chains, becoming an essential part of those production networks, a centrality that was exposed—to great concern—during the COVID-19 pandemic as governments around the world raced to secure personal protective equipment for health care workers. More recently, attention has focused on its control of rare earths (a topic taken up in more detail in subsequent chapters). As Treasury Secretary Bessent and Trade Representative Greer observed, that domination of the rare earths supply chain gives China leverage over trading partners that is every bit as powerful as that US presidents have historically wielded—access to the US market. Moreover, by all appearances, China has studied and copied US strategies and is using them in retaliation for US actions.

This power is an example of what political scientists Henry Farrell and Abraham Newman call "weaponized interdependence," or the use of global economic networks for strategic advantage.[9] While all countries use whatever leverage they have—economic, military, or some other type—to advance national interests, weaponized interdependence is coercion on steroids. The concept isn't new: OPEC used its control of global oil production during the 1973 Yom Kippur War to bend Western governments dependent on its product to take a harder line towards Israel. The United States uses its central role in the global financial system to press other governments to align with its preferred policies, such as sanctions against Iran or North Korea, even when they don't have the veneer of international legitimacy bestowed, say, by a resolution of the United Nations Security Council.

Domination of critical minerals supply chains uniquely positions China to exercise influence over huge swaths of the global economy. Cheng Hui, a researcher at the Chinese Academy of International Trade and Economic Cooperation, explained that China controls rare earths and related technologies because they are key to keeping the country at the top of the global value chain and are also "critical foundations for the independent development of domestic industries like new energy, high-end equipment manufacturing, and defense."[10] Cheng is channeling Xi Jinping, who has repeatedly declared that China, as a matter of national sovereignty, security, and survival, must master the development of new technologies. Xi calls innovation "the primary driving force of development,"[11] and credits advanced technology as key to the West's "sway over the world in modern times."[12] Chinese officials are cautious about revealing such ambitions in public, but many "commentaries and think tank pieces seem to suggest that surpassing the United States in high technology would end its era of global leadership, and presumably, usher in one of Chinese leadership."[13] Xi warned that "the control of core technology by others is our biggest hidden danger" and that allowing foreigners to control core technology "is like building a house on someone else's foundation. Only by winning the tough battle over key core technologies and achieving independent, controllable, and secure supply and industrial chains can we truly realize high-level scientific and technological self-reliance and self-strengthening."[14]

He's right. Mastery of technology—creating it, building it, applying it—will determine who prevails in the global competition for power and primacy. The US National Intelligence Council explained the implications in its 2021 Global Trends report, which argued that "The convergence of seemingly unrelated areas of scientific research and technological applications is making the rapid development of novel applications possible, practical and useful." It goes on to note that "some technological areas appear to offer the potential for transformative change. … advances in these areas will combine with other technologies, such as energy storage, to shape societies, economies and perhaps the nature of power."[15] The US National Commission on Innovation and Competitiveness Frontiers warned in April 2025 that "winning the competitive challenge for the future, defending the current world order, and our global leadership rest with the strength of our ability to innovate with speed and at scale."[16] Or, more simply, "technology is a key driver of any transformation of power at the international level."[17]

The European Union concurs, noting that "those who control digital technologies are increasingly able to influence economic, societal and political outcomes."[18] "A New Industrial Strategy for Europe" embraces the logic that animates the new national security economy when it reasons that:

> The twin ecological and digital transitions will affect every part of our economy, society and industry. They will require new technologies, with investment and innovation to match. They will create new products, services, markets and business models. They will shape new types of jobs that do not yet exist which need skills that we do not yet have.

Significantly, "these transitions will take place in a time of moving geopolitical plates which affect the nature of competition."

This melding of the digital and physical economies has several names. Klaus Schwab, founder of the World Economic Forum, calls it "the Fourth Industrial Revolution"; in Germany it is Industrie 4.0; in Japan, it is Society 5.0, a vision of national organization and rejuvenation that incorporates those technologies into a model that Japan can export to the world. Japan's National Industrial Structure Vision starts from the premise that:

> The rapid pace of technological innovation is shaping the course of the future. This "Fourth Industrial Revolution"—spurred by technologies such as the Internet of Things (IoT), big data, robots, and artificial intelligence (AI)—is the key to a new phase of growth in Japan. By accurately identifying the trends associated with these innovations and making bold changes to its economic and social systems, Japan can play a leading role in ushering in this new interconnected future.[19]

While this new era was launched by the arrival of digital technologies, it is much, much more. Schwab explains that it is:

> not only about smart and connected machines and systems. ... Occurring simultaneously are waves of further breakthroughs in areas ranging from gene sequencing to nanotechnology, from renewables to quantum computing. It is the fusion of these technologies and their interaction across the physical, digital and biological domains that make the fourth industrial revolution fundamentally different from previous revolutions.[20]

Experts expect the transformation of business models, industries, warfare and the very nature of work itself.

These technologies are variously described as "critical," "disruptive," "emerging," "foundational," or "game-changing." Broadly speaking, they are technologies related to biology, robotics, information, nanotechnology and materials, energy, and systems. (A useful mnemonic is BRINES).[21] As foundational technologies, these are the "keys to the kingdom," control of which is essential to leadership in a digital world. Significantly, success in each area will yield disproportionate returns, since "materials and manufacturing are inextricably linked in a longstanding virtuous cycle ... most likely will be accelerated by convergent increases in high performance computing, materials modeling, artificial intelligence (AI), and biomaterials."[22]

The first and perhaps most important is artificial intelligence. The combination of algorithms and data produces an enabling force, like electricity, that can be applied across many types of activity and industry. Eric Schmidt, the former chairman of Google, spoke for many in the US government when he warned that "Leading the world in AI is essential to growing our economy and protecting our security."[23] Money is part of that story: AI is reckoned to have the potential to boost global economic output by up to 15 percentage points over the next decade, growth that would add 1 percentage point to annual growth rates, roughly equivalent to the growth spurt initiated by nineteenth-century industrialization.[24] It is estimated that total economic activity attributable to AI will exceed $13 trillion by 2030.[25] More optimistically, McKinsey estimated that AI would increase global economy activity between $17.1 and $25.6 trillion annually.[26] As important though are ways that governments can use AI to strengthen their power and control their citizens. AI is already used as a tool of authoritarianism, enabling monitoring and surveillance through facial recognition technologies. It's also used to make social media disinformation campaigns more effective, supercharging a tool that can be used to disrupt democracies.

AI should not be considered an emerging or disruptive technology. It is much more. AI is the master tool, mastery of which will yield supremacy in all the other fields identified below. As Henry Farrell explained, generative AI would be a "self-improving system" which would "be capable both of making itself better *and* advancing technological progress on a variety of fronts ... The US would have created a scientific cornucopia, a self-licking ice cream cone of technological advantage that would secure its long-term dominance."[27] Vladimir Putin put it more simply: "whoever becomes the leader in this sphere [AI] will become the leader of the world."[28]

The second key technology is biotechnology. Biotech was developed in the 1970s and commercialized in the agricultural sector in the 1980s, primarily in the form of genetically modified crops to make them pest or herbicide resistant.[29] While genetic manipulation has transformed agriculture, it is more commonly associated with health care, and it is thought that "at least 45% of the current global disease burden could be addressed using science that is conceivable today."[30]

In addition, as much as 60 per cent of the physical inputs to the global economy could be produced biologically, such as wood, cotton, or animals bred for food: biology could "determine what we eat, what we wear, the products we put on our skin and the way we build our physical world."[31] The bioeconomy already accounts for about 5 per cent of US GDP, about $1 trillion a year, and that number could reach $4 trillion in two decades.[32] In 2016, bioscience firms in the US employed 1.74 million people.[33] In 2013, McKinsey estimated that biotechnologies could impact the work of 2.5 billion people employed in agriculture around the world, reduce the 26 million lives lost annually to cancer, cardiovascular disease, or type 2 diabetes, cut $6.5 trillion in annual health care costs or alter the $1.1 trillion global market in wheat, rice, maize, soy, and barley.[34] Biotechnology has the potential to be the most disruptive of all technologies as it can allow us to re-engineer humanity as we manipulate DNA and the building blocks of human existence. As Jamie Metzl explains, we are "beginning a process of hacking Darwin."[35] This is enabled by AI and the exponential growth in computing power.

A third disruptive technology is quantum computing. Quantum computing uses quantum mechanics to exponentially increase the speed and power of computers, solving in a single run problems that might otherwise take years to figure out. Scientists anticipate both unbreakable codes or instant code-breaking with quantum computers. The ability to build and test vastly more complex models, to run more intricate and detailed simulations, promises advances in fields ranging from physical sciences to finance. While the science remains rudimentary, China launched the first quantum satellite in 2016, and several quantum computers, albeit of limited size and capability, have been built. Several governments support quantum computing programs and see progress in this area as vitally important to national security.

Nanotechnology is the manipulation of matter at scales ranging from individual atoms to 100 nanometers. Nanotech provides the prospect of "miracle materials" that can repair themselves, retain or revert to their original shapes, or have completely novel reactions to their environment, becoming superconducting, superstrong, or capable of neutralizing foreign substances (such as cancer cells).

Nanotech-scale manipulation was first achieved in the 1990s, and the first commercial products entered the market a decade later. The global market for nanotechnology was estimated to be $54.2 billion in 2020 and is projected to reach $126.8 billion by 2027, expanding 12.9 per cent annually during that time.[36]

Additive manufacturing, often called 3D printing, could transform manufacturing. Today, its application is limited, and it is mostly used by enthusiasts with few mass-produced applications. The cost curve is sharply declining, and more general use is just over the horizon. Widespread acceptance will transform the production of goods: it will allow immediate alterations of production processes to adapt to a change in demand or a design update; restructure supply chains by enabling local production of products (potentially eliminating nodes in those networks); change the calculus for locations; and redefine "just in time" delivery.[37] With the potential to impact the livelihoods of 320 million manufacturing workers, or 12 per cent of the global workforce, and $11 trillion in global manufacturing revenue, it is truly a disruptive technology.[38]

A sixth disruptive technology is robotics. Robots have seized the imagination as quasi-humanoid objects that are either the servants of humankind or its greatest threat. The global market for robots reached $100 billion in 2020 and is forecast to grow around 26 per cent annually to reach just under $210 billion by 2025.[39] There is no consensus on how wrenching widespread deployment of robots will be: one study concluded that as many as 800 million people across forty-six countries—one-third of the workforce—could be displaced by robots.[40] Another study provides hope with the conclusion that technology has created more jobs than it destroyed over the past 144 years.[41]

Today, most robots are industrial models that bear no resemblance to the fictionalized version of Terminator-type machines that have dominated the imagination for generations, but humanoid robots are increasingly evident in daily life, cleaning rooms or providing companionship. Bank of America anticipates that global sales of those machines will explode, with global sales of just 18,000 units in 2025 expanding to 1 million units in 2030 and 10 million in 2035, an 88 per cent annual growth rate over the next 10 years. By 2060, Bank of America expects there to be 3 billion in operation, or one robot for every three people on the planet.[42]

Other technologies will pop up in the pages that follow, and they should be mentioned here. One is the Internet of Things (IoT), the billions of physical devices that are connected to the internet that have unique identifiers and can transfer data over a network without human-to-human or human-to-computer interaction. As many as 1 trillion "things"—humans, machines, animals, clothing, devices—will be part of the IoT. A critical enabler of the IoT is 5G (fifth-generation) communications technology, which is being rolled out across the world and offers much higher connectivity and download speeds, which in combination with edge computing (in which servers are closer to the end user) facilitates the connectivity that is the essence of the IoT. The mobile internet, which incorporates those two technologies, contributed $6.5 trillion to global GDP in 2024 and is projected to reach $1 trillion in 2030, or 8.4 per cent of global GDP.[43] The US Defense Innovation Board has warned that "the country that owns 5G will own many of these [critical] innovations and set the standards for the world."[44]

Billions of connections will generate massive amounts of data—nearly 2.5 million terabytes of data were collected each day in 2018,[45] and that amount reached 474 million terabytes a day in 2025[46] (increasing nearly 200 times)—which makes big data another area of concern. Those databases and the ability to turn data into information will become a new measure of national strength. This in turn makes data governance a vital concern. Experts believe that 5G is a "winner takes all" sector, where control of the infrastructure equates to control of data.[47] Here perhaps more than anywhere else the cumulative impacts of technological leadership are most evident. The combination of 5G hardware, the data created and supplied by billions of sensors, devices, and connections, and the AI that can sort and analyze it all in real time will create a new and different world. AI is the key: the ability to sift through data at higher speeds and with greater discernment (if done properly—a big if) will accelerate advances in every field. When we can think faster, we can do things faster, and in a world that rewards first movers (in many but not all cases), that speed will separate winners from losers.

A suite of technologies will transform the production, distribution, and consumption of energy. One of the most important of these concerns storage. New battery tech is central to this evolution, and new materials (the product of nanoengineering) and economies of scale are already driving down cost curves. The mounting concern over climate change has instilled new urgency into the drive to create alternatives to fossil fuels, and many of the other technologies already discussed will help in that effort.

It is critical to recognize that these technologies interact. They shouldn't be considered separate or distinct. Just as AI is "the master tool" to accelerate the development of all the others on this list (and to repeat, these technologies can be packaged or identified in different ways), leadership in one field generates spillover effects in others. For example, robotics experts note that when it comes to manufacturing, 70 per cent of the components of humanoid robots are interchangeable with those of EVs.[48]

While the economic impacts of these technologies are plain—trillions of dollars representing, in the most optimistic assessments, around 20–25 per cent of global GDP[49]—there are other, perhaps even more significant, consequences. The importance of mastering biotechnology was evident throughout 2020 as the world struggled with the COVID-19 pandemic. The race to identify the causes of the disease and develop a vaccine had profound implications for international diplomacy and perceptions of global leadership. The development of alternatives to fossil fuels will reduce the influence of oil producers and threaten instability in those states and regions as the $100 billion spent annually on gasoline and diesel fuels is redirected. Other raw materials producers will be equally affected if the prediction that biotechnology will replace 60 per cent of all physical inputs is correct.

Countries that lead in the development of technologies are "first movers," and they enjoy considerable advantages in the competition among states. The companies and countries that get the first product to market create a foundation that others cannot easily build into or upon. So, for example, once a consumer has acquired an Apple product, he or she usually continues to buy more so that the devices can talk to each other. As Eric Schmidt explained, "In critical sectors with strong network effects like telecommunications, a winner-take-all dynamic raises the stakes for rapidly developing leading technology platforms."[50] Many governments installed Huawei equipment in their fourth-generation communications networks, which gives Huawei, and thus China, an advantage as those governments consider their 5G provider. If they don't purchase Huawei—a choice that they are being increasingly pressured to make—then they must rebuild their entire infrastructure, rather than incorporate some 4G elements into the new system.

First movers also gain significant advantages because of the interconnections between these technologies. Robots provide an eye-opening example. It's tempting to dismiss the explosive growth of the industry that has been projected—88 per cent—as hyperbole, but analysts say it's possible if we allow robots to design the next generation of robots. "Robotics systems manufacturing more robotics systems, and with each unit produced the cost will be driven down continuously and the quality will improve, only strengthening their production flywheel. This will repeat ad infinitum."[51] The cost structure rapidly decreases as parts are commoditized and mass production spreads. For those who remain unconvinced, there is precedent for this mind-bending growth: the compound annual growth rate of passenger electric vehicle sales in China reached 90 per cent in the decade between 2013 and 2023. Bank of America analysts warn that "China has the potential to replicate its disruptive impact from the EV industry in the humanoid space,"[52] and industry analysts insist that the impact of this is "exponential compared to their last strategic industry captures."[53]

There are other economic reasons to focus on these technologies:[54] they are increasingly important to trade performance, accounting in the US case for 60 per cent of its exports; they create high-wage jobs, with workers in advanced industries earning 80 per cent more than average, and even workers without college degrees earning 57 per cent more in high-tech industries; and they generate growth and vitality in the larger economy, with 35 per cent of US economic growth occurring in seventy-five intellectual property-intensive industries.

There is more. A vibrant, leading technology sector will attract smart young minds, ensuring a steady flow of talent. Innovation precedes development, which requires the engineering skills that turn ideas into concrete items. Prototypes precede mass production, and that availability encourages local adaption, which in turn stimulates more innovation. As the OECD summarized:

> the capability to innovate and to bring innovation successfully to market will be a crucial determinant of the global competitiveness of nations over the coming decade. ... innovative activity is the main driver of economic progress and well-being as well as a potential factor in meeting global challenges in domains such as the environment and health.[55]

Robert Manning, who has worked extensively on the geopolitical implications of such technologies, concluded, "The hierarchy of nations in the 21st century will be largely measured by their capacity to innovate and/or to adapt and absorb emerging technologies. The degree of success will be the key indicator of their economic growth prospects, their relative weight in the global economic system and geopolitical clout."[56]

Emerging technologies also have critical military applications. Most of the work that addresses these technologies starts with the premise that US military superiority depends on continued mastery of them.[57] A US Council on Foreign Relations Task Force on Innovation and National Security offered an apt summary, noting that "countries that can harness the current wave of innovation, mitigate its potential disruptions, and capitalize on its transformative power will gain economic and military advantages over potential rivals."[58]

The competition sharpens

Ultimately, the case for leadership in emerging tech is this: The US enjoyed technological supremacy for the past half century, a status that buoyed its economy, its society, and its military capabilities. Linda Weiss, a student of the relationship between the US national security state and its technological infrastructure, points out that "all the major advanced industries of the past 60 years have been pioneered in the US."[59] The Export Control Reform Act of 2018 was explicit: "The national security of the U.S. requires that the U.S. maintain its leadership in the science, technology, engineering, and manufacturing sectors, including foundational technology that is essential to innovation."[60] Or, as the US National Commission on Innovation and Competitiveness Frontiers argued, "the convergence of technologies—from artificial intelligence to quantum science and computing, advanced nuclear, robotics, and advanced biology—is accelerating innovation, reshaping traditional business models, spawning new industries, and fundamentally redefining the world order."[61]

Daily there are reports that America's pre-eminence is threatened. The primary challenger is China, which has developed a series of plans, programs, and initiatives to master and dominate those technologies. Building on the 2006 Medium- to Long-Term Plan for the Development of Science and Technology, China has laid out the Innovation Driven Development Strategy which includes Made in China 2025, the New Generation of Artificial Technology Development Plan, China Standards 2035 Plan, the Strategic Emerging Industries Plan, and others.[62] Much attention has focused on Made in China 2025, the country's first ten-year blueprint to comprehensively upgrade China's industrial economy and achieve the goal of becoming a world-leading manufacturer by the 100th anniversary of the founding of the PRC in 2049. The *Made in China 2025 Notice* calls for China to achieve 40 per cent "self-sufficiency" by 2020, and 70 per cent "self-sufficiency" by 2025, in core components and critical materials in a wide range of industries, including aerospace equipment and telecommunications equipment.[63] There are another sixteen "Major Special Projects" in grand-scale technological innovation, each of which has been likened to the Manhattan Project. Fields include quantum communications, 5G systems, biotechnology, high-resolution Earth observation, and manned space flight.[64]

Numerous assessments testify to China's success in these vital areas. The Australian Strategic Policy Institute's (ASPI) covers sixty-four critical technologies that include defense, space, energy, the environment, AI, biotech, robotics, cyber, computing, advanced materials, and key quantum technology areas and offers "a leading indicator of a country's research performance, strategic intent and potential future science and technology capability." Looking back over a twenty-year period, the ASPI analysts conclude that the US led in sixty of the sixty-four technologies from 2003 to 2007. By 2023, however, China emerged as the leader in fifty-seven of these technologies, while the US led in just seven, "marking a significant upheaval in global research dominance."[65]

ASPI's conclusion aligns with that of the Special Competitive Studies Project, a think tank established by Eric Schmidt, which focuses on the national security implications of tech leadership. It considered leadership in twelve critical technologies and found that China led in four—advanced batteries, 5G, commercial drones, and advanced manufacturing—while the US led in five—internet platforms, semiconductors, synthetic biology, fusion energy, and quantum computing—and three were contested: AI, next-generation networks, and biopharmaceuticals.[66]

Analyses of research papers, a powerful indicator of who is on the cutting edge of science, provides yet more revealing data. A 2025 study concluded that "China is on track to lead the world in science," when authorship of scientific papers is taken into account. An analysis of almost six million research papers found that Chinese scientists are taking the lead role in almost half of all collaborations with US counterparts, "a shift that underscores Beijing's growing influence in setting the global research agenda." Chinese-based scientists filled 45 per cent of leadership roles in US–China joint studies in 2023, up from 30 per cent in 2010, and the two are projected to reach parity—the point at which both sides lead an equal share of joint research—in 2027 or 2028. In eight of eleven critical technologies—including artificial intelligence, semiconductors, energy, and materials science—Chinese researchers are expected to achieve leadership parity with the US before 2030.[67] That matches the conclusions of another study which found that "China has entirely closed the gap with the US in terms of contribution to the global top 1 percent high-impact scientific production, specializing in four key fields—engineering/electrical/electronic, materials science, physics, and chemistry."[68]

Given the stakes—technological supremacy and the resulting global leadership, what we call "Tech-Gemony"—the first Trump administration concluded that "the quest for technological supremacy is too important to be left to markets alone to decide."[69] The US argument was simple: "Chinese industrial policy aims to acquire the intellectual property and technologies of the world and to capture the emerging high-technology industries that will drive future economic growth."[70] Or as the 2025 Annual Threat Assessment of the National Intelligence Community concluded:

> China is using an aggressive, whole-of-government approach, combined with state direction of the private sector, to become a global science and technology superpower, surpass the United States, promote self-reliance, and achieve further economic, political, and military gain. Beijing has prioritized technology sectors such as advanced power and energy, AI, biotechnology, quantum information science, and semiconductors ... China is accelerating its S&T progress through a range of licit and illicit means.[71]

Projects like Made in China 2025 and the widespread theft of technology and intellectual property—"the greatest transfer of wealth in history," according to Keith Alexander, former director of the US National Security Agency, which costs the United States "anywhere from $300 billion to $600 billion" a year, as estimated by William Evanina, director of the National Counterintelligence and Security Center[72]—"are *not* being made simply for profit and to drive foreign competitors out of business, but also in support of the PRC's geopolitical ambitions of a so-called national 'return' to the dominant center of the geopolitical system." This strategy is "a mortal threat to non-Chinese technology sectors, even if most of them don't realize it yet." Chinese leaders believe that technological, economic, and military capabilities determine a country's "comprehensive national power," (CNP) which determines a nation's international status. "Building superlative CNP is viewed as the key to acquiring dominant global power ... and acquiring controlling positions across a critical range of cutting-edge technologies is a critical part of the PRC's revisionist geopolitics."[73]

Prior US administrations had sought to check Chinese advances and the encroachment on US (and allied) interests, but Trump's administrations differed from its predecessors in two critical, but related, ways. The first was its focus on the economic dimension of the bilateral competition. This reflected the president's obsession with trade balances and the framing of this era as one of "great power competition" and its recognition that the US must up its game in domains other than that of the military. Both the George W. Bush and Obama administrations had been moving in this direction; the "Rebalance to Asia" was based on the notion of a wide-ranging competition with China that required both a reconceptualization of the geographic space in which the two countries contested (the Indo-Pacific) and emphasis on its nonmilitary dimensions. The Trump administrations took that approach to eleven. Washington imposed unprecedented levels of tariffs on Chinese exports to the US, goaded by Trump's claim that trade wars are "easy to win." Neither the trade sanctions, nor the trade agreement they were supposed to facilitate, had an appreciable impact on the trade balance, however. In July 2020, the US trade balance recorded a deficit of $63.6 billion, $10 billion more than the month before, and the biggest monthly deficit since July 2008.[74] In August, Chinese figures showed the country's trade surplus with the US rose 27 per cent over the previous year: exports to the US increased 20 per cent from 2019, hitting $44.8 billion, while imports of US goods rose only 1.8 per cent, to $10.5 billion.[75] Ultimately, China only bought 58 per cent of the US exports it committed to purchase under the deal, well below import levels that preceded the trade war. Or, as the economists at the Peterson Institute for International Economics concluded, "China bought *none* of the additional $200 billion of exports Trump's deal had promised."[76]

At first glance, Trump's approach across both terms appears to continue down a well-trodden path while attempting to adjust to the peculiarities of Chinese policy, in particular its military–civil fusion (MCF). The State Department describes this as "an aggressive, national strategy ... to enable the PRC to develop the most technologically advanced military in the world. ... a key part of MCF is the elimination of barriers between China's civilian research and commercial sectors, and its military and defense industrial sectors." Personally overseen by Xi Jinping (in his capacity as chair of the Chinese Communist Party's Central Military Commission and the Central Commission for Military–Civil Fusion Development), it is "systematically reorganizing the Chinese science and technology enterprise to ensure that new innovations simultaneously advance economic and military development."[77] Christopher Ford, former US assistant secretary of state for international security and nonproliferation and a longtime student of China, was frank about the implications of this policy, noting that it "makes rather a hash of traditional approaches to export controls which tend to presume a meaningful distinction between civilian and military that is becoming increasingly opaque, or even meaningless, in the PRC."[78]

While insisting that "these changes were hardly what we really *wished* to do," with the administration preferring to continue previous policies, Ford explained that the US has been forced to adopt a restrictive approach because MCF rendered the old logic "incoherent, for *in effect, there is no longer any such thing as a purely 'civilian' export to the PRC*."[79] In other words, the Trump strategy sought to control not only the growth of the Chinese military but that of the most advanced and dynamic parts of the Chinese economy.

This brings us to the second key difference between the Trump strategy and those that preceded it. The United States has developed and sharpened tools to slow—if not halt—China's rise. They are discussed in Chapters 4 and 5, but the point here is the widening scope of measures to block access to technology, which is based on the conclusion that "the economic and security implications of AI and other emerging technologies are all but impossible to disentangle."[80] In some cases, the obstacles are so formidable that the US strategy appears to deliberately threaten a competitor's survival.[81]

After surveying the first Trump administration's steps to punish Huawei, the Chinese telecommunications behemoth, the editorial board of the *Financial Times*, no hysteric, concluded that "the U.S. steps appear part of an attempt to constrain China's rise. … designed to cripple or crush one of the first Chinese tech companies to become globally competitive—and one that relies on American suppliers in both mobile phones and network equipment."[82] Darren Lim of Australia National University argues that this is to be expected: "the logic of economic security expands the scope of legitimate government intervention far beyond the export control of militarily sensitive technologies. Instead, it potentially justifies a broad-based and activist industrial policy—in the name of safeguarding national security."[83]

While the Biden administration reviewed and rejected many of its predecessor's policies, it continued the hard line against China. That reflected a shift in the center of gravity in US thinking toward Beijing during the second decade of the 2000s. While there is acknowledgment that the PRC has a role to play in the resolution of many regional and global problems, there is also an unmistakable embrace of a more skeptical and competitive relationship with Beijing. Views are not monolithic—even within any US administration—but there is consensus in Washington that there is a strategic rivalry between the two countries and that China is a revisionist power.[84] The Biden administration maintained the Trump administration's wide-angle approach to the technology competition, adopting an expansive strategy to slow the pace at which China acquires new technology and deploys its own technology overseas.[85]

The second Trump administration appears to be sticking to that line. Jacob Helberg, undersecretary of state for economic growth, energy, and the environment, explained that "the Trump administration is seeking to establish a "new global consensus between the US and its like-minded friends and partners on the basic fact that economic policy needs to be tethered much more closely in line with US national security."[86] Yet, we have to hedge and only say "appears" because the zigs and zags of US policy under the 47th president are difficult to follow and the underlying strategy almost impossible to discern. The president has threatened and imposed significant sanctions, all the while suggesting that they are in service of an eventual trade deal. Trump has ignored legislation that mandated the sale of TikTok, the popular social media app, and instead facilitated a deal that would allow it to continue to do business in the United States but under new, American, management. His administration stepped up strategic trade controls to block China's access to the most advanced semiconductor and chip-making equipment and then issued waivers to allow the trade to continue. At his October 2025 summit with Xi in South Korea, Trump traded US soybeans for semiconductors: he suspended imposition of trade controls on Chinese businesses—allowing them to continue buying US chips—for increased sales of the US farm export to China.[87]

More than the search for an enemy

Chinese logic is simple: "China must have its own technology, and it must have strong technology."[88] Surveying Chinese history, Julian Baird Gewirtz explained that:

> In the minds of China's leaders, from Mao Zedong to Xi Jinping, technological progress is not only a means to economic and military prowess but also an ideological end in itself—offering final proof of China's restoration as a great power after decades of struggle. Mao, like other Chinese revolutionaries and reformers, blamed the country's having fallen behind partly on its inability to keep up with international technological advancements.[89]

Today, the country's technology entrepreneurs are spurred on by Xi's call to "catch up and surpass." That objective, called *ganchao*, "remains the essential framework for understanding China's ambition to become a technological superpower today, bringing together the legacies of Marxism, Maoism, and the tortuous pursuit of modernization by the CCP."[90] In his report to the 19th CPC National Congress, Xi urged his audience to:

> aim for the frontiers of science and technology, strengthen basic research, and make major breakthroughs in pioneering basic research and groundbreaking and original innovations. We will strengthen basic research in applied sciences, launch major national science and technology projects, and prioritize innovation in key generic technologies, cutting-edge frontier technologies, modern engineering technologies, and disruptive technologies.[91]

Since then, interest and ambitions have intensified: the 14th Five-Year Plan (2021–2025), unveiled in March 2021 during the annual legislative Two Sessions meetings, "described technology innovation as a matter of national security, not just economic development, for the first time. This represents the increasing perception that technology is a battleground for competition with the West."[92] It names "artificial intelligence, quantum information, integrated circuits, life and health, brain science, bioengineering, aerospace technology, deep earth, deep sea and other cutting-edge fields" as focus areas.[93] The 15th Five-Year Plan, released in March 2026, will continue that trajectory and will "first and foremost intensify efforts toward achieving greater self-reliance and strength in science and technology, actively develop new quality productive forces, and deliver substantial progress and new breakthroughs in boosting scientific and technological innovation, fostering new drivers of growth, and promoting economic structural improvement and upgrading."[94] Vice Premier Ding Xuexiang spelled out the implications and goals of this thinking:

> In the period ahead, technological revolution and major power competition will intertwine, and competition in new technologies and new tracks will become even fiercer. We must focus on national strategic needs, pool our strength to make original and leading breakthroughs in science and technology, significantly enhance basic research and original innovation capability, make rapid breakthroughs in key and core technologies in critical areas, increase the number of areas where we run alongside or lead, promote deep integration between scientific and technological innovation and industrial innovation, and see the role of innovation in driving development markedly enhanced.[95]

That technological prowess serves several purposes for the Chinese leadership. Most significantly, it provides the foundation for CCP legitimacy. Economic growth and development confirm the government's most important claim that it has improved the lives of its citizens. Second, as economic historian Adam Tooze explained:

> the successful expansion of renewable power confers output legitimacy. As recently as the early 2000s, China was a chronically power-poor society. Answering that poverty with coal power was an achievement, but it created the world's greatest pollution disaster. Cleaning that up whilst delivering more and more electricity is a spectacular achievement of development. Creating a world-leading new industry to produce solar panels, wind turbines, batteries and EV adds to the success.

That success reflects a "clear and well-resourced industrial policy framework. Backhandedly even Western critics acknowledge Chinese government and party leadership. … China's green energy revolution confirms the CCP's stewardship."[96]

In addition, it girds the executive power of the government. Support for technology development embeds the state in a web of nominally nongovernment activities that extend its authority and power: it directs financing; promotes research, development, and innovation; and affords it a seat in corporate decision-making, for both domestic and foreign companies (seeking to crack the local market).[97] Technological leadership also serves as a springboard for efforts to set international standards, a key part of China's global governance agenda.[98]

Notable successes have occurred. Chinese industrial polices have produced the world's largest high-speed rail system, with trains that travel at speeds between 250 and 350 kph; an indigenous navigation satellite system, BeiDou, as an alternative to the US, European, and Russian GPS systems, and which is said to be the most accurate of them all; the world's two fastest supercomputers, the Sunway TaihuLight and the Tianhe-2; and Huawei, the world's largest telecommunications equipment manufacturer.[99] Other projects underway include: a national big data strategy to build a dynamic digital economy; an all-of-nation strategy to advance next-generation AI development for commercial and military applications that will make China the world's "premier AI innovation center" by 2030; a billion-dollar investment program in biotechnology and genomics with a focus on precision medicine; and initiatives to assume global leadership in quantum technologies with transformative implications.[100] All attest not only to Chinese ambitions but its abilities as well, and it challenges the global order to which we have been accustomed. While global order is most frequently an abstract notion, in the new national security economy these developments should also be viewed as threats.

Japan leads the way

If any country should be ready for the new national security economy, it is Japan. That nation was obliged to adopt a holistic approach to national security because Article 9 of its constitution—the famous "war renouncing" clause—expressly gives up the sovereign right to wage war, the threat or use of force to solve international disputes, and the maintenance of armed forces that would allow the government to do so.[101] Deprived of a military to shape the international environment, governments in Tokyo relied instead on other tools for power projection, most commonly economic instruments. Concomitantly, those governments embraced "comprehensive security," first articulated in the 1970s and which recognized that national security had to consider both military and nonmilitary threats. Japan's rooting of national security in economic concerns, especially the development of its neighbors and regional countries, won it considerable influence and approval among those nations. Not surprisingly, opinion polls consistently show Japan to be "the most trusted major power" in Southeast Asia.[102]

This inclination was reinforced by the reliance of the Japanese government, like that of other countries, on economic performance for domestic and international legitimacy. The country's extraordinary record, building what was formerly the world's second-largest economy from the rubble of World War II with no natural resources other than its populace, yielded a keen sensitivity to economic concerns. The success of its state-led development program reinforced an inclination toward mercantilism and a cultural emphasis on *monozukuri*—production or making of things—that prioritized manufacturing as the foundation of economic activity.

Not surprisingly, then, Japan has worried about disruptions in its production networks. Tokyo was first to experience China's weaponization of its domination of the rare earths supply chain when Beijing cut off exports of those critical minerals in 2010, following the arrest by the Japanese Coast Guard of a Chinese fishing boat captain for fishing in disputed waters and ramming a coast guard cutter when it tried to stop him. That troubling experience found expression in Japan's first National Security Strategy, published in 2013, which put economic security at the very heart of its national security.[103]

But China is a fact of life. It is Japan's neighbor, a historical rival and the nation with which it has a territorial dispute. At the same time, China has been Japan's largest trading partner for nearly two decades, accounting for one-fifth of Japan's total goods trade in 2023; while that share has fallen from a peak of 24 per cent in 2020, China remains the largest source of imports at 22.3 per cent.[104]

This ensures that Japan is extremely sensitive to the geopolitical struggle between the United States and China. The US is Japan's only ally, and because of Article 9 Tokyo must maintain the trust and commitment of its security partner. (Japan's brutal history during the Pacific War makes that partnership even more important.) The US remains the economic partner of choice as well, which means that the Japanese government and Japanese businesses work hard to ensure that there is no daylight between their positions and that of the Washington government on questions of economic security. Both aim to retain access to the US market and avoid any action that would court Washington's anger.[105]

As a result, Japan has been forced to "think about national security from a variety of perspectives, not just force"[106] and recognizes that economic issues are central to the strategic competition defining contemporary geopolitics, and the "tectonic shift" occurring in the development and application of new technologies is critical to success in that struggle.[107] Significantly, there is also a belief "that Japan could not confront a rising China while relying entirely on the market and economy. A form of 'state capitalism,' or cooperation between industry and the political leadership, with the application of administrative acumen, would be needed."[108]

This history has put Japan at the forefront of economic security policy-making. Over the last decade, successive Japanese governments have put the essential elements of economic security—resilience, autonomy, promotion of innovation, and deeper international cooperation—into basic economic policy and have reorganized the bureaucracy to effectuate that intent.[109] Tokyo is pacing government responses to the new national security economy, but even it is coming up short.

2

A WORLD OF INSECURITY

The 2020–21 COVID-19 pandemic provides an exceptional case study of the new national security economy. Sadly, predictably, developments revealed or underscored the many dangers of this world, and just how far individuals, organizations, and societies must go to adjust to these realities and protect against new threats. The pandemic upended the daily lives of individuals and transformed business routines and operations.

COVID-19 had an equally powerful impact on governments. National health care systems are now recognized as a critical part of the national security infrastructure, and governments learned how to surveil their citizens and mobilize them for mass campaigns, while simultaneously seeking to fine-tune the balance between individual rights and social responsibilities. Internationally, the outbreak reoriented and sharpened the competition among nations, with governments racing to procure personal protective equipment, tests, and vaccines needed to protect their citizens. All the while, status and standing have ridden on a government's ability to spur vaccine development and provide it to needy populations within and beyond its borders.

The COVID-19 case study

Fear of infection closed schools, shuttered retail outlets and restaurants, and forced employees to work from home, accelerating long-anticipated moves to virtual schooling, online shopping, and telecommuting. In each case, a reliance on digital technology exposed vulnerabilities in communications networks and a recognition that most citizens don't understand the security issues. Greater use of electronic or cashless payments in the wake of the pandemic exposed the public to cyberattacks. Cybercriminals employed phishing and charity scams through email or by setting up malicious domains to take advantage of people searching for information online about the coronavirus, stealing personal information or implanting malware. Criminals impersonated the World Health Organization to gain access to information in personal computers by distributing a fake "My Health e-book" attachment containing a file with malware.[1] And in a reminder that there are truly ugly and dangerous corners of the internet, the growing number of young people spending time on computers at home during the pandemic facilitated the spread of child abuse networks.[2]

Businesses played catchup as well. The growth of telecommuting exposed vulnerabilities in internal communication networks. Companies rushed to set up virtual private networks to allow increasingly far-flung employees to work together, but many of those systems were compromised.[3] Patrice Caine, head of Thales, a French technology and security company, observed that "The fact that working from home has been. … applied in a very short period of time has clearly increased the number of weaknesses," making it "more simple" to attack companies.[4] Cybercriminals targeted employees to gain control through remote access to corporate networks to steal sensitive information. Google detected more than twelve state-sponsored hacking groups using coronavirus to craft phishing emails and attempt to distribute malware.[5] (The rise of AI has put phishing in overdrive: phishing attacks exploded by 4,151 per cent since the public release of ChatGPT in late 2022, and there is an estimated cost of $4.48 million per breach).[6] One favored form of access is exploiting vulnerabilities of teleconference tools.[7] Criminals shifted toward soft targets, prioritizing poorly defended sectors, such as education, health care, and local government, as well as smaller companies.[8] The problems that plague ordinary telecommuters are evident—and more dangerous—among operators of infrastructure and the industrial Internet of Things.

Businesses reworked fundamentals. Extended supply chains and logistical systems once extolled as cutting-edge were exposed as vulnerable, subject to disruption for technical reasons (hacking), natural reasons (the pandemic), or political reasons (the response to the pandemic). "Just in time" production has been challenged, if not replaced, by "just in case." Resilience has replaced efficiency as a top priority. We've learned that businesses and governments—societies collectively—have little idea how to price security. Most typically, we've privatized profits and socialized losses.

Hackers gave special attention to national organizations and agencies responsible for dealing with the pandemic: as components of national critical infrastructure, they are high-value targets. There were ransomware attacks against hospitals and health care organizations in the United States, France, the Czech Republic, Spain, Thailand, and Germany—a woman in Dusseldorf was reportedly the first known death caused by a cyberattack.[9]

Another target was companies working on vaccines. Victory in the race to develop and deliver an effective COVID-19 vaccine yielded substantial rewards, not just monetary. It demonstrated a mastery of biotechnology, validating the business model and system that developed the vaccine. The achievement bestowed diplomatic status on the government that provided the vaccine to the less fortunate, while accelerating that nation's recovery from the coronavirus-inspired economic downturn. "At times like this, any information about cures or tests or vaccines relating to coronavirus would be priceless and the priority of any intelligence organization of an affected country."[10] Those efforts were facilitated by a COVID-19 vaccine production system typical of contemporary supply chains, with research and development, production, storage, and distribution facilities snaking across the globe, their vulnerabilities visible only in a crisis.[11] The US, UK, and Canadian governments charged Russia, China, North Korea, and Iran with trying to compromise research.[12] Microsoft researchers detected cyberattacks from three nation-state actors against seven leading pharmaceutical companies and vaccine researchers in Canada, France, India, South Korea, and the United States directly involved in researching vaccines and treatments for COVID-19. The majority of the targets were vaccine manufacturers with COVID-19 drugs in various stages of clinical trials.[13]

Hacking did not aim to just steal research. It was also used to compromise—or create the perception of compromised—work. This undercuts confidence in a vaccine, helping to undermine a rival's status in the diplomatic competition. This was part of a broader campaign to weaponize online sources for political purposes. The result was "an epidemic of online misinformation and fake news"[14] in which states and nonstate actors "minimize[d] the infectivity or virulence of the disease," and promoted "false and potentially even lethal 'cures' for the virus, and conspiracy theories about the origins of the virus and its (nonexistent) relationship with 5G phone masts."[15] In some cases, governments used the pandemic to turn against their own citizens. The Syrian government, for example, reportedly used the outbreak to distribute surveillance malware through "watering hole" attacks and third-party app stores.[16] Other governments used the need for disease surveillance to cloak more extensive and intrusive efforts to keep an eye on citizens, and there is a fear that an enduring legacy of the COVID-19 pandemic may be a rebalancing of acceptable (or accepted) limits on privacy.

These are some, but not all, of the threats in the new national security economy. Let's explore each in more detail.

The Soprano state

Our list of threats and dangers begins with cybercrime. A McAfee-CSIS study estimated that the cost of global cybercrime between 2018 and 2020 exceeded $1 trillion, about 1 per cent of global GDP, and registered an increase of more than 50 per cent in two years. Breaking it down, the monetary loss from cybercrime—cash and opportunity costs such as lost sales, reduced efficiency system downtime, brand damage, loss of trust, IP theft, and other items—reached $945 billion; more than $145 billion is thought to have been spent on cybersecurity.[17] (If these revenues flowed to a single organization, it would be the size of the world's seventeenth-largest economy—a genuine "Soprano state.")[18] Global cybercrime is forecast to reach $10.5 trillion by 2025.[19] In 2012, Robert Mueller, then head of the FBI, warned that "there are two kinds of companies out there: those that have been hacked and those that don't know it yet."[20]

There are a wide range of criminal behaviors. While the number of records exposed through data leaks had reached 36 billion by 2020, it's estimated that cybercriminals stole an estimated 33 billion records in 2023 alone.[21] One US law enforcement expert warned that every US citizen should expect that their personal data has been stolen and put on the dark web.[22] Hackers also insert malware to freeze networks and extort payments to liberate them. An industry task force warned in 2021 that ransomware is "a global threat to institutions in all sectors around the world, and while "the nations with the highest percentage of organizations reporting ransomware attacks in 2020 were India, Brazil, Turkey, Belgium, Sweden, and the United States … attacks occur frequently in Russia, Saudi Arabia, China, and nearly every other nation."[23] It's hard to disagree with the conclusion that:

> ransomware attacks present an urgent national security risk around the world. … The immediate physical and business risks posed by ransomware are compounded by the broader societal impact of the billions of dollars steered into criminal enterprises, funds that may be used for the proliferation of weapons of mass destruction, human trafficking and other virulent global criminal activity.[24]

Nor is profit the only motive: the FBI reckons that one in five teenagers who use the internet received unwanted sexual solicitation and estimates that there are between 500,000 and 750,000 sexual predators active on the internet each day.[25]

New-style hacking: Databases to mine

The universe of potential targets has grown drastically for two reasons. First, new vulnerabilities exist—blame IT—creating new ways to wreak havoc on organizations, disrupt operations, and otherwise give managers a reason to pay off extortionists. Second, the motivation for attacks has changed as well, which expands the number of interesting or valuable targets. Companies have to worry about hacks not for criminal motives but for espionage. In its 2020 Data Breach Investigations Report, Verizon found that 86 per cent of breaches were financially motivated and 14 per cent (among businesses with over 1,000 employees) were espionage, although the figure was twice as high (28 per cent) in the manufacturing sector.[26] That tracks with the conclusions of European cyber authorities, who reckon that 20 per cent of data breaches were the result of state espionage, and that 38 per cent of malicious actors are connected with nation-states.[27]

The year 2020 closed with reports of a massive hack of SolarWinds, a company whose Orion network management software is used by over 300,000 customers around the world. Earlier in the year—experts think March, but they're not sure—hackers working for the SVR, the successor to Russia's KGB, slipped a backdoor into an Orion software update, which meant that those customers imported a vulnerability into their systems. This is a textbook "supply chain hack," so-called because it targets a supplier to an organization rather than the organization itself. News of the hack was a stunner. SolarWinds' customers span the globe and include governments, major businesses, and universities. In the US, the list includes the White House, the departments of State, Homeland Security, and Treasury, all five branches of the military, the National Security Agency (NSA, the agency responsible for US government communications security), the national laboratories (where nuclear weapons are developed), as well as 425 Fortune 500 companies, and hundreds of universities and colleges. SolarWinds said that "fewer than 18,000" of its customers installed this malicious update—which means, as cybersecurity expert Bruce Schneier pointed out, "more than 17,000 did."[28]

If the scale of the hack was alarming, so too were its mechanics. The way that access was acquired and the length of time that hackers operated unobserved mean it is impossible to sanitize those networks. The only way to ensure that the hole has been plugged and that there will be no future compromises is to "burn the network down and rebuild it."[29]

Complaints followed news of the hack—teeth were gnashed, fists raised, howls hurled skyward—but the SolarWinds attack wasn't illegal. "This is classic espionage," said Thomas Rid, a specialist in cybersecurity issues, and espionage—acquiring information—is allowed.[30] "By international norms, Russia did nothing wrong. Countries spy on each other all the time. There are no rules or even norms."[31] In a statement about the hack, the Cybersecurity and Infrastructure Security Agency, the FBI, the NSA, and the Office of the Director of

National Intelligence underlined their assessment that SolarWinds "was, and continues to be, an intelligence gathering effort."[32]

This was the reaction too, by intelligence professionals at least, to the 2014–15 hack of the US Office of Personal Management (OPM), in which Chinese hackers exfiltrated the personnel files of 4.2 million former and current government employees and security clearance background investigation information on 21.5 million individuals.[33] Michael Hayden, former head of the CIA, called the data "a treasure trove of information that is available to the Chinese until the people represented by the information age off. There's no fixing it."[34] But James Clapper, then director of national intelligence and Hayden's boss, admitted that "You have to kind of salute the Chinese for what they did. If we had the opportunity to do that, I don't think we'd hesitate for a minute."[35]

Chinese hackers were also behind Salt Typhoon, an attack thought to have been started in 2021 that was orchestrated by China's Ministry of State Security and which penetrated over 200 communications networks in over 80 countries. It has been called "a close cousin" to SolarWinds and the OPM hacks because it aimed to steal information in support of an adversary's intelligence collection priorities, and the access could be actively exploited; experts still aren't sure exactly what information was collected.[36]

The US has probably done as much, if not more. By all accounts, the United States has the world's most advanced, most extensive, and most aggressive intelligence operation. The NSA sets standards for the world's hackers, and it aggressively leverages the US position controlling most of the internet backbone and most major internet companies.[37] President Obama was not exaggerating when he said in 2016 that the US has "more [cyber] capacity than anybody both offensively and defensively."[38] Edward Snowden's revelations pulled back the curtain on some of the NSA's operations and tools that are typically obscured. Equally illuminating—or alarming—is the work of journalists like Shane Harris, who detailed the cozy relationship between the NSA and telecommunications companies and standard-setting organizations, facilitating installation or identification of backdoors or vulnerabilities in networks and equipment to facilitate spying or sabotage.[39] When US officials sound alarms about Chinese capabilities to hack equipment, they speak from experience: their fear is that China will do what the US has long been doing. As one example, Harris notes that "under a secret program called the SIGINT Enable Project, [the NSA] strikes deals with technology companies to insert backdoors into their commercial products. Congress allocated $250 million for the project in 2013."[40] In addition to that sum are millions of dollars to buy "zero-day" vulnerabilities from hackers who find those exploits for a living.[41] Many believe that the United States is responsible for "the most complex, targeted and rigorously engineered cyber attack campaigns to date."[42]

Governments don't have to resort to theft. Unauthorized access to those databases is worrisome, but concerns rise exponentially when AI and data mining are introduced. A digital, connected world is generating massive amounts of data, and "every bit of data, no matter how seemingly trivial, has potential value."[43] Tech companies eye this ever-expanding universe of information with hunger, anticipating the ability to hyper-personalize solicitations or precision-target consumers. Governments will try to "reassemble the data trail" and identify and target individuals.[44]

One piece of data might be innocuous, but combined with other databases it can unlock puzzles and provide startling transparency.[45] That is why the 2018 breach of the Marriott hotel chain's Starwood guest reservation database was so alarming. Hackers got access to information on up to 383 million guests, which included "some combination of name, mailing address, phone number, email address, passport number, Starwood Preferred Guest (SPG) account information, date of birth, gender, arrival and departure information, reservation data, communication preferences, and encrypted payment card numbers." It is believed that "approximately 5.25 million unique unencrypted passport numbers and approximately 20.3 million encrypted passport numbers" were stolen by the hackers.[46] Experts think the attack was a Chinese effort to spy on US government employees and intelligence officers since Marriott is the top hotel provider for the US government and military.[47] In a curious coincidence, in 2016 China's Anbang Insurance Group Company made a $14 billion bid for the Starwood chain but withdrew the offer. Chinese financial media reported that China's insurance regulator was going to reject the proposal because it would put Anbang's offshore assets above a 15 per cent threshold for overseas investments.[48] There was speculation, however, that a more important consideration was the prospect of a US government rejection because of Anbang's control of the customer database.[49] A senior US Treasury official explained, "Any time we see a company that has lots of data on Americans—health care, personal financial data—that's a vulnerability."[50] Researcher Elisabeth Braw was even more expansive, warning that "If we, as liberal democracies, can't be sure that that data remains safely handled and if we have to worry that a foreign government will access it, even the most innocuous tech-based service becomes a matter of national security, ranging from fintech to dating websites."[51]

China scrapes social media platforms, like LinkedIn, the professional connections site, to identify people with access to sensitive information and then targets them to try to obtain it.[52] In 2020, the Australian internet security consultancy Internet 2.0 revealed a database of about 2.4 million people that had been assembled from open-source data found at Twitter, Facebook and LinkedIn by Zhenhua Data, in Shenzhen, China. Included in the data set were former British Prime Minister Boris Johnson, former Australian Prime Minister Scott Morrison, other politicians, members of the British royal family, celebrities, and military figures: influential individuals across a variety of sectors. As a cybersecurity expert explained, "when agglomerated, this data opens up myriad opportunities to conduct targeted influence activities should the need arise. ... This can include dis- and mis-information, inauthentic simulation (deep fakes), straight up bribery and general muddying of the information environment in which democracy operates."[53] Christopher Wray, former director of the FBI, was blunt: "China is using social media platforms—the same ones Americans use to stay connected or find jobs—to identify people with access to our government's sensitive information and then target those people to try to steal it."[54]

While lawful, those penetrations are still dangerous. Even though the original intent may have only been to gain access and observe or acquire information, the transition from espionage to sabotage is quick and easy.[55] Software can be installed that later degrades or destroys networks or modifies data. It is not clear if merely planting "virtual bombs" is an act of war but the possibility is troubling.

Who says crime doesn't pay?

There is a long-time connection between espionage and the economy, which shouldn't be surprising if spying is in fact the world's second-oldest profession. Ancient Greeks complained about rivals stealing recipes at their feasts. A British botanist visited Qing China disguised as nobility to acquire the technique for processing tea leaves for the British East India Company. A Jesuit priest performed a similar task when he stole the technique for making hard-paste porcelain. Francis Cabot Lowell's 1810 tour of English and Scottish factories allowed him to memorize plans so that he could then "invent" the power loom when he returned home to Massachusetts.[56]

Collection of economic intelligence has been an accepted function of intelligence agencies since at least World War I,[57] and those efforts accelerated after World War II as the global economy became more deeply integrated and there was an accompanying rise of economic diplomacy. The acquisition of negotiating positions for trade and other economic talks has been every bit as important as getting those for arms control talks—perhaps more so, since there are far more economic negotiations than military ones. During tough trade negotiations with Japan during the height of tensions between Tokyo and Washington in the 1990s, US Trade Representative Mickey Kantor got daily updates from intelligence officials, including intercepted conversations between Japanese bureaucrats and auto executives. That information helped shape US negotiating positions by providing insight into the pressure their counterparts were under.[58] Twenty years later, US intelligence agencies were in the hot seat after the Snowden leaks confirmed that they routinely spied on trade negotiations, communications of economic officials in other countries, and foreign companies.[59]

Intelligence agencies targeted military technology suppliers and producers to either supercharge their defense development programs or to find weaknesses in their adversaries' equipment. Throughout the Cold War, US and Soviet spies sought insight into the other's weapons programs. In one of the James Bond films, *A View to a Kill*, Bond is rewarded by the Soviet government for thwarting a rogue KGB agent's efforts to destroy Silicon Valley. When Bond's boss asks why he was given the medal—he expected the Soviets to want Silicon Valley destroyed—the KGB heavy replies, "On the contrary … Where would Russian research be without it?" Back on Earth, Chinese hackers have been especially energetic. In 2013, for example, US officials warned that designs for many of the most sensitive advanced weapons systems—among them more than two dozen major weapons programs critical to missile defenses and combat aircraft and ships—had been compromised by Chinese hackers.[60]

It's no longer just defense or military technology companies that have to be on alert. Businesses must worry that their data might make them a target, or that their technology might be of interest, even if they don't have an obvious military application or connection. In 2011, the office of the US National Counterintelligence Executive called foreign economic collection and industrial espionage against the US "significant and growing threats to the nation's prosperity and security," and warned that cyberspace was making that spying much easier. Seven years later, the National Counterintelligence and Security Center served up a reminder that "Foreign economic and industrial espionage against the United States continues to represent a significant threat to America's prosperity, security and competitive advantage." It identified China, Russia, and Iran "as three of the most capable and active cyber actors tied to economic espionage and the potential theft of US trade secrets and proprietary information."[61] The steady expansion of their activities and the increasingly shrill warnings are proof that the threat is undiminished and is instead getting more severe.

While much attention is devoted to "adversaries," when it comes to economic espionage, separating friends and foes can be tough. In 2003, the top twelve states that put economic spies in the United States were China, Canada, France, India, Japan, Germany, South Korea, Russia, Taiwan, Great Britain, Israel, and Mexico.[62] After leaving the CIA, former director Robert Gates observed that "There are probably a dozen or 15 countries that steal our technology in this way," adding that "In terms of the most capable, next to the Chinese, are the French—and they've been doing it a long time."[63]

The US government doesn't do that kind of thing, Gates added, distinguishing the acquisition of economic intelligence for governments from doing so for businesses. "We are nearly alone in the world in not using our intelligence services for competitive advantage for our businesses." Even though the CIA had "an enormous amount of information, either at low classification or unclassified," about other nations that businesses would want and need—details about electrical grids, water supply systems, bridge capacities—he couldn't get other Cabinet officials, commerce secretaries in particular, interested. "This is something we don't do."[64]

Gates' comments are a reminder that threats come from all directions. And one of the more striking recent developments is the proliferation of cyber capabilities. For many governments, cyber is an equalizer: it allows access to information without spy satellites. On the battlefield, it is an asymmetric threat that allows cyberwarriors to inflict extensive damage without spending lots of defense dollars. "It's very inexpensive. It's very efficient," explained John Hultquist, a cyber espionage analyst, and smaller developing countries from Macedonia to Malaysia are building their own capabilities.[65]

Economic espionage can be painful. US official reports put the value of trade secret theft at between 1 and 3 per cent of GDP, or between \$180 billion and \$540 billion for the \$18 trillion US economy.[66] China is the chief culprit, with studies attributing between 50 and 80 per cent of the problem, both globally and in the United States, to China.[67] Former FBI Director James Comey complained that "the Chinese government has blatantly sought to use cyber espionage to obtain economic advantage for its state-owned industries."[68]

Europe is worried. Intellectual property rights (IPR)-intensive industries are core components of the EU's industrial ecosystems and play a major role in the EU economy. They generated 29.2 per cent of EU jobs (63 million) between 2014 and 2016; 38.9 per cent of all employment in the EU (83.3 million) can be attributed, directly or indirectly, to IPR-intensive industries; 45 per cent of EU GDP is attributable to IPR-intensive industries, and is worth €6.6 trillion; and IPR companies offer a wage premium of 47 per cent.[69] Given the value of that property, it follows that 20 per cent of European companies are thought to have suffered a cyber breach.[70]

In addition to the benefits that cyber capabilities bestow on military establishments—both as a way of waging war and improving technology—intellectual property theft can make a tangible difference to the economy as well. One study of commercial espionage during the Cold War showed that East Germany closed its productivity gap with West Germany by 6.3 percentage points by 1989. That analysis concluded that espionage was far more cost-efficient in improving productivity than investing in indigenous R&D. If that same productivity multiplier worked today, Russia would enjoy an annual GDP boost of €12 billion, and China's GDP would increase by €96 billion, the equivalent of a 30 per cent increase in exports to Europe.[71]

The critical infrastructure threat

There are other more conventional ways that governments exploit cyberspace to advance strategic objectives. To prepare for actual cyberwar, they must identify and penetrate targets. The US calls this "defend forward," or establishing persistent presence in an adversary's networks. This is murky legal territory as lawyers insist that surveilling foreign networks is not an act of war or aggression, although it takes only one keystroke to transform reconnaissance and observation into sabotage.

One step removed from those more traditional military activities is the deployment of cyber capabilities against an adversary to disrupt, degrade, or destroy critical infrastructure, the "machinery, facilities and information that enable vital functions of governance, public health and the economy."[72] (This is sometimes referred to as "critical national infrastructure.") If the purpose of warfare is to break an adversary's will or ability to fight, then cyberattacks can be just as if not more effective than the kinetic effects of bombs or bullets (although cyberattacks can be equally destructive and kinetic). What constitutes critical infrastructure has evolved. It first emphasized the physical foundations of national power, but it has since taken on a more expansive concern with the provision of essential services and customary conveniences to the public, part of a "global trend among developed countries favoring society-centric resilience at the system level over security-oriented protection of specific assets deemed at risk."[73] The US has identified sixteen critical infrastructure sectors. They include: chemicals, commercial facilities, communications, critical manufacturing, dams, defense industrial base, emergency services, energy, financial services, food and agriculture, government facilities, health care and public health, information technology, nuclear reactors, materials and waste, transportation systems, water, and wastewater systems.[74] Almost every government has recognized that protection of critical infrastructure—however defined—is vital to national security and is a core component of its economic security.[75]

The National Counterintelligence Strategy of the US 2020–22 warned that:

> Foreign intelligence entities are developing the capacity to exploit, disrupt, or degrade critical infrastructure worldwide. Their efforts likely are aimed at influencing or coercing US decision makers in a time of crisis by holding critical infrastructure at risk. ... they also are targeting the facilities and networks that underpin global energy and financial markets, telecommunications services, government functions, and defense capabilities.[76]

The 2025 Annual Threat Assessment of the US intelligence community warned that "A diverse set of foreign actors are targeting US health and safety, critical infrastructure, industries, wealth, and government."[77] It identified China as:

> the most active and persistent cyber threat to US government, private-sector, and critical infrastructure networks. The PRC's campaign to preposition access on critical infrastructure for attacks during crisis or conflict, tracked publicly as Volt Typhoon, and its more recently identified compromise of US telecommunications infrastructure, also referred to as Salt Typhoon, demonstrates the growing breadth and depth of the PRC's capabilities to compromise US infrastructure.[78]

In 2014, China was charged with having "successfully targeted over 600 corporate, private, and governmental entities within a five-year period," and cyberattacks concentrated on US industrial centers stole "corporate and military secrets and data about America's critical infrastructure, particularly the electrical power and telecommunications and internet backbone."[79] Russia targets critical infrastructure, including underwater cables and industrial control systems, in the United States and in allied and partner countries. The 2019 report warned that Russia could "execute cyberattacks in the United States that generate localized, temporary disruptive effects on critical infrastructure—such as disrupting an electrical distribution network for at least a few hours—similar to those demonstrated in Ukraine in 2015 and 2016. Moscow is mapping our critical infrastructure with the long-term goal of being able to cause substantial damage."[80] Iran "has the ability to conduct attacks on critical infrastructure" while Pyongyang "probably possesses the expertise to cause temporary, limited disruptions of some critical infrastructure networks ... it may be able to conduct operations that compromise software supply chains."[81]

From autos to heart valves: Hacking the Internet of Things

One of the most invisible, and insidious, manifestations of the new national security economy is the proliferation into and penetration of wired devices in daily life. An expanding universe of physical objects—from appliances to toys, heart valves to watches—is embedded with sensors, software, and other technologies to permit the exchange of data with other devices and systems over the internet. This web is estimated to include over 21 billion devices worldwide, a number that will double by 2025. The sensors identify or measure changes in position, location, and so on, and transmit the data to a device or server, which analyzes the data to generate "information" for the user. The promise of an enhanced digital life is based on these systems, usually referred to as the Internet of Things (IoT). The speed and distinctive architecture (low latency, or delays between sending and receiving signals, and edge computing, which moves computation and data storage closer to where it is being used) of fifth-generation (5G) technologies make this promise real—and generates risks of its own. While we are most familiar with IoT applications in the house—"smart" homes and wired toys—the reach of the IoT goes well beyond consumer goods to cars and planes, every business sector, and infrastructure components, including hospitals, telecommunications networks, and power grids.

Cybersecurity expert Bruce Schneier put it best when he explained that "everything is now a computer"[82]—and that means all are hackable. According to the 2020 Nokia Threat Intelligence Report, IoT devices made up roughly 33 per cent of infected devices, up from about 16 per cent in 2019, and the adoption of 5G systems will ensure that those numbers grow.[83] The size of IoT components and the scale of their production almost guarantees that hacks will be massive. In 2019 IBM found a vulnerability in a series of IoT connectivity chips used by the French manufacturer Thales that connect more than 3 billion devices a year.[84] In one depressing exercise, Kaspersky researchers found 27,716 open entry points that could be exploited by a hacker in a single hospital, some of which weren't even medical equipment. Hackers could gain access through lighting systems, air-conditioning units, and printers.[85] These are not technical musings: the 2024 SonicWall Cyber Threat Report noted an approximate 107 per cent surge in IoT malware attacks during that year.[86]

In a July 2015 wakeup call, researchers at IBM hacked a Jeep, taking control of the vehicle through a firmware update vulnerability. That exploit allowed them to hijack the vehicle, controlling its speed and its steering.[87] That bit of derring-do—headlines screamed "Hackers remotely kill a Jeep on the highway—with me in it!"—followed successful breaches into the systems of a Toyota Prius and a Ford Escape.[88] Alissa Knight, an expert on cybersecurity for automobiles, warns that "your car is connected if it was made after 2001" and susceptible to a remote hack and control.[89]

Automotive hacks can be used for espionage too. Most vehicles are now sold with cameras to ease parking or fill blind spots, and ultrasonic sensors allow the detection of objects up to 250 meters away. Radar is available as well. While these features aid human drivers, they will be essential to the safe operation of autonomous vehicles. They are also susceptible to hacking. One automotive cybersecurity expert explained that "Most connected cars have cameras and audio recording capabilities that can be activated remotely. This is by design."[90] And while a hacker might try to direct a vehicle's cameras to take photos of targets, connected cars may store sensitive and private data as well. That's why China in March 2021 banned Tesla cars from entering military facilities over concerns that the cameras in its vehicles could be used for spying[91] and the Israeli Defense Forces has adopted similar measures.[92]

The Holy Grail: Supply chain hacks

This is the ultimate nightmare for electronic equipment manufacturers and users: the surreptitious compromise of a product with a device so small and obscure as to practically require advance notice of its existence to be discovered. This is a "supply chain hack,"[93] so-called because the infiltration is through vendors or products in less secure positions on the supply chain: a subcontractor or via software updates. In each, the customer imports the vulnerability into its product. The most successful (widespread and damaging) cyberattacks have been supply chain hacks: NotPetya, "the most devastating cyberattack in history," which did an estimated $10 billion in damage, was malware installed in accounting software.[94]

Supply chain hacks are a fact of life in a globalized economy based on production networks that span the world. Not only do vulnerabilities emerge from links in those chains, but the means for monitoring those networks introduce weaknesses, too. Indeed, for many years, the prevailing security concern was management of supply chains. More recently, however, there is a growing belief that it is the supply chains themselves that are the chief danger. A World Intellectual Property Organization analysis agreed, concluding that "The multinational nature of business increases the probability of "inadvertent, accidental or willful disclosure of confidential information and trade secrets."[95] A US government analysis went further, arguing that "The globalization of the supply chain for new—and increasingly interconnected—IT products will offer more opportunities for malicious actors to compromise [product] integrity and security."[96] Businesses must recognize that the design and layout of production networks must be assessed for political, rather than just business, risk.

In the early days of his presidency, Biden announced a review of critical US supply chains, saying that "The United States must ensure that production shortages, trade disruptions, natural disasters and potential actions by foreign competitors and adversaries never leave the United States vulnerable again."[97] He added that Americans "shouldn't have to rely on a foreign country, especially one that doesn't share our interests or values."[98] COVID-19 prompted a stampede for national solutions. A typical response was the European

Commission warning that it would consider emergency controls on vaccine production and distribution if needed to deal with the "crisis of the century." That meant, said Commission President Ursula von der Leyen, that the EU could halt vaccine exports to countries that did not reciprocate in allowing shots to flow to the EU. "If the situation does not change, we will have to reflect on how to make exports to vaccine-producing countries dependent on their level of openness."[99]

Other governments were alarmed by the potential consequences of supply chain overextension. An Australian government report estimated that one in twenty Australian imports (292 products worth about A$20 billion) originate from concentrated sources of global supply and might be vulnerable to supply chain disruptions: two-thirds of those vulnerable imports came from China. The main risks that could be "problematic"—the report's word—come from reliance on concentrated imports of basic chemicals or personal protective equipment.[100]

Japan, long sensitized to the need for resilient supply chains, is rethinking corporate strategies. Between 1985 and 2018, 14.3 per cent of global economic losses due to disasters occurred in Japan—the second-largest share of disaster-related losses in the world.[101] The March 2011 "triple catastrophe"—earthquake, tsunami, and nuclear accident—closed factories and shut down ports and airports, with "dozens, if not hundreds, of Japanese manufacturers facing disruptions to their supply chains" as a result of the crisis.[102] Yet even with that history, the COVID-19 outbreak forced Japan to look differently at supply chains.[103]

China is typically seen as sinned against rather than a sinner, but it too has policies that effectively decouple China—but on its terms. Xi provided marching orders in 2020, noting that "to safeguard China's industrial security and national security, we must focus on building production chains and supply chains that are independently controllable, secure and reliable, and strive for important products and supply channels to all have at least one alternative source, forming the necessary industrial backup system."[104] Since then, local governments across the country have created "supply chain chiefs" as part of the national effort to encourage self-sufficiency and self-reliance.

The focus on "supply chain resilience" is a gloss on longstanding efforts to promote indigenous development of critical technologies.

An analysis by the European Chamber of Commerce in China and Merics, a leading China-focused think tank in Berlin, argued that "China's leaders have advanced extensive industrial policies in an attempt to develop self-reliance in critical technologies and seek dominance in high-value-added industries." "Made in China 2025" (MIC2025) "was just the most visible expression of a deeply engrained and extensive support system that protects China's rising stars."[105] The report concluded that the US:

> may have provided the spark, but the powder keg China had created was bound to ignite eventually. … Whether it's the US' Clean Network proposal or measures by Chinese authorities aimed at creating "autonomous and controllable" technology, it is all part of the same slippery slope: the technologies that are defining the future, and which are increasingly integrated into every sector of the economy, are being divided between two of the world's three largest economies, each of which has a growing firewall separating itself from the other.[106]

So, while the diversification of supply chains and the resulting globalization of production was once considered an asset for companies, countries, and the world, this process is increasingly viewed as a liability. More and more frequently, policy-makers around the world think like Michael Wessel, a commissioner on the US–China Economic and Security Review Commission, who urged the US Congress to:

> treat supply chains as integral components of our national and economic security as well as vital to our critical infrastructure. The past years have seen multiple warning shots across the bow of our country ranging from inadequate supplies of PPE to rare earth minerals and magnets to products like grain-oriented electrical steel vital to power transmission. Globalization has shown its limits and its risks and it's time to act.[107]

Rare earths for me and not for thee![108]

Rare earth elements (REE) are often exhibit 1 in the case for supply chain security. In truth, the seventeen elements, including fifteen lanthanides, and scandium and yttrium, with exotic names like ytterbium, cerium, praseodymium, and dysprosium, aren't in fact rare: they are relatively plentiful in the earth's crust but are dispersed, however, and difficult to extract and purify, making their mining an expensive and dirty business. During the Cold War, the United States was the world's largest producer of REE, but operations tapered off and eventually shuttered. By the 2010s, China produced nearly 85 per cent of global REE and 95 per cent of processed REE, and the US was importing 100 per cent of the rare earth metals and compounds it consumed.[109] Dependency ratios in other countries were just as high.

As noted in Chapter 1, that dependence was alarming given the critical role of REE in the production of high-technology applications, some for defense, others for critical industrial sectors. China showed that it recognized the value of its grip on the REE supply chain in 2010, when Japanese authorities arrested a Chinese fishing boat captain who had rammed Japanese coast guard vessels and tried to flee after being stopped for fishing in waters surrounding the Senkaku islands, a group of islets administered by Japan but claimed by China. When Japan refused to release the captain (the usual protocol, but not followed because of the ramming, the flight, and a new, inexperienced government in Tokyo), China retaliated by suspending for 59 days REE shipments to Japan, the world's largest consumer of rare earth materials, which are critical to its advanced manufacturing sectors. Japanese government officials privately described the cutoff "as an 'invisible tsunami' that could threaten the viability of Japan's industrial ecosystem."[110] While that incident should have triggered a priority effort to find alternative sources, handwringing outpaced action, creating the crisis in October 2025 when Beijing announced new trade controls on those exports, and the rest of the world feared that China had a weapon "designed to *demilitarize* and *deindustrialize* the United States and other democracies."[111]

Rare earths are part of a broader group of critical materials that have two distinct features: they are of considerable economic importance for industries that consume them; yet they are also subject to heightened levels of supply risk that can interrupt physical availability and/or affordability.[112] There is no single universal list of "critical materials," but they are "of existential importance for modern, technology-intensive societies," writes Jeffrey Wilson, an Australian researcher. Their critical role in scientific and industrial applications, digital and renewable energy technologies, and defense equipment means that "they pose unique risks to the security and sustainability of an economy."[113]

The transitions to digital and sustainable economies are increasing the importance of critical materials. Advanced materials and manufacturing are central to the creation of new energy generation, storage, and distribution technologies. Computing technologies—AI, big data, cloud computing—are essential to energy distribution. No wonder then that "leaders in technological innovation are positioned to gain the most from the global energy transformation."[114] Those companies will profit financially and the network and first-mover advantages will generate even higher revenues as consumers are locked into particular technology ecosystems. Experts warn of a new geopolitics of Green Energy,[115] with developed economies facing potential critical shortages of metals that would prevent their transition to clean economies.[116]

Supply risks stem from complex value chains, monopolized markets, government policies that distort markets, and the idea that critical materials can be used as a resource weapon.[117] As Wilson details, many of those markets have a high degree of producer concentration, where a single dominant supplier accounts for over half of world supply. Examining thirty critical materials, Wilson found only one market—nickel—in which the top three suppliers accounted for less than half of world supply. China looms large as the world's top supplier of eighteen critical materials and a near monopolist (over 70 per cent market share) in five.[118] As a result, he concluded, "open, transparent and competitive markets for critical materials do not presently exist. Industrial consumers cannot be confident that international markets will provide either reliable or cost-effective access to these essential inputs. Importing governments are also dangerously exposed to diplomatic coercion by monopoly suppliers."[119]

Economic coercion, from dyads to networks

Attempts to use market power for political purposes are not uncommon. Perhaps the most famous incident was the oil embargo imposed by OPEC against countries that supported Israel during the 1973 Yom Kippur War. Their shutoff forced oil prices to skyrocket, rising 300 per cent around the world (in some countries increases were higher still) and transforming the global economy though a massive transfer of wealth to oil-producing countries and permanently altering the thinking about supply chains and dependence on fossil fuels from the Middle East. Ever the student of history, Deng Xiaoping quipped in 1992 that "the Middle East has oil; China has rare earths."[120]

In the last decade or so, Deng's successors in Beijing have been willing to flex their economic muscle to get desired political outcomes. One study identifies 152 cases of China's coercive diplomacy against twenty-seven countries and the European Union, with a sharp escalation in tactics—economic measures, such as trade sanctions, investment restrictions, tourism bans, and popular boycotts, and noneconomic measures, such as arbitrary detention, restrictions on official travel, and state-issued threats—since 2018.[121] Angry that a Chinese dissident won the Nobel Peace Prize in 2010, Beijing banned Norwegian salmon.[122] Angry that the South Korean government in 2017 defied it and agreed to deploy a US missile defense system, Korean celebrities were banned from Chinese TV, the number of Chinese tourists visiting South Korea dropped by 60 per cent, and previously popular Korean products were boycotted.[123] Angry that Australia had called for an international investigation into the origins of the COVID-19 outbreak, China banned imports from Australian abattoirs, imposed anti-dumping and anti-subsidy duties on barley (a major export), called for an anti-dumping investigation into Australian wine exports to China, and threatened tourist and student boycotts.[124] The threat of a ban on its rare earth exports prompted Ryan Grimm, a vice president of Toyota North American, to warn that "they can shut us down in two months, the entire auto industry."[125]

China isn't the only offender. Moscow has used its role as chief energy provider to press European governments. Well before the 2022 invasion of Ukraine, a study identified "at least 15 discrete instances where Russian entities used price and physical volume manipulation of crude oil or natural gas supplies—often amid political tensions—to pressure consumers located in Central and Eastern Europe and the former Soviet countries" and reckoned that the "true number of energy disruptions is significantly higher."[126] The author concluded that even though Moscow's energy weapon had not been widely successful in the former Soviet zone, "its use against Western European countries in various—perhaps more nuanced and sophisticated—forms still constitutes a strategic threat that warrants close attention from policymakers."[127] Or as former Polish Prime Minister Donald Tusk warned, "excessive dependence on Russian energy makes Europe weak."[128]

In 2019, Japan put the squeeze on South Korea when Tokyo announced that it would restrict the export of three specialty chemicals—fluorinated polyimide, hydrogen fluoride, and photoresist—that are vital to the production of semiconductors and other high-end products such as smartphone displays by some of South Korea's largest companies. Officially, the decision was a product of concerns about trade diversion and the proliferation of weapons of mass destruction; others suspected the driver was pique at Korean court rulings over forced labor.[129] South Korea's Supreme Court awarded laborers who had worked at Nippon Steel during World War II 100 million won (around $85,000) each in reparations. The ruling rejected Japan's argument that all such claims were settled "completely and finally" under the 1965 treaty that established diplomatic relations between the two countries.[130] Japan's move was a classic power play: Japanese companies account for about 80 per cent of the global market share of the three chemicals and 100 per cent of top-end supplies.[131]

And there is the United States. The US has used coercive economic measures since the early nineteenth century and embraced them frequently during the Cold War. Deployment of that tool accelerated at the end of the Cold War.[132] During the first two decades of the twenty-first century, there was "a significant upward trend in the number of people, companies, and entities placed on U.S. targeted sanctions lists since at least 2001," with about 600 total additions in 2014, 500 in 2015, 600 in 2016, nearly 900 in 2017, and almost 1,200 in 2018.[133] US leverage stems from its role as the world's preeminent economic and financial power. One analysis identified six pillars of US economic leverage: the global dominance of the US dollar, the leading role of US financial institutions, the size and attractiveness of its market, the role of US companies in global supply chains, the global investment footprint of US investment funds and companies, and financial transparency and the US legal regime.[134]

The US use of such leverage is not restricted to "adversaries," however. Washington has been ready to deploy coercive measures against allies and partners to compel compliance with US foreign policy toward third countries: these are the so-called "secondary sanctions" that punish, say, European companies for doing business with Iran and North Korea—trade that might be permitted by their home governments but which the US would like to stop. Edward Knudsen, a European researcher, warned that "there are no guarantees that the U.S. will confine its secondary sanctions to relatively minor and isolated economies," and then pointed to US restrictions on doing business with Huawei and ZTE as "dramatic reminders of America's willingness to confront allies and adversaries alike."[135]

The central role of the US in the global financial system underscores a shift in international relations that buttresses our claim that a new national security economy has emerged. While governments have long tried to practice economic coercion,[136] they most of the time used market access (or the denial thereof) as leverage. In recent years, emphasis has shifted from bilateral relationships to those defined by networks, which can be created by finance (the SWIFT system for banking), energy, supply chains, or communications (social media), to name a few examples. The first-mover advantage of digital technologies will reinforce this tendency and extend it to new areas. In networks, all countries are not created equal: some nodes are bigger, with more connections, and subsequently have more influence over other members of the network.

This is Farrell and Newman's "weaponized interdependence" and is distinguished from the earlier, crude use of "mere" market access.[137] The leading example is the US readiness to deny other countries access to its financial system and the use of the dollar as medium of exchange. Fear of being cut off stopped banks in 2007 from doing business with Banco Delta Asia, a Macau-based financial institution that the United States charged was laundering money for the North Korean government. Successive US presidents, including Joe Biden, have played the sanctions card. Tufts University professor Daniel

Drezner summed it up well, noting that "to the extent that weaponized interdependence exists, the United States has been the hegemonic coercer."[138]

Trolls and warriors: The social media battlefield

Digital threats take many forms. One that has assumed greater significance in the last decade is digital influence operations, by which adversaries seek to shape politics and social dynamics in a target country, to weaken it or make it more amenable to the acting country's ambitions. Sun Tzu extolled such efforts over 2,500 years ago when he noted that "to subdue the enemy without fighting is the acme of skill." Modern Chinese governments operationalize this logic through the "three warfares" (*san zhong zhanfa*): media, psychological, and legal warfare. Russia has adopted similar logic and tactics, led by General Valery Gerasimov, who argued that "the role of nonmilitary means of achieving political and strategic goals has grown, and in many cases, they have exceeded the power of force of weapons in their effectiveness."[139] The Islamic State's propaganda handbook enthused that "media weapons [can] be more potent than atomic bombs."[140] They've all embraced an updated version of Clausewitz: the internet is politics by other means.

These efforts go by several names: disinformation, malign influence operations, or, more colloquially, "fake news." Whatever the label, the intent is the same: to dilute, distort, or alter perceptions of events or issues, to erode public trust in state institutions, and amplify social divisions and fear. The bottom line is simple: to undermine social cohesion and resilience, to threaten domestic stability and the effective functioning of society. In 2019, the US Director of National Intelligence's Worldwide Threat Assessment sent the same message, noting that "all our adversaries and strategic competitors will increasingly build and integrate cyber … influence capabilities into their efforts to influence US policies and advance their own national security interests. … They are becoming more adept at using social media to alter how we think, behave, and decide."[141] Mounting concern prompted the World Economic Forum to add "the erosion of social cohesion" to its annual Global Risks Report in 2021.[142] It warned that "disinformation and misinformation campaigns can erode community trust in science, threaten governability and tear the social fabric."[143]

Influence campaigns are a form of asymmetric conflict. They soften up adversaries, sapping their will to fight. General Nick Carter, head of Britain's armed forces, warned that "[Our enemies'] goal is to win without going to war: to achieve their objectives by breaking our willpower, using attacks below the threshold that would prompt a war-fighting response."[144]

There is another way to assess influence operations, one that looks at them more broadly. Rather than seeing them as "mere" attempts to protect national interests, these campaigns also reflect the propagation of an ideology fundamentally alien to that of the West, one that posits a completely different relationship between the citizen and the state. The struggle then is not merely between two countries and their respective national interests but between much more elemental questions of social organization and the social contract, the right of individuals to act and think as they wish. "At the heart of this contest is the values-based distinction between democratic and autocratic governance: open and free, defined by diversity of opinion and debate; or closed and controlled, supported by censorship and surveillance."[145] The contest is as much within states as between them.

> Information technologies and architecture—from 5G telecommunications systems to data-driven surveillance both on and off social media—increasingly underpin every aspect of our lives. The architects of these information platforms influence how they are used, who can access the data flowing through them and how, and—for algorithmic platforms—what content is served up to which individuals. As the ability to construct more and more detailed behavioral profiles on each citizen advances, so too do the vectors for influence. This is a contest not just to wield digital tools but also to shape information realities.[146]

In this interpretation, the stakes are not the national interests of particular states but the very structure of the global order and the degree to which its guiding principles are liberal or illiberal values. If this framing is correct, then a focus on electoral campaigns when dealing with social media manipulation misses the larger and more important point.[147]

PART II

GOVERNMENT IN THE NEW NATIONAL SECURITY ECONOMY

3

CHINA REIMAGINES CAPITALISM

In early 2016, Aixtron, a German semiconductor manufacturer, was struggling. The company's success at the cutting-edge of the industry—its equipment deposits atoms-thick layers of chemicals to grow the crystals for chips and light-emitting diodes—had not inoculated it from a weakening market. Its share price plummeted following the last-minute cancellation of a large order by San'an Optoelectronics, a Chinese customer, reportedly because "specific qualification requirements were not achieved."[1] Fortunately, Fujian Grand Chip, a Chinese investment fund, was poised to step in as a white knight and take over the company, a role that Chinese companies were increasingly ready to assume: by 2016, Germany was the largest recipient of Chinese investment in Europe.

Relief turned to concern, however, as regulators explored the deal. Despite first approving the sale, German authorities reconsidered after the US government reportedly raised questions about ways the company's technology could be used to bolster China's nuclear program. When the US government demanded that Aixtron's US subsidiaries be excluded from any sale on national security grounds, Fujian Grand scrapped the deal.[2]

The US decision was met in Germany with equal parts dismay and delight. First eager for Chinese investment, German politicians and businesses were growing increasingly anxious that purchases would facilitate the transfer of technology and knowhow out of the country—a greater likelihood in Germany where companies tend to be smaller than the behemoths that hold desirable intellectual property in the United States. Daniel Bauer, a spokesperson for a German investor protection group, explained that "People fear that foreign investors will just transfer the know-how and have no interest in keeping the facilities."[3]

Investigations into the deal amplified that concern. "A web of relationships" among Fujian Grand Chip, San'an Optoelectronics, and the Chinese state suggested that something more fundamental and alarming was occurring. When the Aixtron purchase was announced, San'an and Fujian Grand Chip had a common investor and a financial relationship. Both had received funds from the same local government investment fund in Xiamen, and a national investment fund had provided funds for the takeover while it held a stake in San'an. Investigators have been unable to prove coordination between the two Chinese companies or any wrongdoing—such as deliberately canceling the San'an order to tank Aixtron's share price and make the company a takeover target—but the discoveries rang alarms. "It is not regular investment that is at work here," said Sebastian Heilmann, president of the Mercator Institute of Chinese Studies, a Berlin think tank. "Instead, we see governmental-program capital working behind the scenes."[4]

The Aixtrom imbroglio followed the successful acquisition of Silex Microsystems, a Swedish company that makes accelerometers, gyroscopes, and other microscopic sensors that are increasingly vital to mobile phones and cars, and are of longstanding use in satellites and weapons systems. After the purchase, NavTech announced plans to build a facility in a state-run industrial park in Beijing that would "rely on Silex's technology" to build "micro-electro-mechanical systems," a core component of China's Made in China 2025 program.[5]

NavTech didn't buy Silex outright. Instead, Silex was purchased by Ruitong Semiconductor Technologies, a newly established subsidiary of China's Integrated Circuit Fund (IC Fund), a state-run program that invests in building Chinese national capacity in semiconductors and related technologies through its subsidiary GAE. NavTech bought GAE, which gave it control of Silex. The IC Fund later invested nearly one-third of the funds required to build the Beijing facility. There were fears that the layering and nesting of investors were an attempt to obscure ownership and connection to the Chinese state and evade controls designed to ensure that such technologies are not used for military purposes. Jerker Hellstrom, an analyst at the Swedish Defense Research Agency, warned that "There is a risk that these acquisitions circumvent the Swedish export control regime ... and Sweden may inadvertently assist the Chinese military in modernizing its capabilities."[6]

On the other side of the Atlantic, another eyebrow-raising set of maneuvers was playing out in a Delaware bankruptcy courtroom. As the court debated how to dispose of the assets of ATop Tech, a California company whose automated designer helped make microchips for a range of products, including high-technology weapons systems, a mysterious purchaser appeared to buy the remains of the company.[7] Avatar Integrated Systems was incorporated in Hong Kong, and its sole director was a locally based businessman who had spent his career in the iron and steel industries, not the high-tech sector. Avatar was set up in March 2017, seemingly for the sole purpose of buying a firm like ATop. The purchase was ultimately approved, despite the objections of a former Department of Defense official who warned that the deal might be intended to build Chinese defense capabilities and could pose a national security risk to the US.[8] But if ATop's technology would have been subject to US strategic trade controls as it scaled, that was not the case when the company declared bankruptcy. As one lawyer who worked in the US Department of Homeland Security warned, "This artful maneuvering of the US legal system to circumvent CFIUS [Committee on Foreign Investment in the United States, which screens foreign investors for national security issues] review is neither new nor uncommon."[9]

By the fall of 2025, some European governments appear to have caught on. In September, the Dutch government took control of Nexperia, a semiconductor manufacturer that was purchased by Wingtech, a Chinese company, in 2019 for 33 billion RMB, the largest semiconductor acquisition in China's history and the first time a Chinese company acquired a leading global semiconductor firm.

Wingtech was put on the US entity list—which controls access to sensitive products—in December 2024 for "aiding China's government's efforts to acquire entities with sensitive semiconductor manufacturing capability." In September 2025, the US revised its export control regulations to include the subsidies of companies it was already targeting, a move that could put Nexperia on that list. Dutch authorities denied that the US prompted them to act.

The Dutch did fear that Zhang Xuezheng, the company's CEO, was stripping the company of its assets—he was closing European facilities and moving production to China—so the Netherlands government removed Zhang from office, citing national security concerns.[10] Using the "Availability of Goods Act," a 1952 Cold War law that had never before been evoked, the Hague replaced Zhang with a Dutch businessman who holds a "deciding vote" and transferred control of almost all of Nexperia's shares to a Dutch lawyer for management.[11] In response, China blocked exports of the company's chips from China, where 70 per cent are packaged, a move that threatened the auto industry, which relies on those chips in its products. A deal was reportedly worked out between US President Donald Trump and China's top leader Xi Jinping at their October 2025 meeting in South Korea to allow the exports to resume.

As the Chinese economic juggernaut has increased in size and momentum, complaints about its practices have been balanced by historical analyses that note this behavior is not unprecedented. Closer scrutiny has yielded a consensus view—outside China at least—that these practices are fundamentally different from those of states that played catchup in earlier days. Today, China is credited with forging a new model of capitalism and its emergence and success have forced other governments to respond. This evolution has contributed to the rethinking of national security detailed in Chapter 1 and is referred to in a variety of ways: techno-nationalism, strategic capitalism, and geoeconomics are but three. Whatever its name—and there's a competition to win the branding battle—it is a growth industry and will continue to be as this new approach fits poorly into the prevailing international business paradigm and the regulatory framework that it has yielded.

Unparalleled, but not unprecedented

There is a rich and detailed history of the ways that countries on the technological periphery have tried to obtain technology from those at the frontier.[12] Before the Industrial Revolution, continental European governments poached scientists and skilled workers from Great Britain. They incentivized the import of cutting-edge technology and tools. They dispatched scientists and engineers to British cities to learn what they could. Britain did the same with weavers from the Netherlands and Venice. Industrial theft and espionage have been a constant in the modernization of societies and economies.

A passive government role—a willingness to turn a blind eye to intellectual property theft from other countries—became more active after World War II, as governments more frequently intervened in the workings of the market. As economist Karl Polanyi explained, "The road to the free market was open and kept open by the enormous increase in continuous, centrally organized and controlled interventionism."[13] Forced technology transfer in particular became a core component of state policy in several Asian states and central to their success. "Japan, the ROK and Taiwan used to tightly regulate inward foreign investment with strict performance requirements, such as those on local technology transfer, local content, and joint venture arrangements and export orientation."[14] Japanese behavior looks like a model for China. "The government granted multinationals access to the Japanese market in exchange for entering into joint ventures with Japanese firms, licens[ing] their technology to Japanese firms, or forego[ing] patent protection on certain products or processes."[15]

Asian governments had no monopoly on this approach. Throughout its history, the United States has balanced two competing strands of thought: government activism, which drew inspiration from Alexander Hamilton's mercantilism, and the restraint captured best by Thomas Jefferson's belief that "the government that governs least governs best." As one economic historian concluded, tension between the two has been "resolved by putting the Jeffersonians in charge of the rhetoric and the Hamiltonians in charge of policy."[16] One student of industrial policy cut through the rhetorical smoke screen to argue that "The American state is no less active than many other states in governing the market; it is however less conventional in its activism and thereby also less visible."[17] Eric Schmidt, former chairman of Google, agrees, writing that "Many of Silicon Valley's leaders got their start with grants from the federal government," and put himself on that list.[18] Another analysis goes further, concluding that "Silicon Valley would not exist as it does today without the Cold War-era tsunami of federal defense contracts. Not only did the US government provide vast sums of money to develop computing technologies across various small 'startup towns,' it also stood as a ready customer long before these technologies were commercially viable." Silicon Valley, ground zero for capitalism, iconoclasm, and entrepreneurism, "was built on a foundation of private defense contracting."[19]

Ironies abounded at the end of the Cold War. First, the presumed superiority of the West's free-market capitalism over the Soviet planned economy was vastly overstated, even though that story became the accepted narrative. Second, rather than basking in that victory—whatever the cause—the US promptly turned its sights on Japan, such that in the early 1990s, commentators could argue that "[w]e are witnessing today a continuation of the same prosaic struggles for national commercial advantage that have characterized the world trading system since the days of Colbert and Clive."[20] In writing that sounds eerily familiar, a team at the Berkeley Roundtable on International Economy warned in 1992 that existing multilateral institutions are "ill-equipped to deal with practices of strategic trade and technology development."[21] Over the intervening three decades, concerns have not diminished. And while commentators "emphasize heterogeneity among market-based economies leading to useful distinctions amongst 'varieties of capitalism,'"[22] there appears to be "a shift toward what can be described as 'state capitalism' [which is] characterized by a system in which national governments promote the interests of leading (sometimes state-owned) companies in key national industries, constrain inward FDI [foreign direct investment], and restrict competition."[23]

While there is a case to be made for continuity, the consensus is that this moment represents something new and novel. There is ample precedent for predatory industrial policies and aggressive mercantilism, but the China challenge is different in degree and kind.

The rise of CCP Inc.

China's trajectory after Mao is an extraordinary story. Deng Xiaoping's decision to open the economy and introduce free-market reforms has improved the lives of hundreds of millions of people in China with ripple effects that have affected billions more around the world. Since that process began in 1979, China has been among the world's fastest-growing economies, with real annual gross domestic product growth averaging 9.5 per cent through 2018, doubling every eight years and lifting an estimated 800 million people out of poverty. The World Bank calls that "the fastest sustained expansion by a major economy in history." China has become the world's largest economy (on a purchasing power parity basis), manufacturer, merchandise trader, and holder of foreign exchange reserves.[24]

While growth continued, the commitment to reform faltered. By the early 2000s, major market-oriented reform initiatives were abandoned in favor of government activism designed to counter slowing growth in a largely marketized economy. "New development strategies focus on identifying and funding 'drivers of growth'—principally, high-tech industrial sectors that are strategic emerging industries, part of the innovation-driven development strategy."[25] The World Bank would conclude that "China has become one of the world's most active users of industrial policies and administrations,"[26] and employs an array of measures to protect its home market, build national champions, and help establish them as global leaders in their sectors. Some of those measures are conventional: tariffs on imports and subsidies for domestic suppliers.[27] In other cases, the government actively discourages imports of technologies for which China is deemed to "possess domestic R&D capabilities."[28] A European analysis argued that:

> through market access and other barriers, China has long managed its interdependence with the world economy in a highly strategic and limited manner: selective coupling where it needed foreign technology or competition—such as in high-speed rail and the financial sector respectively—and remaining uncoupled in sectors reserved for China's, often state-owned, national champions.[29]

Increasingly, China has embraced new and innovative ways to acquire the technologies that will help it prevail in international economic competition. The result is a hybrid innovation system that defies simple categorization and in important ways challenges political and business orthodoxy.

Its first tool is technology transfer. Governments intent on accelerating their country's economic development have often forced foreign investors to transfer technology to local companies to build knowhow and capacity. China has been especially aggressive in those efforts. As a Beijing official explained, "China's offer to the world has been straightforward [...] Foreign companies are allowed to access China's markets but they would need to contribute something in return: their technology."[30] Surveys of foreign businesses operating in China show that isn't idle talk: in a poll by the American Chamber of Commerce in Shanghai, 21 per cent of respondents reported pressure to transfer technology in exchange for market access;[31] that number was 20 per cent in a survey by the European Union Chamber of Commerce in China.[32] The Office of the US Trade Representative in its report on Chinese trade practices, noted that in "many, many private interviews with companies ... *we did not find a single instance in which companies had not felt pressure and in many cases caved into the pressure to share technology*."[33]

If the scale of the pressure is troubling, so too are the methods. Investors are squeezed by local partners who use requirements for technology transfer to play one foreign investor off another and secure the best deal.[34] The playing field is tilted further by the tendency to render China's technology transfer policies and practices more implicit, often carried out through oral instructions and "behind closed doors,"[35] and to not leave a paper trail, which makes questionable behavior difficult if not impossible to prosecute. In addition, China is charged with using discretionary and nontransparent administrative reviews and licensing processes to pressure technology transfer or force the unnecessary disclosure of sensitive technical information.[36]

At each stage of the approval process, in a wide variety of industries, vaguely worded provisions give government officials ample discretion to force technology transfer.[37] China imposes administrative licensing requirements on more than 100 different business activities, such as food and drug production, mining, or telecommunications services, for all enterprises in China. Even if a foreign investment in a particular industry is technically permitted, a foreign invested enterprise must still obtain an industry-specific license to conduct activities.[38] Despite assurances by regulators, companies are not confident that proprietary formulae or designs that they disclose will be protected. For example, environmental impact and energy conservation assessments may require expert panel reviews and sometimes involve a "pre-review" by a separate panel prior to application.[39] As a result, the Office of the US Trade Representative concluded investment approval authorities can work "in a nontransparent way with the local partner to ensure that valuable intellectual property, market channels, and other assets of the foreign investor are made available to the joint venture—often on extremely favorable commercial terms for the local partner."[40]

The counter argument is that foreign businesses choose to operate in China. Any deal is the result of clear-eyed analysis—and they actually do quite well, regardless of the terms of their investments.[41] China's payment of licensing fees and royalties for the use of foreign technology quadrupled between 2007 and 2017 and reached nearly $30 billion; the country was ranked fourth globally in such payments after Ireland, the Netherlands, and the United States, and more than Japan, Singapore, and South Korea combined.[42]

There is growing agreement that in China the line between voluntary and mutually agreed-upon technology transfer and that compelled by the government is blurred, however.[43] Foreign technology owners are deprived of the ability to bargain and set terms for technology transfer that are free from interference by the Chinese government, and many deals are decided on nonmarket-based terms that favor Chinese recipients.[44]

What isn't in doubt is the impact of that transfer. A study from the National Bureau of Economic Research found that joint ventures often generate Chinese companies' most technologically advanced and innovative procedures and products, acquired through technology transfer from their foreign joint venture partner. Chinese firms directly involved in the joint venture—the direct beneficiaries of advanced foreign technology and knowhow—generate positive externalities to other domestic firms operating in the same industry. As a result, these technology transfers make all Chinese firms more productive and competitive.[45]

Another traditional method of government intervention is the operation of state-owned enterprises (SOEs), a tool that China has used with vigor. While the initial reform agenda corralled these dinosaurs—between 1997 and 2003 the state sold off nearly half its SOEs[46]—and restructured the survivors, kindling hopes that they might be on the path to extinction, they have been fortified in recent years. The Chinese government has a constitutional mandate to "play a leadership role" in the state sector[47] and President Xi has embraced that mandate, urging in 2016 that SOEs become "stronger, better, and bigger."[48] Xiao Yaqing, the country's main regulator of SOEs in his position as chairman of the State-owned Assets Supervision and Administration Commission (SASAC) of the State Council, confirmed that framework a year later, insisting that "SOEs should stick to the political principle that all SOEs must be under the Party's leadership."[49] Accordingly, top executives of SOEs are generally CCP members, cycle between corporate and government positions, and are subject to evaluation by the CCP's Organization Department.[50]

Official Chinese government figures identified 150,000 SOEs at the central and local government level. Accounting for as much as 50 per cent of nonagriculture GDP, they dominate vital sectors of the economy, are shielded from competition, are the main sectors encouraged to invest overseas, and dominate listings on China's stock markets. One study found that SOEs made up one-half of the 500 largest manufacturing companies in China and 61 per cent of the top 500 service-sector enterprises.[51] They are sprawling enterprises: the average size of the largest 100 conglomerates in China increased from 500 companies to more than 15,000 between 1995 and 2015; among the top 1,000 conglomerates, the share of subsidiaries that are joint ventures with other firms increased from 30 to 80 per cent.[52] Their size and the protection they are provided at home has fueled their success. China has overtaken the US on the Fortune Global 500 list (124 companies vs 121 from the US), and nearly 75 per cent of these are SOEs. The combined assets for China's 96 largest SOEs total more than $63 trillion, an amount equivalent to nearly 80 per cent of global GDP.[53]

Those vast revenues can be, and often are, deployed in the service of the state. The European Commission noted in 2012 the "[Chinese] government's determination to further develop the dominant role of the state-owned economy, in particular by selectively creating large SOEs, shielded from competition domestically and expanding internationally which would serve the Government's strategic industrial policies rather than focusing on their own economic performance."[54] The 2018 report of the Office of the US Trade Representative agreed, noting that "[m]uch Chinese FDI comes from state-owned enterprises that often have different motives than simply maximizing profits. Rather, their investments often serve strategic state goals."[55] And the problem appears to be getting worse. In the aftermath of the COVID-19 crisis, the role of these companies has been resurgent. With the Chinese government emphasizing "self-reliance over openness and a free, international market," the idea that there will be economic convergence between China and the West "is not on the table at present in Beijing."[56]

While coerced transfer of technology and the use of state-owned enterprises are time-tested tools of state intervention, China has innovated in this area as well. One of China's most creative methods is the extension of state influence to the private sector so that even those actors work to advance government-defined interests. There are a variety of ways that the government has created this convergence of interests: control over access to funds has proven one of the most effective. Many transactions are funded by state-owned entities or banks, often in situations where comparable commercial financing would have been unavailable.[57] And just as foreign companies' investment decisions in China are shaped by administrative and regulatory intervention by the state, so too are Chinese private companies' overseas investment decisions.[58]

Communications between the state and the private sector are facilitated by the presence of Communist Party cells in those companies, and the employment of party functionaries in executive or managerial positions. Article 19 of the Company Law of the People's Republic of China says that an organization of the CCP may be set up in all enterprises, whether it is a state, private, domestic, or foreign-invested enterprise, to carry out activities of the CCP.[59] According to one study, in ninety-five of the top 100 private firms and eight of the top ten internet companies, the founder or de facto controller was either currently or formerly part of the People's Congress or the People's Political Consultative Conference.[60] The CCP's Organization Department reported that just 0.9 per cent of nonstate firms had CCP committees in 1998, a number that had swelled to 16 per cent a decade later. By 2013, committee presence in nonstate firms had more than tripled to 58 per cent, and by 2017 it reached 73 per cent, accounting for 1.9 million firms.[61]

The party and the private sector recognize that a mutually supportive relationship serves both well. The results are complex relationships "that defy simple categorizations as 'state-owned, or 'private.'"[62] Jude Blanchette, one of the most astute observers of the transformation of this business environment, writes that:

> since Xi's accession in late 2012, and picking up speed in 2015–16, important changes to the political, regulatory, and financial architecture that underpin China's state capitalist system have been so pronounced that it is time to replace the existing "China Inc." framework in favor of a new name, which I call CCP Inc. If China Inc. was comprised of mostly state-owned giants, the new CCP Inc. ecosystem is a dizzying mix of public, private, and hybrid-ownership firms.[63]

This evolution has made it "almost impossible" to figure out "with any precision where CCP influence ends and where firm autonomy begins."[64] The *Financial Times* put it bluntly, warning that "it's no longer possible for China's trade partners to assume private enterprises are free agents."[65]

That is how Chinese officials think. Hao Peng, who succeeded Xiao Yaqing as chair of the State-owned Assets Supervision and Administration Commission, insists that "Regardless of whether state-owned or private enterprises, they are all Chinese enterprises."[66] Or as the Office of the US Trade Representative report summarized, "the Chinese government has the means and authority to prevail (and does prevail) on Chinese firms on where to invest, what to invest, and how much to invest."[67] Trey McArver, who offers advice for businesses seeking to understand the Chinese market, explained the phenomenon from the other side of the mirror. "No company, private or state-owned, gets ahead in China without aligning itself with the party's larger goals and strategies. That is more the case than ever in Xi's China."[68]

Xi Jinping's goal is "the great rejuvenation of the Chinese nation." That demands economic strength, which in turn requires mastery of new and emerging technologies. Xi is acutely aware of that prerequisite, arguing that "Core technology is the country's important weapon, and the most critical and core technology must be based on indigenous innovation and self-reliance." And critical to our purposes, he doesn't trust the market to do the job. "Market exchange cannot bring us core technologies, and money cannot buy core technologies. We must rely on own research and development."[69]

The plans developed by the Chinese government, of which the Medium- to Long-Term Plan for the Development of Science and Technology is just one among several (components include Made in China 2025 and the Strategic Emerging Industries Plan, among others), incorporate a three-step strategy to transform China into a world leader in advanced manufacturing. In step one, by 2025, China should "approach the level of manufacturing powers Germany and Japan during the period when they realized industrialization." In step two, China should "enter the front ranks of second tier manufacturing powers" by 2035. In the third and final step, China should "enter the first tier of global manufacturing powers" by 2045, at which point China will have "innovation-driving capabilities," "clear competitive advantages," and "world-leading technology systems and industrial systems."[70]

Step one relies upon a supercharged indigenous technology development effort that is supplemented by an aggressive foreign acquisition program, both of which are guided by the state. The direction and funding come directly from government mandates by bureaucrats, at every level of government—central, regional, and local.[71] As Jude Blanchette explained, "state investors … guide and direct trillions of RMB in cash and capital into sectors deemed strategic priorities by Beijing via government guidance funds and other investment mechanisms."[72] The principal targets are industries prioritized in Made in China 2025, and include seven emerging strategic industries: energy conservation and environmental protection, emerging information industry, the biotech industry, new energy, new energy vehicles, high-end manufacturing, and new materials.[73] A broad range of actors support this effort, including SOEs, state-backed funds, government policy banks, and private companies.[74]

The sums available are astronomical. Government guidance funds channel state funding to domestic companies for R&D and overseas acquisitions: in March 2018, there were an estimated 1,800 such funds linked to Made in China 2025 that were collectively valued at $426 billion.[75] Another study estimates that in the first quarter of 2020 the total number of such funds was slightly smaller—just 1,741—but their registered target size was much bigger: RMB 11 trillion ($1.55 trillion). Even though they had only raised RMB 4.76 trillion ($672 billion), that was still nearly 150 per cent larger than previously thought.[76] One such government guidance fund was a national semiconductor fund of RMB 204.2 billion ($28.9 billion) set up in October 2019; it followed an RMB 139 billion fund that was launched in 2014.[77] An infrastructure initiative revealed in 2018 but given substance in the 2020 government work plan anticipated government spending of roughly RMB 17 trillion for "next-generation information technology infrastructure such as 5G; upgrading and 'smartening' existing infrastructure; and establishing incubators and frameworks to advance scientific research, tech development, and product development."[78] The money continues to flow: in the spring of 2025, China announced the launch of a RMB 1 trillion ($138 billion) government-backed fund to support emerging technologies.[79]

According to the IMF, China has implemented around 5,400 subsidy policies from 2009 to 2022, or about two-thirds of all measures adopted by all the G20 economies combined.[80] Those subsidies are concentrated in priority sectors such as software, automobiles, transportation, semiconductors, and more recently, green technology.[81] Subsidies for electric vehicles and other green technology goods have received widespread attention as exports have surged. China was the world's largest EV manufacturer in 2023, producing 8.9 million EVs (about two-thirds of annual global EV production) and exporting 1.2 million, making it the leading exporter of electric vehicles.[82]

Craig Allen, president of the US–China Business Council, warned that China "is engaged in a techno-utopian quest that is breathtaking in scope. … In a manner that is not constrained by the market, [China is] willing to throw vast sums of capital and unlimited numbers of engineers into developing the products of the future." There is, he writes, "a huge sum of public money to support the New Productive Forces," with estimates ranging from 1.8 per cent of GDP annually to 11 per cent of GDP over five years. These sums are four times larger than those of the next highest incentive supplier, South Korea.[83]

While the primary use of state funds is helping domestic enterprises compete, those monies are increasingly deployed overseas to acquire technology targets. In most cases, when the monies make their appearance in a business deal, they appear to be in the hands of private enterprise; in fact, the state is one or two layers away, its role concealed behind an opaque network of ownership and funding structures.[84]

Helping to guide that spending are more than 140 "science and technology diplomats" stationed in embassies and consulates in fifty-two countries who monitor their host country's technological breakthroughs, identify investment opportunities for Chinese firms, and act as brokers for many deals. They advise Chinese companies "to purchase equity, invest specific amounts of money, or strike patent license agreements with foreign companies,"[85] all in ways that align with the Chinese government's development objectives.[86] Those diplomats "form the outward-facing portion of China's broader technology transfer ecosystem, and monitor scientific breakthroughs, technology enterprises, and other forms of innovation that may be of interest to the Chinese government."[87]

The collection and distribution of science and technology information is a longstanding Chinese national priority. "By 1985, there were 412 major science & technology intelligence institutes nationwide [in China] ... employing ... 60,000 workers ... investigating, collecting, analyzing, synthesizing, repackaging, benchmarking and reverse engineering." The National Internet-based Science and Technology Information Service Systems has 26 million holdings of foreign journals, patents, and reports that are available to the public. "Chinese exploitation of foreign open-source science and technology information is a systematic and scale operation making maximum use of diversified sources: scanning technical literature, analyzing patents, reverse engineering product samples and capturing conversations at scientific meetings."[88]

Searching for a window

Traditionally, attention and outrage focused on China's questionable and in some cases illegal efforts to acquire technology and intellectual property. More recently, however, the spotlight has shifted to the broader and more systematic effort to acquire that knowhow through legal means.[89]

The first instrument is foreign direct investment—investing in or buying companies. According to China's Ministry of Commerce, the country's overseas foreign direct investment (OFDI) flow ranked second and the stock was the third-largest in the world. By the end of 2019, China's OFDI stock totaled $2.2 trillion, less than one-third of that of the United States ($7.7 trillion) and slightly less than that of the Netherlands ($2.6 trillion). Chinese OFDI flows accounted for more than 10 per cent of the global share for four consecutive years and 10.4 per cent in 2019; the stock held steady at 6.4 per cent.[90] By the end of 2023, China's OFDI stock reached $2.96 trillion. China's Ministry of Commerce and the State Administration of Foreign Exchange estimated that China's total outbound direct investment in 2024 reached $162.78 billion, a 10.1 per cent increase over the previous year (or RMB 1.159 trillion, up 11.3 per cent in yuan terms). Chinese investors made nonfinancial direct investments in 9,400 overseas enterprises across 151 countries and regions, totaling $143.85 billion, a 10.5 per cent increase (or RMB 1,024.45 billion, up 11.7 per cent).[91]

China's total investment in US technology (electronics, information and communications technology, biotech, and energy) for 2006–16 totaled $35 billion, and in 2016 alone the amount was about $8.5 billion.[92,93] Between 2012 and mid-2017, China-based investors engaged in tech investments worth $19 billion throughout the US, across 641 deals, many of which reflected the Chinese government's strategic priorities of AI, robotics, and augmented or virtual reality.[94] Between 2013 and 2016, China-based firms leveraged state funding to attempt to acquire or invest in at least twenty-seven US semiconductor firms totaling more than $37 billion.[95] That pace has slowed as US administrations have become alert to the implications of Chinese investments and subjected them to greater scrutiny, as is detailed in later chapters.

In their study of Chinese technology transfer, Michael Brown and Pavneet Singh note that China invests far more in US early-stage technology companies than any other country outside the US, using venture firms, Chinese companies, private equity, special purpose vehicles, and acquisitions.[96] They highlighted Chinese investment in early-stage technology deals: in 2015 Chinese investors participated in 271 deals, with a total deal value of $11.5 billion, nearly 16 per cent of the value of all technology deals in that year ($72 billion), with a substantial increase in these investments during 2010–14, when they ranged between 1 and 6 per cent. They identified more than 500 Chinese-based or affiliated entities investing in US early-stage companies in 2017.[97] According to Brown and Singh, between January 2010 and October 2017, China-based investors participated in 1,201 financings in the US and contributed some $31 billion in venture-backed funding, about 8 per cent of all deals. Chinese investments peaked at 16 per cent in 2015 and then fell to about 10 per cent in 2016 and 2017.[98] Another study found that between 2000 and May 2018 Chinese investors targeted strategic technologies in 78 per cent of all US venture capital funding rounds involving a Chinese investor (out of more than 1,200 funding rounds with Chinese participation).[99] Chinese firms invested in at least fifty-one US AI startups and firms from 2010 to 2016.[100]

Early-stage investments are critical. Brown and Singh point out that Chinese firms are investing in the same technologies as US venture capital firms—the new and emerging technologies that are the cornerstones of the digital economy: AI, autonomous vehicles, augmented/virtual reality, robotics, blockchain, and genetic engineering.[101] And they note that the investment data is incomplete, since investors don't invest in every deal they see.

> It's likely that Chinese investors, in aggregate, have seen *upward of half* of recent US venture financings; Chinese investors have a broad view of US innovation across a range of technologies ... This also means that Chinese investors are learning about foundational technologies *at the same time and at the same rate* that the US does—which precludes any time-based advantage for the US with these technologies.[102]

The US isn't the only target. One analysis concluded that the main objective of most Chinese FDI (primarily by SOEs) in the Netherlands was not to gain access to the Dutch or EU market but instead to access advanced technology and established global innovation networks.[103] An assessment by the Bertelsmann Stiftung found that 64 per cent of analyzed Chinese mergers and acquisitions in Germany between 2014 and 2017 aligned with Made in China 2025 (MIC2025) key areas, especially energy-saving and new energy vehicles (20.5 per cent), electrical equipment (18.8 per cent), biomedicine and medical devices (16.1 per cent), and high-end machinery and robotics (15.2 per cent). And the proportion of MIC2025-relevant investment increased markedly after the announcement of MIC2025 in 2015.[104]

Greater scrutiny of and restrictions on their investment in the United States prompted Chinese companies to focus more on Europe. Chinese FDI in the EU and UK reached EUR 10 billion in 2024, a 47 per cent increase from the previous year and the first increase since 2016. Much of that increase was in EVs, but that industry has reached saturation and analysts anticipate greater attention to the renewable energy sector.[105]

As Scott Kennedy explained, Beijing's leaders encouraged the country's investors to search for "emerging companies that have technologies that may be extremely important … but aren't proven."[106] They are increasingly creative as they do so, skirting the lines that have been drawn and evading the mechanisms designed to enforce them. Providing seed capital or making bankruptcy purchases as in the ATop case at the beginning of this chapter have been two ways "to circumvent the labyrinth of federal regulations designed to prevent foreign investment and technology acquisition that impede US national security."[107] "They have diversified to look for smaller targets," Kennedy said. "Those things typically do not generate a CFIUS reaction. That is part of it."[108] Or as Senator John Cornyn warned, "The Chinese have figured out which dual-use emerging technologies are still in the cradle, so to speak, and not yet subject to export controls."[109] China has "weaponized" its investments in America "to vacuum up US industrial capabilities from American companies," he complained.[110]

A second method to acquire technology is talent recruitment. This includes initiatives that target Chinese and foreign nationals for professional opportunities at Chinese universities and companies in priority sectors.[111] There are hundreds of these programs at various levels of government—national, provincial, municipal—and they are "woven into government and [Chinese Communist Party] organs, state-owned enterprises, defense research and academic institutions, national laboratories, 'private' industry, domestic and overseas 'NGOs,' and global diaspora organizations."[112]

The Chinese government launched Project 111 in 2006 to recruit 1,000 foreign experts in strategic sectors from the world's top 100 universities and research institutes. By 2009, it had recruited thirty-nine Nobel Prize winners and 591 academics.[113] In 2008, the Thousand Talents Program was established; by 2018 more than 7,000 foreign researchers had participated in the project.[114] As international attention focused on the program, the resulting concern prompted the Chinese government to have it absorbed by the High End Foreign Expert Recruitment Plan.[115] While the name has changed, the mission continues.

In China, "returnee parks," "pioneering" parks, tech transfer centers, and technology "incubators"—large physical complexes that collectively number in the several hundred—all provide physical and intellectual space for commercializing foreign technology.[116] Professional associations—"ethnic guilds focused on a technical association or a geographic area"—have memberships that claim to be in the thousands. There are over twenty such groups in the US and another two dozen worldwide. The Chinese-language versions of their websites usually include paeans to China as the country of their ancestors and pledges to support its development. These groups are associated with local PRC consulates, and they in turn broker investment and joint research opportunities.[117]

The government funds networks of entrepreneurs, like TechCode, which is "committed to breaking down geographic barriers and eliminating potential inequalities of international cooperation." It's a global program that uses an online development platform for projects focused on China's development.[118] There are research centers, such as the US–China Clean Energy Research Center that promotes a sustainable environment and clean energy. Chinese firms are doing the same. Baidu established the Institute for Deep Learning in Silicon Valley to compete with local companies for Silicon Valley talent.[119]

Finally, there are Chinese students in US colleges and universities: 328,000, or one-third of all foreign students in 2016.[120] Breaking it down, in doctoral programs, 57 per cent of engineering, 53 per cent of computer science, and 50 per cent of math and statistics candidates were foreign; half of these were Chinese. Forty-five per cent of STEM (science, technology, engineering, and math) undergraduates are foreign and one-third of these are from China, meaning that about one-quarter of the graduate students in STEM fields are Chinese foreign nationals.[121] Many of these are ordinary students: some are not. A 2018 report from the Australian Strategic Policy Institute determined that the Chinese military has since 2008 sponsored more than 2,500 scientists and engineers to travel to universities in the US and elsewhere as students or visiting scholars.[122]

The crime is what's legal!

Responses to this lengthy list of policy innovations and indictments take several forms. Some argue, as noted, that business deals and the concomitant technology transfer are consensual agreements in which both sides weigh their respective interests and proceed (or not) as they deem best. Whatever leverage the Chinese government may use on behalf of its business interests must be acceptable to competitors because they are prepared to proceed on those terms. A second response is that the Chinese behavior is not illegal. Again, that may be true in some cases—although a lack of evidence is not necessarily dispositive—but that is another way of restating the problem. As the saying in Washington goes, "the crime isn't what's illegal, but what is legal!"

A third and more powerful argument is that Chinese behavior is not much different from that of other countries that sought to supercharge their economic development to catch up with international powers of the time. History shows that this claim is largely true and Beijing's behavior is like that of other countries—in particular those that are today's leading nations—fueling countercharges that complaints are hypocritical attempts to either contain China or halt its development, or worse, pure racism.

While Chinese behavior may be similar, its policies can be distinguished from those of its precursors. The most recent and seemingly analogous parallel is the US–Japan frictions of the 1980s and 1990s. Japan's extraordinary economic success, the ascendance of Japanese companies to the tops of global whatever lists, and a wave of trophy purchases—the Pebble Beach golf club, Rockefeller Center in New York City—fueled intense unease in the West (and yes, a chunk of the motivation was racism) amid fears that Japan's rise signaled the end of the American Imperium and the emergence of a new era, one led by Asia, which was following Japan's development model.

In fact, however, this was "manageable disruption." While their policies were mercantilist, Asian challengers were relatively small economies whose political systems fit within the prevailing (accepted) paradigm: they were evolving toward greater openness, liberal democracy, and the rule of law.[123] While there has been a long-scrutinized divergence in Anglo-American capitalism and its continental European cousin, there was never doubt that European practice was rooted in a democratic culture.[124] But, as one legal scholar concludes:

> the same cannot be said for China. Postwar Japan was a working democracy; China is governed by a Leninist clique. Japan is a midsized economy occupying an archipelago. China is subcontinental, with the largest population in the world. ... China's ambition is far more assertive and influential than anything Japan planned near the height of its post-war industrial growth.[125]

Mark Wu, another legal scholar who served in the Biden administration as an advisor to the US Trade Representative, also dismisses the claim that Chinese policies are merely a variant of "state capitalism," a commonly accepted model in which "the state acts as the dominant economic player and uses markets primarily for political gain." Russia, Saudi Arabia, Venezuela, the UAE, Iran, and Ukraine are all considered examples.[126] Instead, he argues that "China's economy is fundamentally different—even unique."[127] That uniqueness is attributable to several factors, and the creativity of the instruments created and deployed among them. But most observers agree that the key issue "is not so much that the Chinese government is willing to subsidize and support such industries, but that the size and scope of such activities goes far beyond the classical conceptualizations of industry subsidies or protection."[128] One damning indictment concludes:

> China has been developing a systematic and coordinated approach of institutional, market, and bargaining strategies to shape its global technological leadership, and bring the benefits back home for its domestic players. ... By focusing primarily on state-led interventionist policies to acquire technology advances that originate in other countries, thereby supporting its national objectives, China has moved beyond traditional conceptualizations of techno-nationalism.[129]

That divergence has important consequences. First, it threatens the way that business is conducted. Chinese protection and assistance are of such a scale that it transforms the international business environment. Companies with state backing "have a competitive advantage over conventional acquirers because of their access to lower-cost state finance and the implicit promise of state support if the acquisition is not successful in income statement terms."[130] Chinese companies backed by the resources of the state have been called "national strategic buyers" (NSBs) because their objective is to further the interests of a nation-state in the pursuit of national industrial policy or national security concerns.[131] Information asymmetries—one side's inability to see the resources and interests that the other brings to the table—create unfair advantages, as properly valuing a deal is difficult if not impossible when the purchaser's motives are not apparent.[132] As a result, "large-scale NSB acquisitions are perceived by other governments as threatening and as violations of the existing order in cross-border mergers and acquisitions (M&A)."[133]

This commitment to public policy objectives by ostensibly private actors also distinguishes China's behavior from that of governments that have sovereign wealth funds.[134] There are codes of conduct for such entities.[135] And, in the case of the most successful of such funds, Singapore's Temasek, the fund is run according to market principles even as its profits are applied to government objectives.[136]

Which brings us to the second consequence: Chinese behavior threatens the legal and regulatory structure—and by extension the business practices that support them—of the global economy.[137] Deals that feature hidden stakeholders and obscured interests erode foundational assumptions of globalization.[138] Wu insists that "WTO rules are not equipped to handle the range of economic problems associated with China's rise." He asks, "how do we determine what entities count as an extension of the state? What if the links run through only the Communist Party, but not the state? What if the links are only informal, and no direct control mechanisms exist?" These questions are outside the purview of existing regulatory frameworks,[139] and thus the "highly distinctive traits resulting from China's political and economic systems" yield an actor that "the multilateral trade and investment regimes that took shape in the post-war period simply do not contemplate."[140] The Federation of German Industries declared that China's hybrid economy is a powerful enabler of a "systemic challenge" to the West.[141] Or as two legal scholars argue, "China's entry into the global M&A market threatens the fundamental assumptions of the current permissive international regime."[142]

Some blame China's rise for the splintering of the global trade system. The inability of institutions like the World Trade Organization to check Chinese excesses has undermined trust in those bodies, but it has pushed governments toward alternatives such as the Comprehensive and Progressive Agreement on Trans-Pacific Partnership (CPTPP, the successor to the Trans-Pacific Partnership). "Countries are using mega-regional trade agreements to define new rules regulating state-owned enterprises. ... one source has acknowledged that the TPP's legal provisions on SOEs were drafted clearly with Chinese SOEs in mind."[143]

The third consequence of Chinese behavior is that it threatens the national security of other countries. China's policy directly ties private business and economic outcomes to national standing and status. "National Strategic Buyers, applying a more directive industrial policy, could capture key inputs and channel them to NSB-home country advantage." Purchases of the scale that China is pursuing would shift the location of innovation, and thus durable economic growth, from the US to China.[144] After all, China's primary goal is the creation of self-sufficiency and cooperation only insofar as it creates dependencies *upon* China. As Xi Jinping explained, "we must tighten international production chains' dependence on China, forming a powerful countermeasure and deterrent capability against foreigners who would artificially cut off supply [to China]."[145]

There is another response to Chinese efforts to direct investment and stimulate the country's economic development: derision or dismissal. In some cases, this reflects orthodox views of industrial policy regarding associated dangers of corruption, capture, and inefficiencies. The views of Gary Hufbauer, economist at the Peterson Institute for International Economics, are typical: "There is not much evidence that government has better insight than private firms on choosing the best frontier technologies."[146] Other observers believe that China cannot innovate or that its attempts to integrate cutting-edge technology will fail.[147] Stated most elegantly, they assert that "China's appropriation of foreign creative resources has been assumed—in the United States especially—to be an Achilles heel that guarantees China will always trail liberal democracies in innovation and in the national power that innovation is believed to support."[148]

One assessment of Made in China 2025 suggests that there is basis for skepticism. That high-profile effort to "transform China into a global leader in advanced manufacturing by 2025" had mixed success. It "largely succeeded" in reducing import dependencies, but the country remains "highly dependent on foreign companies in many critical sectors." While competitiveness improved in many low- and mid-tech sectors, "in most MIC25 sectors, Chinese companies generally lag their foreign counterparts in global revenue, market share and cutting-edge technologies." And even though Chinese companies "made significant strides in closing the gap with foreign firms and advancing toward the technological frontier. ... Chinese firms have yet to reach parity in many MIC25 sectors."[148]

That record is not grounds to dismiss China's efforts more generally. In their analysis of China's strategy to acquire foreign technology, William Hannas and Huey-meei Chang of Georgetown University's Center for Security and Emerging Technology concluded that "China is building indigenous S&T capacity and can innovate on its own terms,"[150] and China uses a different model of innovation, one that reflects its particular culture and history, but dismissing it because it deviates from practice in the West is a mistake. Hannas and Chang argue that "This alternate form of creativity in which China clearly excels accounts for most of what happens in science and should be acknowledged alongside the idealized ('breakthrough') form of creativity." Significantly, they "also believe the West's creative advantage is overstated."[151] The evidence detailed earlier of China's technology successes suggests that they are right.

4

GOVERNMENTS RESPOND

PLAYING DEFENSE

A new mindset has emerged, one that recognizes the enhanced and variegated links between the economy and national security. Governments around the world have been adopting measures to tighten cybersecurity, control trade, scrutinize investment, or check other ways to transfer intellectual property. They are taking vigorous steps to spur domestic innovation, rebuild or shore up manufacturing capacity, and aggressively promote their normative and policy preferences in regional and global forums. These are articulated in policy declarations of all stripes and colors, a list that includes national economic statecraft strategies, cyber strategies, research funding strategies, supply chain resilience strategies, democratic protection strategies, critical materials strategies, critical infrastructure protection strategies, critical and emerging technologies strategies, etc.[1] In other words, governments are taking seriously the challenges identified in Chapters 2 and 3.

Governments have several sets of options. One group is "defensive." These measures are designed to protect home markets and indigenous products and ideas from predation. They include strategic trade controls that restrict access to vital technologies; investment controls that limit opportunities for foreigners to invest in domestic businesses; regulations that reduce access to universities and other research facilities; and various initiatives, focused on particular concerns like supply chains or telecommunications networks, that are intended to reduce vulnerability to foreign interruption or subversion. They also include efforts to counter disinformation campaigns. Those are the subject of this chapter.

A second set is "offensive" and designed to build national capability, rather than merely protect what exists. This group includes strategies to promote innovation, or industrial policy more generally, as well as efforts to create parallel or competing initiatives that give third-country consumers options when considering technology options. Going on the offensive can include aggressive litigation strategies to force change on trade partners, or forward-leaning efforts to block a competitor's access to investment funds. There is also an offensive component to efforts to promote cybersecurity, an approach that is deeply intertwined with notions of deterrence (and is contested on many levels). These approaches are explored in more detail in the next chapter.

These distinctions are not definitive; one person's offense is another's defense. And, importantly, while all are necessary, no single line of effort is sufficient. A multidimensional strategy that integrates various approaches is required. Equally necessary is a multinational strategy that engages like-minded governments to act together to protect and preserve both national interests and a multilateral order that stands atop a foundation of liberal values.

Cybersecurity strategies

As a first step, any government that is serious about understanding and preparing for the new national security economy will develop a national cybersecurity strategy. Most have. In 2012, an OECD assessment of ten national cybersecurity strategies concluded that cybersecurity had become a national policy priority,[2] and by 2024 132 countries had a national cybersecurity strategy, among them the United States, China, and many EU member states.[3]

Those strategies share common features: enhanced governmental coordination at policy and operational levels; reinforced public–private cooperation; improved international cooperation; and respect for fundamental values. They reflect "the consideration that cybersecurity addresses the protection of the society as a whole and requires a whole of government integrated approach."[4] A national strategy is one way of communicating to national stakeholders the main cyber threats that they and their country face.[5] While threat perceptions vary, "all strategies result from the recognition of increased cyber risks, i.e., increased cyber threats, vulnerabilities and potential impact on the economy and the society."[6] Or, to nestle this development squarely in our concerns, central to government thinking has been the rising salience of economic aspects of cybersecurity.[7]

Strategic trade controls

A second frequently used tool is strategic trade controls, or export controls.[8] These historically focused on "dual-use" items—those with civilian and military applications, reflecting the self-evident (and compelling) logic that one shouldn't sell adversaries the means to make themselves stronger. Western governments developed and honed instruments during the Cold War to prevent technology leaking to the Soviet bloc. Having recognized that national efforts would fail if the same technology could be acquired from other nations, multilateral coordination was a priority. The first such instrument was the Coordinating Committee for Multilateral Export Controls or CoCom. It dissolved in 1994, following the end of the Cold War, but was revived a few years later by a larger, looser framework, the Wassenaar Arrangement. Now there is the Australia Group, the Chemical Weapons Convention, the Missile Technology Control Regime, the Nuclear Suppliers Group, and the Zangger Committee. While most efforts concern materials and technology needed for weapons of mass destruction, some focus on conventional weapons.[9]

There is a growing belief that those strategic trade control regimes are no longer fit for purpose. Part of the problem is the lengthening list of new and emerging technologies with potential or suspected military applications; keeping up is tough. But it's not only difficult to identify possible military uses: governments now worry about technology being used for security purposes that are not military in nature. As

European analysts explained, "China's ambition to be a world leader in certain technologies … has led to concerns, including about the potential erosion of Western norms governing the use of emerging technologies."[10] As a result, in June 2021 the US restricted the export of technology that could be used for facial recognition, a tool for surveillance and repression but of limited strictly military utility.[11]

A third problem for the strategic trade control regimes stems from the shifting nature of strategic competition. While military capabilities were once the focus, today "competition in technology has become a central component of the relationship between powerful international actors."[12] Kevin Wolf, former US assistant secretary of commerce for export administration, argued that revisions in US export control authorities in 2018 "stem from the comprehensive Chinese strategy described in the 2015 'Made-in-China 2025' plan. In sum, China's … stated goal is not to join the ranks of the leading high-technology countries but to replace them as such."[13]

In the United States, those shortcomings produced the Export Control Reform Act (ECRA), part of the FY2019 National Defense Authorization Act. ECRA aims to "maintain [US] leadership in the science, technology, engineering and manufacturing sectors, including foundational technology that is essential to innovation."[14] ECRA "significantly expanded" the authority "to regulate and enforce export controls" by focusing on "emerging or foundational technologies" essential to the national security of the United States.[15] It also requires the US to coordinate with allies and international export control regimes to encourage conformity and convergence across lists that limit access to "emerging technologies." It has largely done so, working with multilateral groups as it expanded its own lists or restricted items.

In addition to creating new categories of restricted items, the US has increased reliance on the BIS (Bureau of Industry and Security) Entity List,[16] which identifies foreign persons—a legal term that includes businesses, research institutions, and government and private organizations, as well as individuals—"that are subject to specific license requirements for the export, reexport and/or transfer (in-country) of specified items."[17] Those are the "bad guys" that shouldn't get advanced technologies. The Entity List is "one of the most potent mechanisms of US export control policy,"[18] and its use has expanded throughout the Trump and Biden administrations.[19,20] In September 2025, Jeffrey Kessler, undersecretary of commerce for industry and security, warned that more restrictive measures were coming. The effort to close loopholes and "ensure that export controls work as intended" has been called the "most dramatic expansion of US export control regulations in years."[21]

An increasingly powerful—and controversial—tool is the Foreign Direct Product Rule, which restricts access to items that don't contain controlled US technology but were designed or manufactured with the assistance of US products, software, technology, equipment, or materials. In other words, these are products that have US intellectual property inside but might not have been made by US companies. This is the regulation that China has copied and employed in 2025 to devastating effect for rare earth exports, triggering the outrage by US officials detailed in Chapter 1. The Trump administration also tightened rules for "deemed exports," or trade with foreign citizens who are lawfully living and working in the United States. Previously, foreigners without military affiliations could work in cutting-edge semiconductor or telecommunications sectors without great concern. No longer; now businesses need permission to give them access to this technology as well.[22]

Efforts shifted into overdrive in October 2022, when the US updated and expanded restrictions on exports of semiconductors and chip-manufacturing equipment to China. The Biden administration proceeded alone after months of discussions with Japan and the Netherlands, despite concerns that companies from those countries could backfill—replace the lost US business—which would undercut their effectiveness. Eventually, the two governments agreed to follow suit, and the following summer Japan imposed even more expansive restrictions on those exports to China. The Netherlands stepped up in September. Washington later fine-tuned its policy to align its rules to close gaps among the three governments. Since then, the US has progressively restricted access to semiconductors used for AI and the equipment needed to manufacture them, measures that have been called "a bombshell."[23] Japan continues to tighten its regulations as well.[24]

These changes amount to a "paradigm shift" in US policy.[25] That judgment reflects the belief that the traditional "sliding scale" approach, by which the US would try to stay a few years ahead of adversaries in terms of technological capabilities, should be discarded for one in which the US tried to "maintain as large a lead as possible."[26] Commerce Secretary Gina Raimondo captured the spirit of the new thinking when she declared that "Our national defense is more than guns, missiles, tanks and drones. It's technology, it's innovation, it's working with our allies." Turning to China, she added, "We cannot let them catch up, so we are going to deny them our most cutting-edge technology." More bluntly still, "We can't let China get these chips. Period."[27]

European governments worked with the US throughout the Cold War to restrict Soviet bloc access to key technologies. European Union policy is guided by Regulation 428/2009, which creates a common list of dual-use items, destinations, and guidelines for member states.[28] Each national government must create its own export control regime and issue its own export licenses. From 2013, the EU has debated modification of its strategic trade controls, concerned about the proliferation of information and communications technologies. But unlike the US, European discussion centered on the protection of human rights and ways that those technologies can be used for surveillance and repression.[29] In November 2020, the EU Council and the European Parliament agreed on a draft revision that "explicitly acknowledged human rights implications associated with the export of cyber surveillance items;" it became law in Dual Use Regulation 2021/821, which established "cyber surveillance items" as a new category on dual-use lists. The EU has also used export controls against Russia following Moscow's invasion of Ukraine. Individual states are also acting, with the Netherlands, home to chip-equipment manufacturer ASML, announcing in early 2025 that it would expand export controls on advanced semiconductor equipment.[30]

For all those efforts, Europe is being squeezed as it balances competing demands from its ally and partner against those from China, which has pressured European governments to ignore Washington, pursue strategic autonomy, maintain trade with China, and enjoy the economic opportunities it creates.[31] Thus far, Europe has rebuffed

Chinese entreaties. European Commission President Ursula von der Leyen laid out EU thinking in 2023, explaining that:

> there are some areas where trade and investment poses risks to our economic and national security, particularly in the context of China's explicit fusion of its military and commercial sectors. This is true for certain sensitive technologies, dual-use goods or even investment which comes with forced technology or knowledge transfers. ... We need to ensure that our companies' capital, expertise and knowledge are not used to enhance the military and intelligence capabilities of those who are also systemic rivals.[32]

China has its own export controls. While it remains outside most multilateral arms control regimes,[33] it has restricted the export of sensitive products, with a strategic trade control law going into effect on 1 December 2020. This export control regime has been called "a strategic pivot for the country as it matures in its view of its own national security and its place in the world as a competitive producer of sensitive goods."[34] The law gives the Chinese government policy tools and justifications to deny and impose terms on foreign commercial transactions—inside and outside China—on the grounds of China's national security and national interest. Critics note the term "national security and interests" is undefined and its scope unknown; it "seems almost intentionally vague ... to allow Beijing to retaliate against countries or companies it deems as having broadly violated its export controls."[35]

The law also authorizes reciprocal measures if "any country or region abuses export control measures to endanger the national security and interests" of China.[36] It appears to be "aimed at creating a Chinese policy counterweight to the US government's use of export control authorities to restrict the transfer of US dual-use technology to China, including provisions for retaliatory action and extraterritorial jurisdiction."[37]

Shortly before passing the export control law, China updated its list of technologies prohibited or restricted from export. New items included: biotechnology, pharmaceuticals, and medical equipment; 3D printing; construction, petroleum, and power equipment; machine tools; high-speed wind tunnel design; aerospace bearings;

unmanned aerial vehicles; space-related remote sensing image acquisition, measurement instruments, and data transmission; vacuum technology; mapping; information processing technologies (e.g., personal interactive data algorithms, speech synthesis, artificial intelligence-based interactive interface, voice evaluation, and intelligent scoring); and cryptographic and cyber-related technologies. Many of those items are on the US list of emerging technologies, and all are prioritized on China's national development plans.[38]

In another effort to beef up its muscle, China created its version of the BIS Entities List. It is intended "to identify and respond to entities that endanger China's sovereignty, security, or development; violate 'normal' market transaction principles; and cause serious damage to the legitimate rights and interests of Chinese companies, organizations, or individuals." Observers warn that justification for inclusion on the list appears to be much broader than that of the US.[39]

Subsequently, in June 2021, responding to US sanctions, the National People's Congress passed legislation that creates a legal framework for retaliation against foreign countries that impose sanctions on Chinese companies or individuals, allowing Chinese firms to seek compensation in response.[40] While many features of the law are unclear, three stand out. First, it is potentially sweeping in scope, extending even to individuals who write articles that back sanctions if they "violate international law or basic principles of international relations and interfere in PRC domestic affairs." Read that again: it's very open-ended. Second, it prohibits companies operating in China from complying with EU or US sanctions and exposes them to civil lawsuits for damages caused by such compliance."[41] Third, it signals a downward spiral of actions and reactions between China and other governments.[42] As explained by the *Global Times*, a nationalist tabloid, the legislation "aims to safeguard China's rights and dignity. It also aims to protect Chinese institutions and individuals from the suppression and bullying of foreign forces. ... It enables China to stand with all countries and peoples of the world who suffer from hegemony."[43] It also means that individuals and businesses can be squeezed from two directions: damned if they trade with entities sanctioned by one government and damned if they don't by China.

This need to be alert to vulnerability from opposing—and unexpected directions—is also evident in the incident described in Chapter

2 in which Japan cut off exports of key chemicals to South Korea. While the toxic dispute over history was driving the controversy, Japan has insisted all along that the measures were "undertaken in order to properly control certain exports and transfer of technologies with potential risk of being diverted for military use."[44] The complexity of global trade and international supply chains means that strategic trade controls will be an ever more salient consideration for governments and businesses in the new national security economy.

Investment screening

A third tool that governments use with increasing frequency is investment screening. One of the pillars of the postwar economic order is the belief that the free movement of capital promotes prosperity and growth for both the recipient and the investing country. This preference has co-existed with a recognition that foreign investment is not unvarnished and some sectors of the economy deserve (or demand) protection from foreign investment. These so-called "strategic sectors" are deemed so essential to national sovereignty and security that international law accepts that they can be denied to foreign purchasers.[45] The definition of "strategic sectors," like that of "national security," varies; traditionally, however, the assumption has been to allow access. That is no longer the case.

In the last decade or so, there has been a "gradual and then steep increase in government attention to threats to essential security interests that occasionally comes with international investment."[46] Blame "geopolitical and geoeconomic changes in recent years—especially the emergence of new actors or foreign investors, a more assertive stance of some emerging economies, and new technological developments"—for the new sensitivity.

The COVID-19 pandemic accelerated this tendency. Many of the changes address health infrastructure and the need to prevent future supply shortages such as those that occurred during the first phases of the pandemic. In mid-2021, critical health infrastructure was or would soon be subject to scrutiny in twenty-one OECD countries that screen inward FDI for security reasons, an increase from fourteen just a few months earlier. Acquisitions of biotechnology or medical devices companies were being added to review authorities in twenty-one OECD countries, up from eleven countries before the pandemic.[47] In other cases, governments introduced or strengthened FDI screening mechanisms to prevent acquisitions in any sector where companies might be rendered vulnerable by conditions created by the pandemic.[48]

Inclusion of health infrastructure is part of a more general trend. Governments now examine investments across a much broader section of the economy, have lowered the threshold at which scrutiny is required, and rules are more detailed and sophisticated and geared toward routine applications. The list of industries now subject to screening should look familiar: strategic industries and critical infrastructure, such as energy production and supply, water supply, transport, telecommunication, mineral resources, or media. Investment screening has been extended to foreign acquisitions of domestic technology and knowhow, such as AI, robotics, semiconductors, cloud computing, 5G, quantum technology, computing hardware, nanotechnologies, biotechnologies, or satellites and aerospace—all emerging technologies. Also of growing interest is foreign investors' access to sensitive data of domestic citizens, an issue that undermined the initial public offering in New York of Didi Chuxing, the Chinese ride sharing service, in the summer of 2021.[49]

The OECD observed that "a transformational change ... has emerged over the past decade in many countries, deepening markedly from 2016."[50] Until then, open investment was the norm. In 2016, however, language "became subtly but steadily more assertive and references to reciprocity considerations, to the 'end of naivety', and to 'strategic' enterprises—rather than 'sensitive' ones—entered the discourse ... Whereas foreign investment was previously associated with opportunities, it is now occasionally associated with sell-offs and the loss of critical assets and technology."[51]

The result is "unprecedented policy making activity" in this area. By 2025, there were more than fifty investment screening regimes, and over 100 jurisdictions had some form of investment screening rules.[52] This includes more than 80 per cent of OECD members, up from 42 per cent a decade ago, and twenty-four of twenty-seven EU member states.[53]

New screening criteria reveal a considerable stretching of "national security." For example, Japan will weigh up the "smooth operation of the economy," the government in London will consider the "stability of the UK financial system," while Canada will examine the "net benefit," as Chinese authorities study "national steady economic growth" and "the R&D capacity of key technologies." A growing number of jurisdictions take into account the potential social costs of foreign investment: Portugal is considering the "security of supply in services fundamental to the national interest"; Poland will consider "the needs of the population to protect health and lives" and "environmental protection"; Finnish authorities will look at the "fundamental interests of society"; and South Korean law will weigh "Korean morals and customs."[54]

Those authorities are not just for show. In Europe, there were 3,136 requests for authorization or cases initiated by the authorities themselves, up from 1,444 in 2022. 41 per cent were formally screened, and the remainder were deemed ineligible or did not require formal screening.[55] In the US and EU as a whole, screened transactions represented about 20 per cent of the total number of FDI transactions in 2022. This share is close to 50 per cent when greenfield investments are excluded.[56]

That scrutiny is taking a toll. UNCTAD tallied at least twenty planned foreign takeovers with a value exceeding $50 million each that were blocked or withdrawn for national security reasons between 2016 and September 2019. The aggregate value of those transactions topped $162 billion.[57]

The US sets the pace

As always, the United States has played an important role in this evolution. Despite being an architect of the liberal and open global economic order, and continuing to press and prioritize access of capital to foreign markets, Washington has been leading efforts to promote greater scrutiny of foreign investments and, inter alia, insisted that declarations of national security interests are nonjusticiable and hence beyond review. This approach is well captured by the logic explained above by the OECD and reflects both the "great power competition" that guides US strategic thinking and the rising importance of emerging technologies for US status and standing.

The primary instrument of investment review in the US is the Committee on Foreign Investment in the United States (CFIUS), established in the 1970s in the aftermath of the oil shocks that produced massive shifts of global wealth and facilitated a wave of foreign investment in the country.[58] Chaired by the Treasury Department, CFIUS includes representatives from all major cabinet agencies to assess various dimensions of foreign investment with an eye to national security concerns. It rarely convened until the 1980s when Japanese companies began to make waves with high-profile purchases in the US. Another important shift occurred after the September 11, 2001, terror attacks when Congress expanded the committee's ambit to include critical infrastructure. That new authority along with rising sensitivity to foreign purchases ballooned its workload. Notice to CFIUS of transactions more than tripled between 2009 and 2019, increasing from 65 to 231; the number of investigations into foreign acquisitions more than quadrupled, swelling from 25 to 113.[59]

China has been a top concern. Between 2015 and 2019, acquisitions by Chinese companies constituted 22 per cent of all CFIUS submissions, and the highest number of notices concerned Chinese purchasers from 2015 to 2018. The strategy described at the beginning of the last chapter sparked worries in the United States that CFIUS was not fit for purpose. The Made in China 2025 plan in 2015 crystallized concern in the US security community about US preparedness for this new competition as Chinese investments in US tech startups jumped from $2.3 billion in 2014 to $9.9 billion in 2015.[60] The mood in the US was captured by a widely circulated report whose title was intended to sound the alarm: "Chinese Investments in Emerging Technology Enable A Strategic Competitor to Access the Crown Jewels of US Innovation."[61]

Rising anxiety prompted passage of the Foreign Investment Risk Review Modernization Act (FIRRMA), another part of the FY2019 National Defense Authorization Act. It is intended "to address growing national security concerns over foreign exploitation of certain investment structures which traditionally have fallen outside of CFIUS jurisdiction." FIRMMA sought to close loopholes espied and exploited by foreign companies who feared that outright purchase of a US company might trigger alarms under old CFIUS authorities. China isn't called out by name in the FIRRMA legislation, but it's there. An analysis notes that "Without mentioning specific countries, FIRRMA allows CFIUS potentially to discriminate among foreign investors by country of origin and transactions tied to certain countries in reviewing certain investment transactions, pending specific criteria defined by regulations."[62]

In most years since FIRRMA was enacted, there has been an increase in transactions reviewed. In 2023, CFIUS reviewed 342 filings, clearing 83 CFIUS declarations, and requested parties to submit a subsequent notice in 18 per cent.[63] Even after the reforms, China remained number one, even though the number of notices involving China dropped to 13 per cent between 2020 and 2024 and 12 per cent in 2025.[64]

Concern that CFIUS, even with the FIRRMA reforms, was not keeping pace with the new geopolitical environment prompted the Biden administration to issue, in September 2022, the first-ever presidential directive defining additional national security factors for CFIUS to consider in evaluating transactions.[65] They are: 1) supply chain resilience, 2) US technological leadership, 3) aggregate investment trends, 4) cybersecurity, and 5) US persons' sensitive data. While the impacts of the new national security economy are evident in factors 1, 2, 4, and 5, number 2 stands out. CFIUS must now consider a transaction's effect on US technological leadership in sectors vital to national security, which it identifies—including but not limited to— as microelectronics, artificial intelligence, biotechnology, quantum computing, advanced clean energy, climate adaptation technologies, and parts of the agricultural industrial base with implications for food security. The list will be periodically updated.

Reflecting their own concern and pressure from Washington, US allies followed with their own investment review revisions. In February 2019, the European Commission approved a bloc-wide mechanism for screening FDI to build on national review mechanisms already in place in twelve member states, and the EU three months later adopted a regulation that established a framework for screening FDI; guidance on implementing that regulation followed in March

2020.[66] In October 2020, the EU introduced a new foreign investment screening framework to harmonize regulation throughout the Union. European action was driven by fears that strategic assets would be purchased by foreign countries with hostile intent, that supply chains could become vulnerable, and that the COVID-19 pandemic had stressed European industrial and corporate assets and exposed them to hostile foreign takeovers. "Economic sovereignty" is a key concern.[67]

Individual countries are following suit. Germany, which absorbed 14 per cent of Chinese FDI in the EU between 2000 and 2019, trailing only Britain during that time, has strengthened the Berlin government's authority to review foreign direct investments in future and key technologies. In May 2021, the National Security and Investment Act became law in the UK. It requires advanced notice for foreign acquisitions in seventeen fields including AI, quantum technologies, communications, defense and energy, and the London government can block any deal that it considers a risk to national security.[68] Canada instituted new investment review guidelines in March 2021; they call for increasing attention to the national security implications of foreign transactions, with special focus on "sensitive technology areas," a list of which will be familiar to readers still awake.[69] Australia updated its investment review procedures on 1 January 2021.[70]

Japan deserves special mention. In November 2019, the Diet (parliament) amended the Foreign Exchange and Foreign Trade Act to impose more stringent controls on foreign investment, lowering the threshold for regulatory approval for foreign investments in strategically sensitive sectors from 10 per cent of a company's shares to 1 per cent; the law also obliges them to look at investments that give foreign investors say over the appointment of directors or proposals to sell certain operations.[71]

The new law is designed to shore up an investment regime that has been somewhat hostile to foreign investment in general yet largely indifferent to the national security risks of such inflows. The revisions should remedy that blind spot and ensure that Japan is working with the US, its only security ally and a vital partner in the development of technologies. Indeed, one of the motivations for the changes is better alignment of Japanese policy with that of the US to facilitate future tech collaboration.

Critics charge that the 1 per cent threshold is too low and the scope of the regulation is too sweeping. The Ministry of Finance, which administers the new law, has identified 518 companies as core industries and subject to the new regulations. While weapons, aircraft, and nuclear power plants are covered, so too are travel companies, *onsen* (hot springs), and baseball stadium operators, prompting charges that the law's purpose is as much old-fashioned protectionism and discouraging shareholder activism as national security.[72] The government has tried to assuage concern about its intent, but the Japanese case is a reminder that laws intended to protect cutting-edge national security concerns can be easily abused to promote old-style protectionism.[73]

Meanwhile, the US is spearheading efforts to "create a process for sharing information on potentially troubling investments with foreign allies," working through initiatives like the Quadrilateral Security Dialogue (whose members include Australia, India, Japan, and the US) and the US–EU Trade and Technology Council. This aligns with the FIRRMA mandate to "establish a formal process for the exchange of information ... with governments of countries that are [US] allies or partners."[74] Cooperation extends beyond mere information exchanges: the US and Japan, for example, are reportedly jointly monitoring Rakuten, the Japanese e-commerce powerhouse, after China's Tencent became a major shareholder, raising questions about data security.[75]

China has its own foreign investment national security review mechanisms, although they have been slow to take shape. While the first law on inward foreign investment was passed in 1979, the country did not address national security concerns until 1995 with "The Interim Provisions on Guiding the Orientation of Foreign Investment," which prohibited foreign investment that "harmed national security or impaired public interest."[76] Since then, the government has provided content and clarity about the means and mechanisms to ensure the protection of those interests.[77] At the end of 2020, the Chinese authorities adopted the Measures of Security Review of Foreign Investment, which went into effect the following January. The law was designed to align Chinese investment screening procedures with those of other major countries (like those listed above). It expands the scope of review from national defense to critical infrastructure and emerging technologies and personal data; it covers new types of transactions, including those that don't affect ownership; and it is intended to ensure more routine implementation of administrative processes.[78] For the most part, the new law is "a formalization and elaboration of existing screening procedures."[79]

Given the complaints detailed in previous chapters about Chinese efforts to tilt the domestic playing field against foreign investors, the need for such regulation is unclear. Since the Chinese authorities already have "substantial latitude for imposing conditions on or rejecting foreign investments" and "there is little evidence of foreign investors acquiring or building de novo controlling positions in sectors critical to national security as defined by China," observers are inclined to see the measure as "a response to the tightening of restrictions of Chinese investment in such jurisdictions."[80]

Protecting university research

The centrality of emerging technologies underscores the importance of research to the new national security economy. Concern focuses on three issues: collaborative research that can support illiberal governments; protection of intellectual property; and reciprocal access to research.

The first concern is that foreign researchers have helped build the repressive capacity of authoritarian states. This is a widespread phenomenon. A 2018 study by the Australian Strategic Policy Institute found that the leading countries for People's Liberation Army (PLA) collaboration, as measured by peer-reviewed papers co-authored by PLA scientists and overseas researchers, were, in order, the US, the UK, Canada, Australia, and Germany, while Singapore, Sweden, and the Netherlands are other major destinations for PLA scientists.[81]

The second concern has been generated by outright theft of intellectual property as well as Chinese initiatives, such as the Thousand Talents program, that encourage researchers to join its research cadres. The UK Centre for the Protection of National Infrastructure has warned, for example, that "Hostile state actors are targeting UK universities to steal personal data, research data and intellectual property, and this could be used for their own military, commercial and authoritarian interests."[82] While China has been the primary offender in recent years, other nations have misbehaved as well.

Governments have responded to this threat by creating or expanding national security screening programs. The US developed rules to protect its science from theft by foreign spies and established guidelines that universities and researchers must follow when reporting foreign financing and collaborations to US science funders.[83] The British government worked with Universities UK, an umbrella body for 139 universities, to create guidelines on managing security-related risks within overseas collaborations and set up the Research Collaboration Advice Team to offer confidential security advice to researchers before entering international collaborations to protect research assets from "hostile actors."[84] The Canadian government has launched initiatives to raise researchers' awareness of the risks of research theft.[85] Nearly three-quarters (72 per cent) of Japanese universities have established export control departments to keep sensitive information from being transferred out of the country as of April 2020, up from 58 per cent in February 2018.[86] The government created regulations requiring university researchers to report any financial contributions from foreign and other external sources when applying for public funds.[87] In December 2025, the Japanese government unveiled guidelines to screen researchers' backgrounds and restrict access to sensitive information in certain high-risk, advanced-technology programs.[88]

Western governments are taking collective action. There is regular discussion among the Five Eyes, as well as with other allies that share concerns about foreign interference. At their June 2021 Cornwall Summit, G7 nations agreed on guidelines to prevent leaks of sensitive research data to protect joint projects amid growing concerns about risk of theft by China. They also set up a Working Group on the Security and Integrity of the Research Ecosystem, which "will develop a common set of principles to protect the research and innovation ecosystem across the G7 from risks to open and reciprocal research collaboration, and preserve the principles of open science and research freedom and independence."[89]

While the need to protect research and the resulting intellectual property seems straight-forward, it can be a tricky proposition as much of the most important work is collaborative, and international teams are increasingly the norm. A British government unit tasked with protecting domestic research explained, "The threats to science and research in particular—primarily the theft, misuse or exploitation of intellectual property by hostile actors—are growing, evolving and increasingly complex."[90] Moreover, not all participants in theft are willing. "Threats can … include the manipulation of students through foreign interference" warned a spokesperson for Canada's public safety minister.[91] Japanese law enforcement authorities also warn that "there may be instances where they are forced to cooperate in exchange for the safety of family in China."[92] Finally, the growing dependence of Western universities on funds from Chinese students makes them reluctant to adopt overly tough restrictions. As a British report concluded, rapid growth in joint research and Chinese student numbers had created a "strategic dependency" on Chinese partnerships, exposing British universities to "poorly understood" risks, including financial dependency, intellectual property theft, and exploitation by an economic rival.[93]

A third concern is access to research. The relationship between China and the West is considered unbalanced. Jean-Eric Paquet, the EU's director-general for research and innovation, spoke for many when he complained that "Reciprocity is very difficult to find." There is, he noted, "essentially, full access to Europe but very cumbersome and formally limited access to resources on the Chinese side," adding that "China is willing to cooperate in areas where Europe leads, but is much more reluctant to do so in the areas China excels."[94]

This imbalance could be exacerbated by Chinese rules to tighten controls on data generated by its scientists. Regulations passed in 2018 require all scientific data generated within China to be submitted "by the lead program entity" to "state data centers" for review and approval before publication.[95] Ye Yujiang, director of the basic research department at China's Ministry of Science and Technology, explained that "In recent years, China has seen major scientific development and explosive growth in the amount of scientific data. However, it has lacked a national-level regulation to govern the data, and its management has lagged behind developed countries." Control of that work product is essential since "Scientific and research data are valuable strategic resources for a nation's scientific and socio-economic development," Ye added.[96]

Critics worry that the new regulations have expansive reach, extending to scientific data held by anyone outside China not directly funded by the Chinese government but funded by a mixture of government and nongovernment sources. Elsa Kania, an authority on Chinese work on emerging technologies, concluded that "The Chinese government is evidently seeking to ensure that innovation 'Made in China' will stay in China." She noted that "The policy purports to promote open access to and sharing of scientific data within China while creating ambiguous new restrictions, that, depending upon their implementation, could render future cooperation asymmetrical in its benefits."[97]

While vigilance is needed, it must be tempered by realism and humanity. Countries concerned about China must ensure that their efforts are properly focused. The China Initiative, the US Department of Justice effort launched in November 2018 to "confront Chinese malign misbehaviors," was criticized for targeting Chinese researchers in the United States. According to the Department of Justice, some 80 per cent of its economic espionage prosecutions allege conduct that would benefit China, and there is at least some nexus to China in around 60 per cent of all trade secret theft cases. The initiative took more than eighty cases to court and yielded a number of convictions—as well as controversy, with an increasing number of charges related to lack of candor on forms—and the criminalization of honest mistakes—and a substantial number of cases being dismissed for lack of evidence. Since all but a few of the individuals involved were researchers of Chinese descent, a former prosecutor warned that "a program focused on the Chinese government has morph[ed] into a people-of-Chinese-descent initiative."[98] That resulted in the shuttering of the program in 2022, although officially, the cause was "unjustified drain on DOJ national security program resources, its chilling effect on research collaboration, and unintended harms to public confidence."[99] The focus and tactics have reportedly been continued in several initiatives of the second Trump administration, however.[100]

If moral concerns aren't persuasive, then there are the practical consequences of profiling. US leadership in science and technology, and its ability to innovate, depends on the country's openness and its ability to attract and retain talent from around the world. The US needs about one million more science, technology, engineering and mathematics (STEM) graduates to maintain leadership in emerging technologies, and China has been the main source of students to fill the gap.[101] US universities awarded 66,690 doctorates to Chinese students in science and engineering fields between 2000 and 2017, and their five-year and ten-year stay rate is 83 per cent and 90 per cent, respectively, the highest of all countries. The Silicon Valley Leadership Group reports that the Silicon Valley workforce was 38 per cent foreign-born in 2018, and Chinese immigrants were among the top three largest groups; 67 per cent of Silicon Valley's new tech talent aged 25–44 was Asian, with the majority from China and India. Continued success demands that we not alienate or exclude them.[102] That was one of the motivations for ending the China Initiative, which was replaced with "a broader strategy aimed at countering espionage, cyberattacks and other threats posed by a range of countries."[103]

Supply chain resilience

Supply chain vulnerabilities loom increasingly large in government and business consciousness. First noted in 2010 when questions arose about access to rare earth elements, the problem assumed new dimensions after the March 2011 triple catastrophe in east Japan shut down component makers in a variety of industries. As the impact rippled through global production networks, companies began to think about adding slack and redundancy to hitherto lean and tight supply chains. The issue exploded during the COVID-19 pandemic when countries were unable to procure essential supplies of personal protective equipment. Since then, regular incidents have underscored a growing sense of vulnerability: a storm in Texas took down the state's power grid;[104] a ship stuck in the Suez Canal blocked that major trade artery;[105] COVID-19 infections in Taiwan undermined semiconductor production;[106] unvaccinated sailors threatened shipping;[107] and a fire shut down a Japanese plant that makes automotive chips.[108]

This has yielded an expanding set of initiatives to promote supply chain resilience. In the initial stages of the COVID crisis, several governments resorted to crude self-help measures, such as forbidding the export of PPE[109] or engaging in bidding wars on airport tarmacs as shipments of medical equipment passed through.[110] As befits the country that invented the just-in-time (*konban*) delivery system, created an extensive global production network, and experienced massive supply chain disruptions, Japan has been leading efforts to address this problem. In April 2020, Japan and ASEAN agreed to an economic resilience action plan that focused on supply chains.[111] Tokyo set aside roughly $2.2 billion to help Japanese firms shift production out of China by either bringing manufacturing home or moving facilities to Southeast Asia.[112] In September 2020, Japan, Australia, and India announced that they would develop a joint effort to address this issue,[113] which took concrete form with the launch of the Supply Chain Resilience Initiative in April of the following year.[114] In March 2021, the US joined those three governments to announce at the first Quadrilateral Security Dialogue leaders' summit that they would convene dialogues in critical technology supply chains.[115]

The Biden administration hit the ground running on this issue. A month after his inauguration, Biden announced a 100-day review across federal agencies of supply chain vulnerabilities in active pharmaceutical ingredients, critical minerals, semiconductors, and advanced packaging and large-capacity batteries, like those used in electric vehicles. The executive order also called for a more in-depth, one-year review of US supply chains in six key sectors: the defense industrial base; the public health and biological preparedness industrial base; the information and communications technology industrial base; the energy sector industrial base; the transportation industrial base; and supply chains for agricultural commodities and food production.[116] The 100-day review was completed on time—a testimony to the urgency of the issue—prompting the creation of a Supply Chain Disruptions Task Force to address near-term bottlenecks that can affect the economic recovery; its announced focus was "areas where a mismatch between supply and demand has been evident: homebuilding and construction, semiconductors, transportation, and agriculture and food."[117]

Europe has been worried, too. The medical equipment shortages of the first months of 2020 prompted a systematic assessment of the bloc's vulnerabilities. An analysis of trade data found that 137 products out of 5,200 originated from "sensitive ecosystems for which the EU is highly dependent," mainly in energy-intensive and health industries. Another thirty-four products were identified as being "potentially more vulnerable given their possibly low potential for further diversification and substitution."[118] Those results prompted the European Commission to update its Strategic Policy and seek to enhance the EU's "strategic autonomy" in areas such as raw materials, batteries, active pharmaceutical ingredients, hydrogen, semiconductors, and cloud and edge technologies. Central to that effort are efforts "toward diversifying international supply chains and pursuing international partnerships to increase preparedness."[119]

Two items get special attention: semiconductors and rare earth elements. Semiconductors are the DNA of technology, critical to every sector of the digital economy, and virtually every industrial sector experienced shortages in 2020 and 2021. The European Commission acknowledged that "the pandemic triggered wider awareness of the need to analyze and address strategic dependencies, both technological and industrial,"[120] but governments and businesses around the world were worried even before the COVID-19 outbreak. As the Semiconductor Industry Association, the industry trade group, explained:

> The US semiconductor industry relies on its deep global supply chains and access to overseas markets, but there are significant risks for the United States. ... All countries, particularly key US allies, are interdependent in this integrated global value chain. ... However, some large foreign nations do not currently operate on an entirely free market and have significant restrictions and incentive targets for their own domestic production.[121]

In some cases, vulnerabilities are a function of geography, not geopolitics. As the Semiconductor Industry Association explained in another study, "There are more than 50 points across the value chain where one region holds more than 65 percent of the global market share. These are potential single points of failure that could be disrupted by natural disasters, infrastructure shutdowns, or international conflicts, and may cause severe interruptions in the supply of essential chips."[122]

Invariably, there are calls for cooperative solutions. Typical is the language of the US Supply Chain Review, which concluded that "The United States cannot address its supply chain vulnerabilities alone." Therefore, the US should expand "multilateral diplomatic engagement on supply chain vulnerabilities, particularly through groupings of like-minded allies such as the Quad and G-7."[123] At first, "friend shoring" was all the rage. The first substantive agreement of the Indo-Pacific Economic Framework for Prosperity was the Supply Chain Agreement, which went into effect on 24 February 2024 for the first five countries—Fiji, India, Japan, Singapore, and the United States; the other nine members will join when they ratify the deal—and established three supply chain bodies—the Supply Chain Council, Crisis Response Network, and Labor Rights Advisory Board—"for deeper collaboration to prevent, mitigate, and prepare for supply chain disruptions."[124]

Still, for all the talk of cooperative solutions, governments look first to build national capacity. The EU announced plans to spend some 140 billion euros ($166 billion) to ensure that Europe-based companies hold at least a 20 per cent share by value of the global market for semiconductors below 5 nm in size by 2030, a doubling of its share over the decade.[125] US administrations have aggressively courted chip makers to set up operations in the US. The Supply Chain Review called for US government investment of at least $50 billion to support the development and production of semiconductors. Major manufacturers, like Taiwan Semiconductor Manufacturing Company (TSMC) and South Korea's Samsung, announced that they would be building facilities in the US, and Biden offered subsidies to encourage more chipmakers to follow their lead.[126] Japan is also in the game. METI established a ¥200 billion fund to court foreign manufacturers, especially those from Taiwan, to set up production in the country, and it has had some success with TSMC announcing that it would open two facilities in the country.[127]

Rare earth elements are another focus. Major consumers have had rare earth strategies for some time: the European Union's Raw Materials Initiative was launched in 2008; Japan's Ensuring Stable Supplies for Raw Materials was articulated a year later; the US Department of Energy created a Critical Materials Strategy in 2010 and in 2019 the Department of Commerce issued the Federal Strategy to Ensure Secure and Reliable Supplies of Critical Minerals; and Australia's Critical Materials Strategy was published in 2019.[128] They have had limited impact. For all the verbiage, China still controls about 70 per cent of global mining and 90 per cent of the supply of processed rare earths.

Western governments and REE consumers consider this troubling exposure, and with good reason as was driven home in 2025. To counter their vulnerability, national strategies have included criticality studies, emergency stockpiles, measures to develop alternative sources, and financing to build new facilities for mining and processing.[129] The US is reinvigorating domestic supply chains, reopening a shuttered mine and planning to build several separation and processing facilities.[130] It's stepping up R&D to diversify supply chains, developing critical minerals substitutes, and improving reuse and recycling.[131] The EU launched the Critical Raw Materials Alliance in September 2020 to identify barriers, opportunities, and investment cases to build capacity throughout the critical minerals value chain.[132] Japan has had a strategy to secure supplies of critical minerals since 2007. The government has provided funds to help companies that experience shortages, promoted R&D to reduce critical minerals use, sought alternative sources of supply, and developed and expanded stockpiles.[133] Multilateral initiatives are also popular. At the 2021 G7 summit, the leaders agreed to "consider mechanisms and share best practices to address risks to the resilience of the critical global supply chains, in areas such as critical minerals and semiconductors."[134]

While China is viewed by Western governments as a threat to their supply chains, Chinese government and industry increasingly see the West as an obstacle to their own growth and survival. Supply chain independence was one of the eight priorities of the 2020 Central Economic Work Conference, a high-level CCP meeting led by President Xi and Premier Li Keqiang that sets national economic goals and priorities.[135]

In rare earths and semiconductors, China has been more focused than the West on building capacity and for longer. Beijing first recognized the importance of rare earth elements in the seventh National Five-Year Plan for Rare Earth Industry (1986–90), which prioritized "research and production of advanced rare earth applications and new materials for domestic consumption and export." "New Materials" is one of the targets of the Made in China 2025 program, and the National Mineral Resource Plan for 2016–20 to ensure steady supplies of rare earths and other critical minerals.[136] China has decreed that "Rare earths are important strategic resources that are not renewable," and is implementing a series of measures to ensure steady, safe, and sustainable production of a vital component of the twenty-first, century economy.[137] Xi highlighted the importance his government attached to those resources during his 2019 visit to Jiangxi, and since 2023, Beijing has progressively tightened restrictions on the export of them.[138]

As a Chinese observer of the semiconductor industry explained, "We need to enhance supply chain security from a holistic perspective ... Chips, raw materials, equipment, design tools ... every single part of the supply chain is critical to China's pursuit of semiconductor independence."[139] Beijing launched its march to that independence in June 2014, when the State Council announced the establishment of a national investment fund dedicated to the development of semiconductors. As detailed in Chapter 3, hundreds of funds have been established to support the industry, while Chinese venture capitalists invest heavily in semiconductor startups.[140] Boston Consulting Group forecasts that China, with a tailwind of some $100 billion in government subsidies, will lead the world in chip production with a 24 per cent share in 2030.[141]

The disinformation threat

Authoritarian governments have been quicker than their democratic counterparts to recognize the importance of controlling information flows in cyberspace and writing the narrative to frame international discussion and debate on issues of key concern. For China and Russia, "cybersecurity and information security are two sides of the same coin, an approach that enables the control and manipulation of information at multiple levels."[142] These governments are more efficient at controlling cyberspace and information flows since their decision-making is much simpler: they don't have to balance individual rights against state power.

China has built the most comprehensive apparatus to control cyberspace; this is part of a policy in place since the founding of the People's Republic of China to control the flow of information and limit certain criticism of the government or the ruling Chinese Communist Party.[143] The Beijing government asserts that "within Chinese territory, the internet is under the jurisdiction of Chinese sovereignty."[144] The Chinese government has prioritized close coordination across all state bodies to ensure that foreign elements don't pollute the cyber and information space, to create a "harmonious Internet" by censoring dissent, and to limit foreign suppliers of technology to China. Officially, this is the "Golden Shield Project": unofficially, it is the Great Firewall. Chinese leaders trumpet the effort as correctly balancing between "freedom and order" and between "openness and autonomy." Lu Wei, who heads the government's efforts to manage the internet, calls it "cyber-governance with Chinese characteristics."[145] The OpenNet Initiative calls it "the most sophisticated content-filtering Internet regime in the world."[146]

While predisposed to Chinese thinking about internet sovereignty, Russia was slow to turn that mindset into policy. Only in 2016 did Moscow adopt the Doctrine of Information Security, which decreed that "protecting the information sovereignty of Russia" was essential to national stability.[147] The "sovereign Internet" law, passed three years later, allows the Moscow government to centralize and control domestic internet traffic and creates chokepoints that enable it to isolate the national internet from the rest of the world. For Russia, these policies are defensive in nature as many in the country are convinced that the United States actively works to undermine their government and a critical tool in this effort is support for Russian dissidents.

Europe has recognized the threats posed by disinformation to its democracy and democratic processes and set up a framework for action. Efforts began in 2015, which yielded a framework for action, "Tackling Online Disinformation: a European Approach," in April 2018; a voluntary "Code of Practice on Disinformation" for private companies published in September 2018; a "Progress Report" on implementation of the initial recommendations published in December of that year; and the "Action Plan against Disinformation," jointly prepared by the EU Commission and European External Action Service was also published in December 2018.[148] If followed up, those principles and action items "would mark the end of Europe's admiring the problem of disinformation and the beginning of its credible efforts to contain and reverse it."[149] One US assessment deemed EU progress in framing solutions to this challenge to be "impressive."[150]

The United States has been slow to the game. The 2018 National Cyber Strategy noted the threat of information operations and authoritarian challenge to an open internet but focused on a traditional view of cybersecurity that is limited to the functioning of networks. By law, the State Department Global Engagement Center is on point to "direct, lead, synchronize, integrate, and coordinate efforts of the Federal Government to recognize, understand, expose, and counter foreign state and foreign non-state propaganda and disinformation efforts aimed at undermining or influencing the policies, security, or stability of the United States and United States allies and partner nations."

Observers fault the US for not investing sufficient resources in the fight against disinformation,[151] unclear lines of authority,[152] and the failure to articulate an integrated national strategy outside the military domain for competition in the information space.[153] Blame politics. The Biden administration's National Security Strategy Interim Guidance, published early in its term, highlighted threats posed by misinformation.[154] The US intelligence community updated its Defense Intelligence Strategy to make countering Russian and Chinese disinformation a priority.[155] Efforts to fight disinformation, such as the Department of Homeland Security's Disinformation Governance Board, imploded under the weight of partisan complaint, however.[156] Republicans see US government initiatives to fight foreign interference as a subterfuge to silence conservatives.[157] Trump insists that claims that Russia interfered in the election in ways that favor him are a hoax and an attempt to delegitimize his administration.[158]

The most effective efforts to counter disinformation will be multilateral. Governments should be sharing information on threats—who they are and how they act—to better identify and respond to them. NATO has worked with the European Union, the United Nations, the G7, and civil society to counter disinformation.[159] At its

June 2018 Summit, the G7 produced the "Charlevoix Commitment on Defending Democracy from Threats," which included general language about cooperation to share lessons and information to counter disinformation.[160] The Five Eyes intelligence alliance has reportedly stepped up cooperation with democracies such as Germany and Japan to counter foreign interference emanating from China.[161]

One important counter to disinformation is programs to promote digital and media literacy among citizens. European countries are credited with generational investments in building media and digital literacy programs. Eléonore Caroit, a senior French official whose work focuses on disinformation, warned that "this is one of the major threats that we're facing in terms of national security sovereignty." Defending the information space is "at the very core of democracy," she explained. "It has an impact on elections, an impact on how citizens see their institutions, trust their representatives, and it is of the utmost importance in particular for younger generations."[162] By all accounts, governments around the world have awakened to the scale of the problem but their responses are, by and large, haphazard and ad hoc.

It has also become clear that purely defensive strategies are not by themselves sufficient to protect national equities. Governments are increasingly going on the offensive to do that.

5

GOVERNMENTS RESPOND

PLAYING OFFENSE

A growing number of governments think that "a good offense is the best defense" when trying to protect their national security and economic interests.[1] Keith Alexander, a former head of the National Security Agency, warned that "If we are under attack, you can't just try to catch every arrow. You have to take care of the person shooting the arrows at you."[2] In the new national security economy, going on offense means deploying trade and investment tools to promote innovation, employing development assistance to promote connectivity, and adopting cybersecurity strategies that look a lot like provocations.[3]

In many national security contexts, deterrence is the guiding principle: manipulate an adversary's cost–benefit calculation of behavior so that s/he believes that the cost will exceed expected benefit. There are many problems with deterrence, not least of which is an accurate assessment of that calculation: can we get inside the head of the adversary, weigh factors as they do, appreciate how they interpret our signals? In the digital space, problems multiply. Attribution is the most famous—can we be certain of the identification of the source of the disruption? That difficulty is magnified by the secrecy that shrouds all activities in this arena. General Michael Hayden, former head of the NSA and the CIA, noted that in the US until the early 2000s the phrase "offensive cyber operations" was classified—not the targets, the techniques, or the technologies but the very phrase itself.[4] But offensive capabilities deployed for deterrence won't work if the adversary is ignorant of their existence. Stanley Kubrick got it right when Dr. Strangelove tells the Russians that their doomsday device—which promises retaliation for a nuclear attack even if the leadership is decapitated—is not the ultimate deterrent but is instead useless if no one knows about it.

In addition, there are many motivations for cyberattacks—theft, extortion, activism, terrorism, espionage, preparations for war, sabotage, tests, information operations—each of which requires its own solution; even when the motives of attackers are ostensibly the same—say, extortion or espionage—the response must be crafted to fit each set of circumstances. In other words, "one cannot speak generally about 'cyber deterrence.' Rather, different kinds of malicious cyber activity demand different, tailored deterrence strategies."[5] Perhaps most important are differences between deterrence in the case of armed conflicts between states and deterrence of behavior that is not intended to be an act of war.[6] Our focus is the latter.

The growing menace of cyber threats is proof that offensive cyber capabilities are proliferating. Governments are generally reluctant to acknowledge possession of those skills, and fewer still attempt to explain their strategies for using them. Most discussion is framed by grand strategic interests, or how those capabilities fit within a broader set of national security concerns. The US is an exception. It has been a bit more forthcoming than most governments, despite the inclination to secrecy that General Hayden described.

The US defense community debated the role of offensive capabilities in the late 1990s, although legal authorities and directives remained opaque.[7] The FY 2012 National Defense Authorization Act authorized the Department of Defense to conduct offensive cyber operations although restraint remained the guiding principle. Then Secretary of Defense Chuck Hagel articulated US thinking in 2014 when he explained that the military "will maintain an approach of restraint to any cyber operations outside the US Government networks. We are urging other nations to do the same."[8] Growing frustration over the efficacy of that approach prompted in 2018 a shift in

Pentagon thinking "from a defensive strategy to a more active approach that is focused on imposing costs on malicious actors who violate accepted norms for responsible state behavior."[9]

"Offensive" operations assume a slightly different meaning when the phrase is used to characterize the campaign to "name and shame" hackers believed to have attacked US networks and the governments that support them. This effort was evident in May 2014 when the US indicted five members of the PLA for alleged cybertheft of intellectual property. The next year, Obama and Xi agreed during Xi's September 2015 state visit that neither government would "conduct or knowingly support cyber-enabled theft of intellectual property ... [for] competitive advantages to companies or commercial sectors."[10]

The strategy is reported to have worked—for a while. Chinese IP and trade secret theft declined after the PLA indictment and the 2015 commercial espionage agreement,[11] and Russian hacking activity appeared to slow after Obama issued a warning to Putin.[12] A National Security Agency official later acknowledged that Chinese attacks had dropped "dramatically" since the deal, but he complained that "it's clear that they are well beyond the bounds today of the agreement."[13] Frustration with Russian activity grew as well, and after the Biden administration's 2021 warnings seemingly fell on deaf ears—the SolarWinds hack was especially galling—the US imposed sanctions on six Russian companies involved in cyber espionage and thirty-two people and entities involved in election interference.[14]

Offensive capabilities can be used in other, creative and threatening ways. In 2017, Dutch hackers penetrated Russian networks and watched and filmed Russian hackers as they tried to penetrate US networks.[15] US investigators managed to track down and identify the private key for the wallet that received more than $2 million in cryptocurrency paid in ransom for the 2021 Colonial Pipeline ransomware attack and retrieve the funds.[16] Being able to put names, faces and addresses on hackers, or having the ability to track payments, provides governments with a precision-targeting mechanism, one that should make a hacker hesitate about proceeding. The team responsible for the Colonial Pipeline hack reportedly went out of business after that attack, perhaps in recognition of the potential exposure and danger that its members faced. After all, terrorist groups that have been no less disruptive have been the targets of drone strikes.[17]

The US National Cyber Strategy calls for cooperation with like-minded nations to impose "swift, costly and transparent consequences" when it or its partners are harmed by "irresponsible" and "malicious" cyber activities.[18] The US has many options. In 2017, US intelligence officials estimated that more than thirty countries were developing "offensive cyberattack capabilities."[19] Most Western governments have made significant investments in those skills, although the bulk of their effort goes into "hard" military programs, and there is little clarity about the resources devoted to offensive projects for other national security interests, such as protecting infrastructure or preventing theft of intellectual property. The US engages with like-minded nations through the Cyber Deterrence Initiative.[20] Central to that effort are the Five Eyes partners: the UK, which declared that it had offensive capabilities in its cyber quiver in 2013, although it's believed it had that ability some years earlier;[21] Australia, which acknowledged for the first time in 2016 that it had offensive cyber capabilities;[22] and Canada, which went public with its possession of those skills a year later.[23] (New Zealand, the other Five Eyes member, has not publicly said that it has offensive capabilities).[24]

NATO governments have declared that they are "determined to employ the full range of capabilities, including cyber to deter, defend against and counter the full spectrum of cyber threats."[25] Germany and Spain are also ready to wage offensive operations, but only after they have been attacked first—they "hack back"—unlike the US, the UK, and France, which are prepared to take preventive actions against potential adversaries.[26] Obviously, the Dutch have these capabilities as well.

Readers will note that unlike previous discussions, this one has focused on the US and has a seemingly benighted view of its policymakers. This stems from several considerations. First, for all the limitations, the US has been among the most transparent of governments to discuss offensive cyber capabilities. Second, there is little if any evidence that governments not mentioned here—China, Russia, North Korea, Iran, for example—are using offensive capabilities to deter attacks by other governments, which is our focus. They have those skills—and use them—just as the US does when it launches offensive operations against its adversaries. In other words, the first group (the US and its allies) is acting more conventionally, attempting to damage or degrade a threat, while the second group—either the government or its agents—is using those capabilities not to deter but for criminal purposes—to steal, extort, or disrupt society. Or as one analysis explained, "For both China and Russia, what the West calls 'offensive cyber' is just the technical component of a wider information-operations capability. It is just one means of controlling their own information space, and subverting those of their adversaries, in what they see as an ongoing conflict of ideas with the West." This likely means that the two governments devote fewer resources to this capability and therefore have "blunter" tools.[27]

Moreover, when it comes to espionage, those other governments are unlikely to publicly complain about US actions—unlike the US, which sees a benefit to doing so. Finally, US behavior is not without criticism. The US is the only Five Eyes member that is not a signatory to the "Paris Call for Trust and Security in Cyberspace," in which stakeholders commit to working together and to adopt a set of nine nonbinding principles for a secure and responsible cyberspace.[28] The US also does not explicitly commit to adhering to the UN norms which it promotes. As the US Cyber Command explained in its 2018 Command Vision:

> The Command makes no apologies for defending US interests ... in a domain already militarized by our adversaries. To the maximum extent possible, we will operate in concert with allies and coalition partners. We will also explain to oversight entities and the public the nature of threats in cyberspace, the threatening conduct of our adversaries, the limitations of passive defenses and our scrupulous regard for civil liberties and privacy.[29]

Trade and investment tools

The US has other tools it can use to go on the offensive to protect its economic interests. To stop foreign trade practices that constitute a national security threat, it has Sections 232 and 301 of US trade law. The Trump administrations have been especially activist in their resort to those authorities, powered by the president's belief that "trade wars are good and easy to win."[30] Section 232 of the Trade

Expansion Act of 1962 allows the president to impose restrictions on imports if they are "being imported into the United States in such quantities or under such circumstances as to threaten to impair the national security."[31] Since it became law until 2020, it was used to launch thirty-one investigations. Prior to Trump's election in 2016, the last Section 232 investigation was in 2001 and a president last acted under Section 232 in 1986. The first Trump administration launched eight Section 232 investigations into imports of: steel, aluminum, automobiles and automotive parts, uranium, titanium sponge, electrical transformers and certain grain-oriented electrical steel components, mobile cranes, and vanadium. While the cases involving uranium and vanadium imports might address issues of the new national security economy, most investigations tackled run-of-the-mill charges of unfair trade. In March 2018, Trump applied 25 per cent and 10 per cent tariffs, respectively, on certain steel and aluminum imports and, in February 2020, expanded the scope of products subject to the additional tariffs.[32]

While President Biden debated whether to continue those tariffs, he did promise to combat "unfair foreign trade practices," especially those that contributed to the erosion of supply chains. Biden declined to impose sanctions after inheriting a Trump-launched investigation of vanadium imports and also passed on imposing tariffs despite a Commerce Department ruling that imports of neodymium-iron-boron permanent magnets met the statutory definition of threatening national security.[33]

The second Trump administration has been especially aggressive in its resort to tariffs to effectuate policy. Nine months into his second term, the US had imposed 232 tariffs on steel, aluminum, automobiles and their parts, and copper and had launched investigations into exports of semiconductors, pharmaceuticals, critical minerals, timber/lumber, and heavy trucks, among other things.[34] The range of affected products suggested that the authority was being stretched well beyond traditional definitions of national security.

US presidents have a second tool: Section 301 of the Trade Act of 1974 gives the United States Trade Representative authority to enforce US rights under trade agreements and respond to certain foreign trade practices. It allows the US to sanction countries that "deny adequate and effective protection of intellectual property rights" or "deny fair and equitable market access to United States persons that rely upon intellectual property rights." There have been 130 cases under Section 301 since it went into force. They primarily targeted the European Union, accounting for about one-third (30 per cent) of all cases, and most concerned agricultural trade. The last case that resulted in retaliation and the imposition of tariffs occurred in 2009 and involved Canada. During the first Trump administration, the Office of the US Trade Representative initiated six investigations against China, the EU, France, a group of ten trading partners, and two against Vietnam.[35] In the second, China's semiconductor policies, its shipping industries, foreign digital services taxes, international seafood, and Brazil's trade practices had triggered investigations, a list that again suggested that the tool was being applied in an expansive manner.[36]

The Special 301 provisions have been an especially powerful cudgel against China. They were used to investigate Chinese trade practices and produced a damning report that readers will remember from Chapters 2 and 3.[37] Its findings were used to validate five major tariff actions in 2018 and 2019, which subjected about three-fourths of US imports from China to increased Section 301 tariffs, ranging from 15 per cent to 25 per cent.[38] In what is largely considered a political calculation, Biden let the tariffs continue.

The second Trump term has witnessed the use of yet another trade tool, the International Emergency Economic Powers Act (IEEPA). It authorizes the president to "regulate" certain economic transactions, including imports, in response to declared emergencies concerning "unusual and extraordinary" threats to national security, foreign policy, or the economy. Trump has extensively flexed that muscle since returning to the White House in 2025, using it to impose tariffs on trading partners around the world. His justifications—declaring "the influx of illegal aliens and illicit drugs" or large and persistent trade deficits and a lack of reciprocity in trade relationships a national emergency—have been challenged, and multiple courts ruled that the administration exceeded its authority when doing this. As we go to press, the Supreme Court ruled that IEEPA could not be used, but the Trump administration has vowed to continue while using other authorities.

The US president has a second set of tools, one that is made effective by the central role of the US in international finance: he can ban US investment in Chinese companies. Trump used that authority as his first term in the White House ended, issuing an executive order in November 2020 to prohibit any US citizen or investment fund from making new investments in companies identified as supporting the Chinese military. They would also have to sell any existing shareholdings. Biden expanded the order to include companies that sell surveillance technology, calling the businesses a threat to US interests and values,[39] and officials expect the list of banned companies to lengthen.[40]

While US trade and investment authorities have been used most aggressively against China in recent years, their history is a reminder that it isn't just Chinese businesses that should be concerned. Both Trump administrations have levied tariffs against a dizzying number of countries, even if they did grant exemptions to their application. Despite the comity that surrounded the creation and launch of the EU–US Trade and Technology Council, Biden administration officials were soon after warning the EU against pursuing "protectionist" technology policies that target US companies. The EU's Digital Markets Act was a source of particular concern, with the US National Security Council complaining that it appeared to "unquestionably target only the five biggest US firms," which raised questions about Europe's intent and "make[s] regulatory co-operation between the US and Europe extremely difficult and send[s] a message that the [European] Commission is not interested in engaging with the United States in good faith to address these common challenges in a way that serves our shared interests."[41]

There is a new twist in efforts to regulate foreign investment: a "reverse CFIUS" that will regulate investment originating in the United States and heading to other nations (outward foreign direct investment, or OFDI), usually "countries of concern," a list that China typically tops but includes Russia, North Korea, Iran, Cuba, and Venezuela. An executive order that focuses on semiconductors and microelectronics, quantum technologies, and AI was issued in October 2024 to "help address the advancement of key technologies and products by countries of concern that may use them to threaten US national security."[42]

The European Commission has studied the problem since June 2023 and released a "White Paper on Outbound Investments" in January 2024 which focused on the security risks arising from OFDI and calls for monitoring such investments in the fields of advanced semiconductors, AI, quantum technologies, and biotechnologies. The Commission will use those studies to recommend additional steps by member states. Germany, Italy, and Spain have already acted, as has the UK.[43]

Europe has taken a page from the American playbook with creation of the Foreign Subsidies Regulation (FSR), a tool that gives the European Commission the power to investigate financial contributions granted by non-EU governments to companies active in the EU. If Brussels finds that the aid creates distortive subsidies, it can impose measures to redress their effects. The FSR has been called "one of the EU's sharpest enforcement tools" and is "gaining prominence as geopolitical tensions reshape global investment and trade." Within two years, there had been 179 FSR filings, an average of eight a month.[44]

While the US has relied on trade and investment tools, other governments have resorted to nontariff attacks when they go on the offense. This is "a new category" of trade weapon that "erodes the global competitiveness of targeted firms through regulatory fines, digital service taxes, forced localization requirements, and operational constraints. They are framed as legitimate domestic policies, but their function is to weaken US technology leadership in favor of domestic champions." Adopted by governments from Brazil to the UK, nontariff attacks "are structured to chip away at the advantages of a few specific, successful US firms by draining cash, eroding scale economies, and constraining product evolution. ... This systematic pattern—revenue extraction, precision targeting, competitive undermining, and strategic exclusion—reveals these measures as coordinated attacks on US technology leadership rather than legitimate regulatory responses to market failures." They are painful: fines alone have resulted in losses to US companies of over $30 billion over the last decade.[45]

Just connect

Infrastructure inequities are stunning.[46] Annual global infrastructure demand is forecast to be between $2.9 trillion and $6.3 trillion until

2030, with the shortfall reckoned to be about $350 billion to $370 billion a year.[47] In 2017, the Asian Development Bank estimated that Asia would need $26 trillion in infrastructure investment through 2030.[48] Latin America and the Caribbean need an additional $120 billion to $150 billion per year in investment to close their infrastructure gap.[49] These shortfalls matter. Countries with inadequate infrastructure under-trade by 30 per cent and fall short of potential FDI by 70 per cent.[50]

The digital dimension is especially important. In 2019, the G20 countries recognized "that improved connectivity and broadband access is a necessary condition for the development of the digital economy, as well as a powerful enabler of inclusive growth and sustainable development."[51] In 2024, however, about 2.5 billion people were not online.[52] Ensuring their access to broadband will cost about $428 billion.[53] This is an infrastructure investment challenge since some 2.6 million 4G base transceiver stations and 700,000 km of backbone fiber transmission infrastructure will have to be installed on top of existing broadband network capabilities.[54] Meeting that need is one way that governments can go on the offensive to shape the international environment, tilting the economic competition in their favor and recalibrating the geostrategic balance.

Conscious of those benefits (and others), China seeks to fill that gap. In 2013, President Xi Jinping announced the launch of the "One Belt, One Road" initiative, now known as the Belt and Road Initiative (BRI).[55] Having become associated with Xi himself, it was enshrined in the Chinese Communist Party constitution in 2017. The BRI's ambitions are breathtaking. The project is intended, with the Silk Road Economic Belt and the New Maritime Silk Road, to evoke the glories of imperial China and its place in the ancient trade route that millennia ago linked Asia, Europe, and East Africa. The original Silk Road was a network that produced not just fabulous wealth and prosperity but also promoted the exchange of ideas and people that helped grow the civilizations that it connected.[56]

Since its launch, the BRI has expanded to recapture that history. It initially focused on Asia, Europe, and Africa, but it has since gone global. As of the summer 2025, China had partnerships with more than 150 countries and over thirty international organizations.[57] BRI countries include 65 per cent of the world's population, half of global GDP, 75 per cent of known energy reserves, and a quarter of all cross-border goods and services trade.[58] By the start of 2020, 2,951 BRI-linked projects—with a total value of $3.9 trillion—were planned or under way across the world, covering energy, transport, mining, information technology, "smart cities," and special economic zones.[59] By 2025, two-way investment between China and participating countries surpassed $380 billion, while cumulative BRI engagement reached $1.3 trillion, of which some $775 billion was in construction contracts and $533 billion was in nonfinancial investments.[60] The BRI includes at least six overland corridors, a maritime Silk Road, and a Polar Silk Road that traverses Arctic shipping routes.[61] There is also the "Digital Silk Road," the "Belt and Road Space Information Corridor," the "Health Silk Road," and the "Green Belt and Road."

Narrowly defined, this is an infrastructure project. But a closer reading reveals a more ambitious design. The BRI aims not just to link China to those distant places but to integrate them into a cohesive network that facilitates the spread of Chinese ideas, interests, and influence. China's leaders tout the the BRI as "win-win" cooperation that promotes peace and development.[62] While foreign analysts warn that nodes of the BRI are strategically located and could be used to facilitate a military presence and power projection,[63] Vice Foreign Minister Zhang Yesui countered that "in pursuing the Belt and Road Initiative, China will not resort to outdated geopolitical maneuvering"[64] and Chinese officialdom repeats that the country has no intention of building overseas military bases.[65]

That doesn't mean the BRI has no security dimension. Most basically, as noted in Chapter 1, economic development has always been a way for the Chinese Communist leadership to realize its goal of "the creation of a rich, strong, powerful state."[66] Even Xi concedes that "peripheral diplomacy" seeks to maintain stability on its borders so that China can establish a new regional economic order through the BRI.[67] Kerry Brown, one of the UK's foremost China watchers, argues that this effort aligns with longstanding Chinese efforts to assert what it believes to be its legitimate rights over issues of concern and its claim to more strategic space. China aims "to create a vast zone built on economic commonality ... the construction of interconnectivity through human capital, finance, infrastructure, and trade."[68]

The project to create a "vast zone built on economic commonality" makes the BRI relevant to our discussion. The BRI and the tools of its construction are some of the most visible and effective examples of ways that governments can secure their interests and extend their influence in this new world.[69]

The easiest way to win friends and influence people is to make them money. The BRI will do that. Its size and scope could boost global gross domestic product by as much as $7.1 trillion by 2040 and reduce global trade costs by up to 2.2 per cent.[70] A World Bank study estimates that once completed the BRI will reduce travel times by up to 12 per cent, with an average decrease of 3 per cent, which will produce a growth in trade for BRI corridor economies between 2.8 and 9.7 per cent and between 1.7 and 6.2 per cent for the world. It also forecasts a 7.6 per cent increase in foreign direct investment in low-income countries because of new transport links. Real income could increase between 1.2 and 3.4 per cent. Overall, BRI transport projects could help lift 7.6 million people from extreme poverty (those earning less than $1.90 a day) and 32 million people from moderate poverty (those earning less than $3.20 a day).[71]

The linkages make those countries more dependent on China. China's trade with BRI countries has been growing more rapidly than trade with non-BRI countries. In 2019, China's total trade with BRI partners was $1.34 trillion, 7.4 per cent higher than the aggregate growth in trade overall.[72] Linking economies to China and making countries depend more on Chinese finance, the BRI gives Beijing additional geopolitical leverage. It increases those countries' reliance on access to the Chinese market and Chinese finance for economic growth.[73] It also creates an advantage for Chinese companies. Nearly 90 per cent of the contractors in Chinese-funded BRI projects are Chinese, 7.6 per cent are local (companies headquartered in the country where the project was taking place), and just 3.4 per cent are foreign.[74] Given the volume of the BRI projects and the role that Chinese companies play in them, it should be no surprise that the fifth- and seventh-largest global contractors are Chinese, and no US company has cracked the top twenty.[75]

The strategy seems to have paid off. One exhaustive poll showed that 79 per cent of leaders surveyed consider Beijing to be actively supporting their countries' development, and 38 per cent selected China as their infrastructure partner of choice, surpassing all other major powers.[76]

Critical to the BRI's success is technology. The Digital Silk Road "not only promotes the development of the digital service sector, such as cross-border ecommerce, smart cities, telemedicine, and internet finance, but also accelerates technological progress including computing, big data, Internet of Things, artificial intelligence, block chain, and quantum computing."[77] This sweeping ambit has prompted a belief that the Digital Silk Road may well be "a brand for virtually any telecommunications or data-related business operations or product sales by China-based tech firms" in a BRI country.[78]

Then there's the Belt and Road Space Information Corridor, which uses space technology to support development of the BRI and tighten participant states' links to China. It relies on China's BeiDou satellite system, which serves as the "digital glue for the roads, railways, ports, and industrial parks" being developed under the BRI. Components of the corridor include navigation, remote sensing, weather, communication, data-relay satellites, and ground stations and data centers.[79]

The spread of its technologies to BRI recipient states gives China a crucial advantage in the race to set national and international standards. As one analysis explains, "the sheer scale of Chinese activity abroad puts it in a very strong position to establish its technical standards as 'global defaults' in a number of fields."[80] Those efforts could also lock states into Chinese tech ecosystems. Once technology is imbedded in a country's network, it is hard to switch to other providers: mixing and matching components is tough, if not impossible. The only alternative is often to rip out existing systems and replace them root and branch, an expensive and time-consuming proposition.[81] Not surprisingly, then, there is also a Chinese Action Plan for Standards Connectivity for the Joint Construction of the Belt and Road which calls for uniform technical standards across the BRI. As of September 2019, China had concluded ninety bilateral agreements on technical standardization cooperation with fifty-two countries and regions.[82] These facilitate Chinese standardization efforts in international forums such as the International Organization for Standardization and the International Electrotechnical Commission.[83]

The proliferation of Chinese digital technologies helps China in other ways. It gives China access to vast amounts of data, the "oil" in efforts to build artificial intelligence technologies (such arrangements are often discussed during project negotiations); it could give China access to sensitive data that could be used for intelligence purposes (these arrangements might or might not be negotiated);[84] and most worrisome, it could give China a "kill switch" that allows it to disrupt or disable a network during a crisis.

This last point takes on ominous proportions as China promotes energy connectivity in the BRI. By 2020, nearly two-thirds of Chinese spending on completed BRI projects went into the energy sector,[85] and Beijing's planned investment in power sources and grids in BRI countries will reach $27 trillion by 2050,[86] with the goal of creating a huge, multinational smart grid that uses Chinese technology and communications networks to allocate energy and resources. These capabilities would give China intelligence-gathering opportunities and the ability to manipulate or deny power in a crisis. A Chinese energy official referred to global energy interconnectivity as "the ICBM [intercontinental ballistic missile] of the power industry."[87]

Critics warn that Chinese investment and finance will lead to corruption,[88] unsustainable debt obligations, and long-term financial dependencies.[89] The evidence is not determinative, partly because many deals are shrouded in secrecy.[90] There is one high-profile case of China seizing property because of excessive debt—Hambantota port in Sri Lanka—but governments have since renegotiated deals to get more favorable terms. There is also evidence that Chinese debt is no more onerous than that of Western creditors.[91]

Despite the many criticisms of the BRI—and there is a steadily expanding cottage industry—it is important to recognize what a far-sighted and strategic initiative it is. BRI is designed to redraw and reconfigure the geopolitical map in ways that benefit China. "Beijing's BRI strategy bolsters its technological, economic, political, and security interests and affords it increased rule-setting power."[92] These efforts "seem to lead toward a regional or perhaps global ecosystem that would disadvantage the United States and other of China's competitors."[93] And driving much of that project is mastery and domination of the high technologies that are included in the mundane descriptor "infrastructure."

The West responds

Western governments have been studying the Belt and Road Initiative and are alarmed about its implications. Those worries were not matched by action, however. US economic engagement with the Indo-Pacific has been anemic. It has not joined the various regional trade and investment projects, famously withdrawing from the Trans-Pacific Partnership despite playing a key role in its negotiation. (There remains hope in Asian capitals that Washington will reconsider and rejoin what is now called the Comprehensive and Progressive Agreement on Trans-Pacific Partnership, or CPTPP; there is no indication that the US will be swayed, however.) Lending to and investment in the region have been limited and declining. Washington has not used its influence in multilateral development banks to get them to fill regional infrastructure needs. One analysis flatly blames "US inaction as much as Chinese assertiveness" for the shift in influence in the Indo-Pacific. "It has not met the needs of the region."[94]

In 2018, the US passed the Better Utilization of Investments Leading to Development (BUILD) Act to boost private sector investment in poorer countries, but it lacked funding, staff, and visibility.[95] Throughout President Trump's term, a series of initiatives seemed to lay the ground for more economic engagement with Asia, but, as one observer sniffed, "There was a lot of rebranding, but there weren't additional resources given to these initiatives."[96]

Other Western governments haven't been indifferent to regional developments. Japanese Prime Minister Abe Shinzo in 2015 launched the Partnership for Quality Infrastructure, which provided $110 billion (a 30 per cent increase) for quality infrastructure in Asia over five years. The following year, the sum was increased to $200 billion. Abe and his successors sought to differentiate their investments from those of China by stressing "quality" as the path to sustainable growth.[97] India and Japan announced an Asia Africa Growth Corridor in November 2016.[98] The EU launched a connectivity plan in 2018 and signed a partnership with Japan in 2019, but neither produced flagship projects in third countries.

The headline effort was the Blue Dot Network, announced by the US, Japan, and Australia in the summer of 2019. Wilbur Ross, then US commerce secretary, called it "a multilateral approach to fostering sustainable economic growth by promoting excellence in infrastructure development and supporting alternatives to predatory lending." More simply, the program was a confidence-building measure that would certify that a project met infrastructure standards and was thus a good investment. Matthew Pottinger, Trump's deputy national security advisor, likened a blue dot to "a Michelin-star rating for your restaurant."[99] Whatever the intent, the program languished; it was pointed to whenever there were accusations that the West had no counter to BRI, but it was starved of funds. Unlike the BRI, the Blue Dot Network had no funding authority. It merely showed private lenders where they could invest safely.

A senior European diplomat lamented, "we are trying to counter Belt and Road mostly with buzzwords and lofty policy papers. But unfortunately there is no real geopolitical strategy or plan which is consistent and coherent."[100] Biden sought to fix that. On his first overseas trip as president, to the G7 meeting in Cornwall in the summer of 2021, the US unveiled a newly revived Blue Dot project. As the initiative was endorsed in the G7 statement, over 150 global executives from ninety-six countries and responsible for some $12 trillion of assets under management were meeting in Paris to launch an OECD-led Executive Consultation Group that would provide high-level guidance for the effort.[101] Focusing on transparency and sustainability, the Blue Dot Network is intended to galvanize private sector investment in infrastructure in emerging markets by providing "a high-standard certification to give investors confidence and move one step closer to infrastructure becoming an asset class."[102]

Europe appears to be awakening from its slumber as well. In the summer of 2021, the European Commission acknowledged the need to "take a geostrategic approach to connectivity" and change the narrative of a sleepy West failing to rise to the China challenge.[103] That shift in thinking yielded the Global Gateway Initiative. Announced in December 2021, the EU initiative will "mobilize" up to €300 billion ($339 billion) between 2021 and 2027 for infrastructure and other projects. European Commission President Ursula von der Leyen said the plan will "support smart investments in quality infrastructure, respecting the highest social and environmental standards, in line with the EU's values and standards. The Global Gateway strategy is a template for how Europe can build more resilient connections with the world."[104] While the word "China" never appears in official Global Gateway documents, von der Leyen pulled no punches when asked. "Countries (had) their experience with the Chinese investments—they need better and different offers," she said.[105] The US has called the plan a complement to Build Back Better World (B3W), a domestic infrastructure program—Build Back Better—that the G7 adopted in 2021 as a global initiative (hence the W) and was rebranded at the 2022 G7 summit as the Partnership for Global Infrastructure and Investment (PGII).[106]

Doubts persist. Jonathan Holslag, a longtime observer of European politics, cautions against Western governments who assume that they are considered "a more honest-and-responsible partner, while China fumbles with credits and megaprojects that benefit no one." China has learned lessons from the first stage of BRI and upped its game.[107]

The big issue is money. The EU plan "mobilizes" up to €300 billion. That is €60 billion a year, an impressive sum—until compared to the €100 billion that China was spending each year before the pandemic. But the EU, along with the G7, also aims to "catalyze" investment with a mix of loans, grants, and guarantees, distributed through public, private, and development finance; just over 50 per cent will be in the form of direct funding. Ultimately, only money on the table counts. Reinhard Bütikofer, chair of the European Parliament's delegation for relations with China, noted the obvious: "The Chinese made inroads because they had something to offer that we did not."[108]

An effective counter calls for creative diplomacy. Australia deserves credit for innovative thinking. In 2024, Canberra concluded an impressive trifecta of diplomatic agreements with Tuvalu, Nuaru, and Papua New Guinea. All promote national development in new ways. The third deal, with PNG, uses that country's love of rugby as a hook, promising entry into Australia's National Rugby League.

Most experts agree with analyst Derek Grossman, who argued that Australia, Japan, and the US "cannot possibly match China's BRI dollar-for-dollar."[109] More significantly, the BRI has raised awareness of the importance of software and "operating systems." As a European diplomat explained "We now understand that it's not just about infrastructure but ... a chance to set standards," adding that "[whoever] writes the rules rules the world."[110]

"Those who set the standards gain the world"

Connection demands uniformity and consistency. Trade, even commerce, is not possible without standards. Qin-era China unified and codified weights to facilitate fair exchange. Without standard railroad gauges, trains must be reloaded at every border. Global standards are the foundation of any technology regime. Standardized USB ports allow data transfers among disparate devices. They guarantee functionality, interoperability, safety, and reduce costs.[111] In 2014, there were an estimated 500,000 global technology standards. With some 96 per cent of trade affected by technical standards and associated technical regulations,[112] they are "the glue that makes technology compatible between both countries and the rest of the world."[113]

The entities that set standards—governments, businesses, and nongovernment organizations—project influence as they do so. Over a century ago, the German engineer and industrialist Werner von Siemens noted that "he who owns the standards, owns the market."[114] That continues to be true, as a Chinese regional official declared a few years ago: "Standards are the commanding heights, discourse power, and the power to control. Those who set the standards gain the world."[115]

For much of the modern era, the West held those heights. While it is fashionable and easy to criticize the European Union for punching below its weight in many areas, it has been the dominant power when it comes to setting global standards.[116] The size of its market and an extensive institutional architecture have created "the Brussels effect," the "unprecedented and deeply underestimated global power that the EU is exercising through its legal institutions and standards."[117] As one analysis detailed:

> Few Americans are aware that EU regulations determine the makeup they apply in the morning, the cereal they eat for breakfast, the software they use on their computer, and the privacy settings they adjust on their Facebook page. And that's just before 8:30 AM. ... EU regulations dictate what kind of air conditioners Americans have and why

their children no longer find soft plastic toys in their McDonald's Happy Meals.[118]

The ability to set standards confers a range of benefits. When a company's intellectual property is included in a standard, it is paid for each use. This "standard essential patent" provides a royalty stream to the patent holder that is often significant. Ericsson, the Swedish technology company, earned €5.2 billion by licensing technology in 2017, more than 20 per cent of the company's revenues.[119] In the same year, Nokia generated $1.86 billion, about 7 per cent of its revenue, from standard essential patents, and it is estimated that Qualcomm receives 5 per cent of the profits from every 3G handset sold.[120]

Setting a standard gives a company a leg up on future sales since many IT networks build on legacy systems. If, for example, a company has installed a Microsoft operating system, it is more likely to upgrade that software than to buy an entirely new system as it modernizes. This "lock in" and the revenue stream it yields help fund research and development, promoting a virtuous cycle of innovation that can keep a company's products at the technological frontier. Conversely, companies and countries that must switch to adopt to new standards incur significant costs to make that transition.[121]

A third benefit is transmitting the standard setter's values. Technology is not neutral. It conveys and disseminates the priorities and the values of the originator. For example, while WIFI is today the standard for local wireless networks, it was challenged by a rival technology, WAPI, that was developed in China. WAPI performed better but had worse privacy protections. The challenger was ultimately rejected for procedural reasons, but "whether intentionally or not, by rejecting WAPI international standards bodies prioritized privacy over performance, shaping what consumers and manufacturers around the globe can expect from [this] technology."[122]

Ultimately, then, while setting a technology standard seems to be a politically neutral act, the work of experts and technocrats, in reality it "is an intensely political process, even if the politics may be hidden beneath a veneer of technical rhetoric."[123] The ability to shape or set international technical standards confers credibility and status by signaling technological prowess. Not surprisingly, it has become an arena of geopolitical competition.

China has long understood this. "Interest in standards ... is rooted in long-held aspirations for Chinese technology and the belief that through technological development, China can reclaim a position of wealth and power lost to technologically superior countries over the course of the past 150 years."[124] President Xi is guided by that mindset, arguing that to become a leading nation, China will have to become a rule-maker and that standards are the "first chess move" of global expansion. He insists that China strengthen its "leadership over standardization work" to promote the "hard power of China's standards."[125] Even more weighty among Chinese considerations are the implications of a failure to seize this opportunity: "If the key technologies and main standards of the Internet of Things are in the hands of Western developed countries, and [China] has no independent intellectual property rights, then China will have no chance of achieving its peaceful rise and national rejuvenation."[126]

Chinese efforts have proceeded on two tracks. The first is prioritizing the setting of standards at home and then exporting them through the activities of Chinese businesses. In 2006, China's State Council made explicit its intention "to make our country's technology standards international standards."[127] In 2016, Xi declared that China would "actively implement a standardization strategy" to "accelerate the promotion of China's international discourse power and rule-making power in cyberspace."[128] In March 2018, Beijing launched the China Standards 2035 project, when evolved into the National Standardization Development Strategy Research in January 2020.[129]

Government ministries with portfolios ranging from information technologies to construction urge companies they supervise to exploit "first mover advantage" and promote Chinese standards.[130] Those companies are supposed to "construct a brand around China standards" and integrate them into "engineering contracts abroad, major equipment and infrastructure exports, and foreign construction assistance projects ... to propel the going out of China's products, technologies, equipment, and services."[131]

The second track includes aggressive efforts to send personnel to international standard-setting organizations. Chinese experts have been comfortably ensconced in the top positions in those bodies. A Chinese national was elected president of the International

Organization for Standardization (ISO) in 2015, another was elected president of the International Electrotechnical Commission (IEC) in 2019 (after having served as vice president for five years), while another Chinese official began his second term as secretary general of the International Telecommunication Union in 2019 as well, after having served as deputy secretary general from 2007 to 2015.[132] China has also stepped up its presence in standards drafting technical committees. From 2011 to 2020, the number of Chinese-occupied secretariat positions in technical committees or subcommittees in the ISO increased by 73 per cent; they increased 67 per cent in the IEC. This growth occurred as the number of secretariat positions occupied by other major participants like the US, Germany, and Japan remained relatively flat.[133] PRC bilateral and multilateral standardization cooperation has made progress: as noted, by September 2019 China had concluded ninety bilateral agreements on technical standardization cooperation with fifty-two countries and regions.[134]

The export of Chinese standards, and the corresponding expansion of Chinese influence, has been a pillar of the Belt and Road Initiative. "China Standards" provide the essential technical connective tissue for the BRI and the Digital Silk Road.[135] The PRC has adopted a practice of offering "package deals" that bind the granting of loans for major infrastructure projects to the adoption of Chinese technical standards.[136] And since Chinese construction companies perform the majority of work on BRI projects, Chinese standards provide the default guidance.

Alarm bells ringing

The EU is well aware of the influence bestowed by "the Brussels Effect." There is a steadily expanding library of reports from EU institutions and think tanks that call on the EU to pay more attention to China's challenge to that longstanding role and to be more aggressive in shaping international norms and standards in a way that reflects European values and interests.[137] European officials believe that the EU can retain its influence over global markets even if European businesses can't go toe to toe with the tech conglomerates in the United States and China.[138] As part of that effort, the European Commission in February 2022 revealed its strategy on standardization,[139] one that will "support a more assertive stance of European interests in standardization (in the EU and globally) and in working openly with others on certain areas of mutual interest (e.g., the US and Canada on lawful and ethical use of AI)."[140]

US alarm bells have been ringing too. An October 2018 report to the US–China Economic and Security Review Commission warned that Beijing's effort to influence and set international standards was "a critical part of China's ambitious state-directed plan to achieve dominance."[141] It urged the US to become more active in standards-setting bodies as "[Beijing's] efforts may lock in Chinese preferences for standards in IoT and supporting infrastructure sooner rather than later, as nascent IoT and 5G standards exist in a fragmented and complex standards-setting environment."[142] The Biden administration responded with a National Standards Strategy for Critical and Emerging Technology, which noted that strength in standard-setting was instrumental to US global technological leadership, underpinning prosperity and fortifying its leadership in the industries of the future. "Bolstering US engagement will strengthen US national and economic security."[143]

5G first movers

Beijing puts emphasis on emerging technologies, areas where global standards are being set, and where China can leapfrog incumbents and assume a leading position.[144] A focus of this effort is fifth-generation or 5G telecommunications networks. 5G not only supercharges the capabilities of the handheld devices of daily life, but it also facilitates an ever-thicker weave of communications, such as those that empower driverless vehicles and the smart home. The sheer number of devices that are part of the 5G ecosystem means that companies that own the standard essential patents will enjoy abundant revenue flows as well as substantial first-mover advantages as those networks are built out. 5G is considered a standard of standards—a system that will empower a cascading set of technologies, capabilities and standards, and will define the larger information technology ecosystem.[145] Even if it is not a "winner take all" game, success will likely empower a "virtuous cycle" as "5G and related applications attract talent and capital while huge datasets generated by applications running on top of 5G networks spur further innovation."[146]

Among Chinese strategists, technologists, and politicians, the stakes could not be higher. "Whoever masters the model, architecture, and standards of 5G technology first has the right to speak in the future mobile network and the first-mover advantage of the industry chain. They can occupy a strategic leading position in future economic trade and military competition."[147]

No company better symbolizes China's ambitions in this field than Huawei. It quickly became a telecommunications giant, with global revenues that grew from $28 billion in 2009 to over $118 billion in 2024 as industry stalwarts saw their bottom line shrink over the same period.[148] In 2019 it was "the only company that can produce 'at scale and cost' all the elements of a 5G network, with its closest competitors Nokia and Ericsson unable to offer a viable alternative."[149] Huawei has aggressively promoted its 5G technologies, aided by government financial support that allows it to price its equipment considerably lower than the competition, and diplomatic support in BRI contracts along with other forms of assistance and "encouragement" from Chinese officials who help negotiate deals with foreign buyers.[150] That strategy paid off: in 2024, Huawei ranked first in both revenue and shipments and had over 42 per cent of the market of global 5G core networks.[151]

Not only has Beijing pushed Chinese 5G technology, but it has, consistent with the strategy outlined above, aggressively pushed its preferences on standards-setting bodies. Chinese companies participating as voting members in the Third Generation Partnership Project (3GPP), the body responsible for setting 5G standards, have more than doubled in the past few years, numbering 110 in January 2020, more than twice the fifty-three US voting members.[152] Chinese representatives held nine of the fifty-three 3GPP chair or vice-chair positions (about 17 per cent) in December 2012, a number that grew to eleven of fifty-eight positions (about 19 per cent) in December 2017.[153] Meanwhile, one-third of 5G technology is believed to consist of standardized patented technology, and the share of Chinese standard essential patent declarations as part of 5G is about 40 per cent, compared to the roughly 7 per cent in the previous generation of mobile technology (4G/LTE).[154] Huawei holds the largest number of standards-essential patents required to make 5G work, followed by Europe's Nokia and South Korea's Samsung, with Qualcomm in sixth place.[155]

If China is pleased with Huawei's performance, Western governments and companies are not. Apart from economic advantages conferred by (what they allege are) predatory practices and protected domestic markets,[156] they worry about the security consequences of a Chinese presence in their national telecommunications infrastructure. James Lewis, a former diplomat and one of the leading analysts of the geopolitics of digital technologies, analogizes to a situation in which the person who builds your house decides to burgle it. "They know the layout, the power system, the access points, may have kept a key and perhaps even built in a way to gain surreptitious entry." An operator could even install a perfectly clean and secure system and compromise it later with an ordinary software update. The operation and its customers would never know.[157] This fear prompted the US, its allies, and like-minded governments to ban Huawei from their national telecommunications grids.[158]

Western governments are promoting 5G systems such as O-RAN, or Open Radio Access Network as an alternative, which allows companies to "mix and match" components using different vendors, reducing or eliminating the "lock in" effect.[159] The Open RAN Policy Coalition includes US, European, Japanese, and South Korean telecommunications and information technology companies to advance this concept. At their 2021 summit, Biden and Japanese Prime Minister Suga Yoshihide agreed to commit $4.5 billion to "next generation and beyond" telecommunications systems that would build on O-RAN, with an emphasis on helping third countries.[160] The Trump administration launched the Clean Network to keep untrusted vendors out of national telecommunications networks. An expansive effort that includes all facets of telecommunications infrastructure, from submarine cables to smartphone apps, the Clean Network explicitly identifies Huawei as one of the "untrusted 5G vendors." Over sixty countries, representing two-thirds of global GDP and many of the world's leading developed economies, are members.[161]

Industrial policy

Long a favorite of developing economies, politicians and pundits throughout the developed world increasingly view industrial policy as the best way to respond to the challenges posed by the transformation of the economic order. Governments "from Washington to Beijing and New Delhi to London, are rediscovering the joy of subsidies and singing the praises of economic self-reliance and 'strategic' investment."[162] The European Commission has proposed a "fresh approach to industrial policy"; Japan is exploring a new "Japan Inc."; India has launched a "Made in India" strategy; China is advancing "Made in China 2025"; and US politicians have been openly calling for a new industrial policy.[163] At least fifty nations operate formal national innovation foundations that regularly produce national innovation strategies.[164] A 2024 study by IMF economists recorded over 2,500 new industrial policies worldwide, over a third of which aimed to address strategic competitiveness concerns,[165] and subsequent research by the same team concluded that the COVID-19 pandemic and supply chain dependencies were "important triggers in the shift in the use of industrial policies worldwide."[166]

The new vogue stems from unease in Western countries as they navigate the new economic landscape and sense that they are falling behind as China challenges Western pre-eminence—a challenge that has been powered to a large extent by industrial policy. This anxiety has been fueled by the economic difficulties that followed the Global Financial Crisis, which shook faith in economic orthodoxy, the COVID-19 pandemic, and intensifying geopolitical tensions. Economic historians also note that downturns invariably trigger attempts at some form of state-led economic revival, and industrial policy is typically employed either to manage the effects of macroeconomic flux or to achieve some other political objective—promote development of leading-edge technologies, resilient supply chains, vaccine, or PPE production, to name three of the most pressing recently.[167] The many uncertainties surrounding the transition to "digitalized, data-driven and technology-rich economies seem to have strengthened the case for state intervention."[168]

As is to be expected among the newly converted, the concept is fluid without an agreed definition. Traditionally, it has been used to support "industry"—manufacturing more specifically—and has invoked protection for domestic industries against foreign competition. In its crudest form, it tries to pick "winners"—companies or industries—that deserve state support. That version still exists, but today industrial policy seeks to address systemic and market failures, creating "an environment (or stable seedbed) out of which 'winners' may arise."[169] It is sometimes referred to as "competition" or "innovation" policy. This enthusiasm reflects the recognition that markets often undersupply the critical resources needed to make and sustain that environment—from higher education to basic scientific research to digital infrastructure—and governments must fill the deficit.[170] Self-aware policy-makers and analysts acknowledge that all societies do that: "Consciously or unconsciously, actively or passively, successfully or unsuccessfully, governments are continuously engaged in designing, executing, and fine-tuning what are effectively industrial policies."[171]

Funding is the easiest metric for industrial policy, and in recent years gross R&D investment levels have been a lever to increase productivity, technological innovation, and other benefits.[172] According to the US Congressional Budget Office, "Economic studies have shown that federal support for R&D—particularly early-stage research—has long been very important in promoting innovation," adding that "economic analysis suggests that the benefits from R&D would justify much higher spending on R&D and, in particular, much higher spending on basic research."[173]

The US once dominated global R&D, accounting for almost 70 per cent of the total in 1960; by 1995 its share fell to around 40 per cent, and by 2020 it was close to 25 per cent. Meanwhile, over the last two decades total global R&D investment tripled, topping $2.2 trillion per year. So, as the world spends more on R&D, the US share is falling, and other countries are catching up. The chief rival is China, whose R&D spending is almost equivalent to that of the US after growing its R&D expenditures by an average of 18 per cent annually since 2000; the rest of the world accounts for half of global R&D, or more than $1 trillion spread among a mixed group of countries.[174] One analysis projects China will spend $42 billion on basic research in 2026, which would approach 2024 US levels. However, the Trump administration's proposed 34 per cent funding cuts means that the US will fall behind China for the first time since World War II and the Cold War.[175] That reversal is one of the reasons that the Council on Foreign Relations Task Force on Innovation and National Security concluded that China is closing the gap with the US and will soon be one of the leading powers in emerging technologies.[176]

The US rediscovers its past

"Geopolitical rivalry is driving the new fashion for state intervention in the economy,"[177] but the US has ample experience with industrial policy. The founding fathers made the case for state action on behalf of its industries, and the government has often acted to protect its economic interests. Since then, the transition from an agrarian to an industrial economy was eased by federal government spending on transportation and creation of a national bank. After World War II, the "federal government made strategic investments in emerging technologies including microelectronics, telecommunications and biotechnology." The economic logic behind that thinking was simple: "Markets—on their own—will not make investments in the technologies and infrastructure that would benefit an entire industry," Brian Deese, director of the National Economic Council under Biden, said.[178] As Jake Sullivan, Biden's national security advisor, put it, "industrial policy is deeply American."[179]

From 1950 to 1980, the US federal government financed between 47 and 66 per cent of national R&D spending; that contrasts with about 20 per cent in Japan and Korea and around 30 per cent in Europe over the same time.[180] Support yielded national security payoffs and helped shape society more broadly. According to a 2019 study, starting in the 2010s nearly one-third of patented US inventions relied on federally funded science. Touch screens, GPS, and internet technologies central to the smartphone are all products of Department of Defense research. Research grants from the Department of Energy contributed to the development of Tesla batteries and solar panels, shale gas hydraulic fracturing, light-emitting diode (LED) technology, and 4D and 5D seismic imaging. Grants from the National Science Foundation helped build the internet, the

Google search engine, and magnetic resonance imaging (MRI) machines. Funding from the National Institutes of Health drove research that supported the sequencing of the human genome, advances in prosthetics, and cancer drugs. Between 1988 and 2010, $3.8 billion of federal investment in genomic research generated an economic impact of $796 billion and created 310,000 jobs.[181]

US policy-makers recognize that the country will continue to lose ground to China on key technologies, such as 5G and solar panels, "if Washington continues to rely so heavily on private sector research and development."[182] One of Biden's economic strategists was blunt: "Strategic public investment to shelter and grow champion industries is a reality of the twenty-first century economy."[183]

Congress has acted. Animated by the belief that "the PRC has chosen to pursue state-led, mercantilist economic policies,"[184] it passed the Chips and Science Act (CHIPS Act), an omnibus bill with a hodgepodge of China-related measures that was signed into law in August 2022. Its provisions included a $52 billion subsidy program for US semiconductor firms and another $174 billion in scientific research and development in areas such as AI, quantum computing, robotics, biotechnology, advanced energy, and materials science—the new and emerging technologies that will define success in the future economy.[185] A key element of the legislation encourages restructuring of supply chains to help the US produce more of its important needs at home, rather than import them from abroad. Biden's Indo-Pacific czar Kurt Campbell called the act "an effort to make clear that the new ramparts of competition, that will define American leadership, increasingly will be in technology, and we're determined to take the steps to run faster."[186] More simply, the US is "shifting the focus from defense to offense."[187]

The second Trump administration is skeptical of the CHIPS Act; the president called it "a horrible, horrible thing" and threatened to end the subsidies.[188] Yet despite denunciations of Democratic plans as socialist or communist, the administration has aggressively inserted itself into corporate decision-making. For example, it used CHIPS Act funds to purchase a 10 per cent stake in Intel, the struggling US chip manufacturer. More broadly, the administration continues to support the reshoring of manufacturing to the United States, but it is primarily using the stick of tariffs rather than the carrot of subsidies to influence business decisions.

Europe tries to claim its future

Europe is as worried, if not more so. While EU governments have been more accepting of—or at least more open about accepting—industrial policy, following the end of the Cold War they moved closer to the Anglo-Saxon approach to economic policy, lifting the government's hand and giving markets more influence over economic outcomes.[189] But those governments agree that economic orthodoxy—and resulting policy—needs updating in the wake of the 2008–9 Global Financial Crisis, the sluggish recovery, and the COVID-19 pandemic. One analysis adds challenges posed by climate change, the stagnation of southern bloc economies, and a pervasive concern that the continent has not adapted to the transition to the digital economy and is living off its twentieth-century successes. Only one of the world's top twenty largest tech companies is from Europe, while there are fifteen from the US and two from China.[190] No wonder that "there has been a growing perception among European leaders that for several aspects regarding the ongoing economic and technological transformation, market forces alone might not be sufficient to ensure that Europe stays afloat of the current."[191]

Like the US, Europe is also concerned about China. Publics across the EU are skeptical of Beijing's intentions, and member states "view China pragmatically as a partner, but also almost everywhere simultaneously as a rival."[192] "The rise of China as a 'strategic rival' in key industries and technologies and the success of its state capitalist economic model are major challenges to the EU's own perception as an innovative and efficient high-tech economy."[193] There is a broad consensus that the EU is not "equipped to tackle the challenges posed by the country's state-led authoritarian model and state-capitalist economy."[194]

Europeans are also worried about being caught in the crossfire between the US and China. David Miliband, the former UK foreign secretary, acknowledged that "Europeans don't see equidistance between China and America in values and outlook. But they don't want to be forced—and they don't think it's strategic to be forced—to disengage with China."[195] That underpins the European Union's commitment to "strategic autonomy" as a guiding principle and technological resilience to ensure that it retains the capacity to act on its own.[196] The European Commission warned that "Europe's economic sovereignty is at stake. Other global powers are combining geopolitical and economic interests to increase their influence in the world. This includes protectionism, export control and the international role of currencies." This is the backdrop to the new Industrial Strategy, first articulated in 2020 and updated a year later after COVID hit.[197]

The strategy argued that Europe needs "a new industrial way," one that is driven by "the ability of Europe's industry to lead the twin transitions [green and digital] and drive our competitiveness. It cannot afford to simply adapt—it must become the accelerator and enabler of change and innovation. Our industrial policy must help make this ambition a reality."[198] It is an industrial innovation strategy at its heart.[199] Its key domestic features include: strengthening EU capacity in critical digital technologies such as 5G, AI, and metadata analytics; boosting innovation; leveraging public–private partnerships to help develop technologies to meet industry goals; financing investment and innovation; revamping competition rules, including those regarding state aid; transitioning toward a circular economy; and upgrading skills available to European industry. These will be complemented by efforts to deepen the EU industrial base in strategic arenas, such as critical digital technologies, defense, space, and pharmaceuticals; and adopting legal and regulatory measures to rebalance global competition, including a screening mechanism for foreign investment, the enactment of a carbon border adjustment tax, and reinforcing customs controls.[200]

For the EU, two tools stand out: industrial alliances and the Important Project of Common European Interest (IPCEI) framework for funding projects. The EU identified industrial alliances as key enablers of change, to "help Europe lead the green and digital transitions, maintain industrial leadership and benefit European businesses and society while reinforcing Europe's resilience." Alliances include the European Battery Alliance, the Plastic Alliance, the European Clean Hydrogen Alliance, the European Alliance for Industrial Data and Clouds, and the European Raw Materials Alliance.[201]

The IPCEI platform was established in 2014 to "give Member States a tool to address market failures in financing large transnational projects of a strategic importance."[202] Its first tranche of funds, released in December 2019, included €3.2 billion for a pan-European research and innovation project in all segments of the battery value chain, which was followed by a second outlay of €2.9 billion thirteen months later for another pan-European research and innovation project along the entire battery value chain.[203] "It is a well-established tool with the ability to deliver short-term results."[204]

Yet for all the attention, there is near unanimity that the EU must do more. The most pointed criticism came in a September 2024 report to the European Commission issued by Mario Draghi, former prime minister of Italy and former president of the European Central Bank. He warned that Europe faced "an existential crisis"; the continent's slow growth was once "an inconvenience" but is now "a calamity." Fixing this, he concluded, requires "a new industrial strategy for Europe," one that includes further consolidation of the single market and $800 billion annually in new funding.[205] Taking stock a year later, the European Commission applauded creation of the "Competitiveness Compass," noting that 90 per cent of its flagship initiatives are directly inspired by Draghi's recommendations, that more than half have been delivered, and over €1 trillion in capital has been mobilized to ensure that "Europe remains competitive, sustainable and resilient."[206] Draghi countered that as the challenges have grown more acute, Europe's "growth model is fading. Vulnerabilities are mounting. And there is no clear path to finance the investments we need."[207]

Japan ups its game

Japan wrote the book on industrial policy and for good reason. In the second half of the twentieth century, Japan enjoyed unprecedented economic growth, with GDP rising from $44 billion in 1960 to $5.45 trillion in 1995.[208] Tokyo used strategic and corrective industrial policy to promote the development of targeted sectors and fix market failures,[209] transforming the economy and helping domestic companies capture significant market shares of key global industries.[210]

The bursting of the bubble economy in the early 1990s and the resulting stagnation shook the foundations and credibility of Japanese economic policy. Despite periodic recoveries and glimmers of hope, events—the Global Financial Crisis, COVID-19—conspired to crush any nascent turnaround.[211] Authorities across the government continue to produce society-wide schemes to promote or facilitate a national transition. Society 5.0, discussed in Chapter 1, is the most sweeping and recent of these visions.[212]

Central to the new policies is a mindset that closely integrates economics and security.[213] Even before the COVID-19 outbreak, the government had recognized a need to think differently since "the limitations of advancing science and technology innovation policies as an extension of past measures have been revealed. Japan is pressed to reconsider its socioeconomic system with its organizations, structures, enterprises conducts, customs, and ways of working."[214] That analysis, like that of the updated Integrated Innovation Strategy which followed, notes Japan's declining competitiveness and its position in digital platforms, R&D funding, venture capital investment, unicorns, (and other measures) relative to the US and China, and is invariably found wanting.[215] METI has insisted that Japan "focus on measures to 'develop' domestic capacity to advance those critical technologies in which we have an advantage while reducing our current technological overreliance on other countries in certain areas."[216]

That logic is evident in the New Growth Strategy endorsed by the government of Prime Minister Suga Yoshihide in the summer of 2021. It has a strong focus on enhancing domestic development and production of cutting-edge semiconductors, a vulnerability that became clear during the COVID outbreak. Most semiconductor factories in Japan make low-end products, and the country imports 64 per cent of the chips it consumes annually from abroad. The strategy will "build a system which ensures a steady semiconductor supply" and reduce dependence on foreign manufacturers by offering research funding and encouraging chipmakers to build factories in Japan.[217] Green tech is yet another priority, with the growth plan establishing a ¥2 trillion ($18 billion) fund.[218] The Economic Security Promotion Law is another element of Japanese efforts to promote development of key technologies, with many of those funds going to subsidize construction of semiconductor facilities.

Japan has also embraced "moonshot" projects, following economist Mariana Mazzucato, who argues that a "challenge-driven, mission-oriented" policy can deliver results.[219] In 2019, the Japanese government launched the Moonshot Research and Development Program, which "aims to create disruptive innovations from Japan and promotes challenging R&D based on revolutionary concepts that are not simply the extension of existing technologies, i.e. moonshots."[220] Its ¥100 billion budget will be used in pursuit of seven goals: overcoming limits of body, brain, space, and time; ultra-early disease prediction and intervention; co-existence of AI and robots; fault-tolerant universal quantum computing; sustainable resource circulation to stabilize the global environment by 2050; sustainable global food supply by 2050; overcoming major diseases by 2040 and allowing for a full productive life to the age of 100.

The Japanese government makes no bones about its role. A pillar of the 2020 Integrated Innovation Strategy is "strengthening the control tower function of government," which guides and coordinates action throughout the government.[221] Less than two weeks after taking office in October 2025, Prime Minister Takaichi Sanae launched her government's economic strategy headquarters that would direct strategic investment (and other policies) in seventeen key industrial sectors, with AI and semiconductors among them.[222] Policy-makers in Tokyo remain wedded to a political-economic model that affords a leading role to government, and the need for greater sensitivity to the security dimensions of economic policies reinforces those instincts.

China sets the pace

Driving much of this is China. Its ability to develop national capacity and in many cases surpass established leaders has forced other governments to reassess their policies and rethink orthodoxy—especially when this extraordinary record is the product of "national strategies." Jimmy Goodrich, a China expert who focuses on new and emerging technologies at RAND, points to the recommendations issued by the Party Central Committee in October 2025 that called for "extraordinary measures" to secure the country's "long-term science-technology autonomy." He noted that *China Daily*, official outlet for the government, said that the report "signals 'profound mobilization' to overcome intensifying containment from the West and win major-power competition by 'seizing the commanding heights of scientific and technological innovation.'"[223]

While the West invariably references Chinese policies and progress to justify its actions, considerably less attention is given (in the West) to the ways that China sees Western behavior as forcing it to act as it does. While self-reliance is hard wired into the Chinese psyche, that inclination is reinforced by decisions of other governments that are interpreted as hostile to China. Typical is the argument of Li Zheng of the China Institutes of Contemporary International Relations, who insists that "The United States views technology as the 'last barrier' to constrain China's challenge," and the US seeks to "systematically and comprehensively curb the rapid rise of China's technology industry."[224]

This escalating competition, an industrial policy arms race, has yielded a "strategic capitalism" in which the twin forces of great power competition and the vulnerabilities arising from market-led economic interdependence push governments to intervene to ensure that market transactions do not imperil the bases of national economic power. The result is a surge in policies designed to shield "strategic assets," "critical infrastructures," or "emerging technologies" from the free interplay of market forces—and to block the transfer of control over these assets to foreign "strategic competitors" or "systemic rivals" through market mechanisms.[225]

PART III

BUSINESS IN THE NEW NATIONAL SECURITY ECONOMY

6

THE NEW NORMAL FOR BUSINESS

Globalization's reckoning dominated the annual gathering of elites at the May 2022 World Economic Forum. The Davos conclave normally convenes in January, but the first in-person meeting in two years was delayed because of Russia's invasion of Ukraine and COVID-19 pandemic concerns, two developments that hung menacingly over discussions that sought to capture the political and business zeitgeist. Tellingly, the theme of the 2022 forum was "History at a Turning Point: Government Policies and Business Strategies."[1]

Gideon Rachman, chief foreign affairs columnist at the *Financial Times* and a Davos mainstay, explained that "For 30 years, from the end of the Cold War to the start of the pandemic, geopolitical rivalries took a back seat to globalization—and nowhere was this more evident than here. The informal motto of Davos was 'make money, not war.'"[2] A long-time observer of globalization's ebb and flow, Rachman in his 2010 book, *Zero-Sum World*, examined how globalization might unravel if only markets converged and not their accompanying political norms and values. With that fragmentation underway, partially because of the failed political convergence, geopolitical concerns have returned to the driver's seat.[3]

Observing this transition, José Manuel Barroso, who served as Portugal's prime minister and later as president of the European Commission, commented, "Tension between the US and China was accelerated by the pandemic and now this invasion of Ukraine by Russia—all these trends are raising serious concerns about a decoupling world ... [Globalization faces] friction from nationalism, protectionism, nativism, chauvinism if you wish, or even sometimes xenophobia, and for me, it is not clear who is going to win."[4]

Though the debate continues, these observations by Rachman, Barroso, and others are the essence of the new national security economy: how the complexities of connectivity, technology, and geopolitics collide and impact companies operating in this new environment.

In October 2021, then HSBC CEO Noel Quinn spoke at Sibos, an influential financial conference organized by SWIFT, the organization that sets standards for facilitating interbank transfers. Drawing on HSBC's long history as a bridge between East and West, Quinn noted that:

> Our clients still want to be international. They still have needs on a global basis and they're asking us to help them with those global needs ... My job is to help this organisation navigate those geopolitics, be very focused on helping our clients navigate their business model needs and remaining focused on the long term, not just the short term.[5]

As head of one of the world's largest banks, Quinn understood the complexities and challenges that business leaders encounter. Before becoming chief executive in 2020, Quinn led HSBC's Global Commercial Banking business. In December 2018, HSBC was thrust into the middle of a complex crisis when Canadian authorities arrested Huawei CFO Meng Wanzhou on behalf of US law enforcement seeking her extradition. The US Department of Justice claimed that Huawei and Meng had defrauded HSBC and violated sanctions against Iran through its relationship with a subsidiary company called Skycom. For over two and a half years, Meng challenged her extradition in Canadian courts before reaching an agreement with US prosecutors that allowed her to return home.[6] Meng arrived at Shenzhen Baoan International Airport in September 2021 to a hero's welcome that was widely covered in Chinese and global media.

Leading up to and throughout the international legal drama that connected Huawei, HSBC, China, the United States, and unwittingly,

Canada, Huawei faced intensifying scrutiny triggered by concerns about its 5G telecommunications infrastructure hardware, which was expected to be widely adopted around the world. As we've noted in Chapter 5, 5G is at the heart of the next wave of connectivity and plays a critical role in the emerging generation of transformative technologies. Huawei is one of the few companies that supports 5G wireless networks and was positioned to dominate the market, even in Western countries. The United States and other countries were concerned that a dominant market position would result in the possession or control of critical infrastructure, data, and other forms of vital information by a strategic competitor. Meng was arrested amid this increasingly fraught technological, economic, and strategic struggle.

The Huawei-HSBC-China-United States saga represents in microcosm the new national security economy, which is transforming global business. This phenomenon reflects the confluence of four factors: 1) deep connectivity within and between societies, which has yielded benefits like globalization but also introduces acute risks related to interdependence; 2) geopolitical competition largely between the West, led by the United States, and China, in which 3) a primary arena is advanced technologies, AI among them; and 4) the private sector serving as a primary actor in this battlespace.

For business leaders, Quinn's framing of the problem was a reminder and a warning. Understanding and mastering this interplay between short- and long-term, or tactical and strategic choices, is foundational in this new business environment. A seemingly tactical decision can create long-term, structural outcomes that may prove difficult, if not impossible, to change or reverse.

What has changed?

While businesses have always had to navigate a fluid operating environment, the current moment is unique. The complexity, gravity, impact, and frequency of today's challenges mean that most business leaders traverse unchartered territory. Blame a once-in-a-lifetime global pandemic or protests, natural disasters, trade wars, actual wars, and other variables beyond the influence and control of businesses. Whatever the cause, there is a growing awareness of and concern about the frequency and impact of nontraditional business risk, or the type of risks that exist outside a typical business function. For example, a financial institution may understand credit risk well, but it is likely not prepared to account for the impact that, say, an election result or pandemic will have on its business and customers.

"Investors today are confronted with a dramatic increase in the political risk profile of what were supposedly low-risk, or indeed risk-free, developed markets,"[7] warned Duncan Wales, former CEO of Exotix, an investment bank focused on emerging markets. "A dozen years ago it seemed geopolitical risk was a thing of the past and that the apparent dominance of the liberal economic and political model had heralded the end of history. Now geopolitical risk has returned as a great shadow on the horizon of all significant investment decisions."[8] Many leaders are not ready for this shifting landscape.

The new classics

In the 1980s, Margaret Thatcher's support of a free-market economy was embodied in the slogan, "There is no alternative," later known as the TINA doctrine. And indeed, by the end of the Cold War, there was no alternative. The world seemed to be on the brink of a new era, one that was defined by democracy, free markets, and globalization, all of which were shaped by Western ideals, values, and norms. The zeitgeist of this post-Cold War world was captured by three best-selling and influential books that were published after the Soviet Union collapsed.

The first was Francis Fukuyama's provocatively titled *The End of History and the Last Man*.[9] Published in 1992, it argued that the fall of the Soviet Union and defeat of communism meant that Western liberal democracy had won and the contest for ideas was settled once and for all. The book cast its author, then a professor at Johns Hopkins University's School of Advanced International Studies, into the leading ranks of global intellectuals. The West's victory in the Cold War, he concluded, marked "the end-point of mankind's ideological evolution and the universalization of Western liberal democracy as the final form of human government."[10] The timing and power of his argument—and the collapse of the Soviet Union—meant that this perspective was widely accepted in classrooms, boardrooms, think tanks, and halls of government around the world.

A few years later, *New York Times* columnist Thomas Friedman produced two international best-sellers that took Fukuyama's thesis further and argued that there was seemingly boundless opportunity in an interconnected world facilitated by free markets, technology, and open flows of ideas, capital, and people—what came to be known as globalization. He first explored this thesis in his 1999 book, *The Lexus and the Olive Tree: Understanding Globalization*,[11] and then expanded the analysis six years later in *The World Is Flat: A Brief History of the Twenty-first Century*.[12] Both were quickly added to university syllabi and became required reading for businesspeople with even the slightest of international aspirations.

Historically, globalization is a natural outgrowth of increasing connectivity between people, organizations, and events through technological advances that produce greater interdependence and reciprocal impacts. Globalization, in its simplest, most basic form, is a process. The evolution in communication from written letters to telegrams to phones and to instantaneous, electronic communications is one representation of this phenomenon.

Globalization largely mirrors advances in technologies and the corresponding rise of international commerce. Harvard professors Joseph Nye and John Donahue expanded on this in *Governance in a Globalizing World*, describing globalization as "the process by which globalism becomes increasingly thick."[13] When they published their book at the beginning of the twenty-first century, globalization was not a new phenomenon, but increasing connections—what they called "thick globalism"—were transforming the landscape. "The general point is that the increasing thickness of globalism—the density of networks of interdependence—is not just a difference in degree from the past. Thickness means that different relationships of interdependence intersect more deeply at more different points."[14]

Friedman, Donahue, Nye, and many other commentators were attempting to describe the same phenomenon: a foundational shift in global business and politics that marked the beginning of a new era. To put it succinctly, for roughly two decades, from the end of the Cold War to the early 2010s, the global environment that business leaders anticipated (and often operated in) rested on two pillars: Western liberal democracy and globalized free markets. CEOs internalized these norms while in school, and they provided the structure for the resulting mental paradigm of the business landscape. This model was augmented by several key assumptions:

1. The United States and the West more broadly enjoyed economic, strategic, and technological primacy and would continue to do so.
2. China was a low-cost, low-skilled, low-technology manufacturing hub that was the world's factory but not a source of innovation.
3. Technology was a powerful tool to facilitate connectivity through communication, helping to create and expand markets to sell goods.
4. A rising tide lifts all boats, and as authoritarian or closed countries participated in the global economy, free markets would lead first to economic development and then political liberalization; this was hoped to be particularly true for China. Typical was the comment of former US President Bill Clinton, who rhapsodized that:

   > China is not simply agreeing to import more of our products; it is agreeing to import one of democracy's most cherished values: economic freedom. The more China liberalizes its economy, the more fully it will liberate the potential of its people—their initiative, their imagination, their remarkable spirit of enterprise. And when individuals have the power, not just to dream but to realize their dreams, they will demand a greater say.[15]

5. While politics and business are never entirely separate, the end of the Cold War meant that businesses could largely ignore geopolitics as countries worked together to open markets and abide by universally accepted norms (e.g., the creation and expansion of the World Trade Organization).

This world view and its epistemological pillars are crumbling. A quick look at those assumptions reveals how much has changed.

1. The United States is no longer an undisputed, dominant power. Its unilateral moment has seemingly passed, and it now faces direct or near-peer competition from China across strategic, economic, and technological domains.

2. China has raced up the manufacturing value chain, has transformed itself into a hub for innovation, and is now home to a sophisticated, nuanced domestic market that is among the most important and influential globally.[16]
3. Technology no longer just supports traditional business; instead, the high-tech industry itself has supplanted many traditional industries and is powering another wave of change that will fundamentally alter the complexion of society.
4. The emergence and superiority of Western-style democracy and free markets are no longer a given. China has thrived under its own rules, while Russia pursues its own path without embracing Western norms. China's successes have demonstrated there are alternatives to the Western blueprint.
5. The global norms, and resultant stability, once certain to shape economic relationships seem distant, while various, but ever-present, political, social, environmental, and health shocks are becoming a permanent part of the operating environment for most businesses.

The mental model that once guided business leaders is in tatters. Though they may be conscious of a changing operational landscape, they are either in denial or are just beginning to process the implications of this transition. This upheaval creates new risks that businesses and their leaders are struggling to understand and address.

Risk in the new national security economy

The new national security economy demands that businesses revisit their understanding and perception of risk and how to manage it. Traditional business risk management focuses on mitigating the risk related to a specific function or area. For example, a company may monitor compliance risk, credit risk, legal risk, operational risk, and so on. These longstanding forms of risk management and mitigation remain relevant, but they do not provide meaningful guidance on how to navigate the current storm. Traditional risk categorization lacks the holistic perspective and contextual awareness that organizations and their leaders need in this new business environment.

Guidance is emerging. Harvard Business School professor Robert Kaplan and his co-authors endorse an expanded type of risk management. Writing with Anette Mikes in the *Harvard Business Review*, Kaplan calls for risk management that focuses on three types of risks: preventable risks, strategy risks, and external risks.[17]

They define preventable risks as those emanating from compliance-based risk prevention. These risks are avoided by following certain rules, processes, and procedures. Preventable risk is solely focused on minimizing or eliminating the risk of a bad thing happening. Consequently, it is important to note that preventable risk measures do not lead to advantages associated with a business taking risk to achieve additional gain. For example, a factory may have a rule that a certain machine needs to be powered down at the end of the day otherwise it may not work the next day. This practice is intended to prevent a bad outcome from occurring but does not provide additional benefit to the factory beyond what is normal or expected. Similarly, a person may be very concerned about capital loss. Thus, the rule-based risk prevention technique that is employed is to only invest in a government-insured bank account. By following this rule, the person will likely avoid significant monetary loss (not accounting for inflation), but following this rule will also require the person to forego any potential gains that could have been made by investing in a different asset class that offers more upside along with more risk. Ultimately, preventable risk is about avoiding self-inflicted harm by complying with guidelines.

Preventable risk, however, is just a first step. After addressing such risks, successful organizations need to incorporate strategic risk into their plans. Strategic risks are calculated risks taken for a potential benefit. For example, despite concerns about losing money, a bank may decide to make a loan having judged that the potential benefit outweighs the likelihood and magnitude of a negative outcome. Strategic risks can be ameliorated through rules-based risk prevention along with prudent risk management practices. In the bank example, preventable risks would include following the proper loan evaluation process, while strategic risk management would add to that by implementing some type of hedging to offset the risk of loan default.

Well-run businesses can manage preventable and strategic risks. They may not fully mitigate these risks, but they generally understand them. This is not the case for their third category, external risk. Kaplan and Mikes define external risk as those risks that originate outside the organization that are usually difficult to anticipate let alone control. External risk overlaps with many of the variables we consider part of the new national security economy. This category, however, is general in nature and does not provide substantive guidance for identifying and engaging the issues that dot the new business landscape.

The vast, inchoate, and somewhat conceptual nature of external risk obliged Kaplan, Mikes, and another co-author, Herman Leonard, to introduce the idea of "novel risks" in a subsequent article.[18] They define novel risks as risks that are unforeseeable because they seem distant, remote, and otherwise unlikely. Novel risks include "black swans" that are beyond the realm of experience, imagination, or are distant in experience or geographically.[19]

Novel risks encompass scenarios where risk emerges unexpectedly, swiftly, and on a massive scale causing figurative, and in some cases, literal destruction. They can be triggered by tsunami-type events like the 2011 Tohoku earthquake, which caused the tsunami that led to the disaster at Japan's Fukushima nuclear facility or the COVID-19 pandemic. Or they can be triggered by bounded thinking. For example, before 2008, it was universally assumed that home values in the United States would only rise. The decline in the housing market that precipitated the 2008 Global Financial Crisis was a type of novel risk. Novel risks tend to be associated with a perfect storm scenario in which multiple failures of routines and processes collectively yield a catastrophic event.

While Kaplan and his collaborators offer useful perspectives for many issues that organizations contend with, their framework does not provide a comprehensive approach to the new national security economy and how businesses should respond. Executives and their organizations encounter an existential challenge as the assumptions, norms, and standards that have guided them for a generation are being torn apart by the convergence of connectivity, technology, politics, and strategic competition. They must understand and then maneuver through this environment to survive and thrive. In the remainder of the chapter, we highlight key concepts that business leaders must monitor and address to succeed in the new national security economy.

1. Geopolitical risk as a strategic input

Businesses have always had to deal with political considerations like policy shifts, regime change, or even armed conflict. But today, increased connectivity, rapid technological evolution, and global operations have amplified the extent to which politics now pervade nearly every aspect of business from intellectual property protection to hiring. Businesses and their leaders must be better prepared for domestic and international political dangers.

Ian Bremmer recognized the potential significance of the shifting landscape as the Cold War drew to a close, and in 1988 founded the Eurasia Group, a political risk consultancy. As former Soviet and centrally planned economies transitioned from communism to capitalism, Bremmer understood that geopolitical factors would play a decisive role in a world of dynamic transformation. Eurasia Group was, Bremmer claimed, the "first firm devoted exclusively to helping investors and business decision makers understand the impact of politics on the risks and opportunities in foreign markets."[20] The interest in such work and its importance has continued to grow since the firm's founding.

That case was echoed in "The CEO as Chief Geopolitical Officer," a 2017 report by Eurasia Group and KPMG, the global accounting firm.[21] The title made clear its chief message: geopolitics should be a strategic consideration for companies and a priority for the C-suite and board. The authors argued that

> only by moving politics to the forefront of strategy can a CEO adequately steer today's global businesses. Such thinking starts with an acceptance that global companies are assumed to be political entities and CEOs de-facto political players. It means recognizing that a different approach is required, one that elevates politics in the boardroom and puts it on a par with other strategic challenges.[22]

John Chipman, Executive Chairman of the International Institute for Strategic Studies, one of the world's leading think tanks, insists that companies need to formulate their own foreign policy to manage increasing geopolitical volatility.[23] While few companies have gone that far, many companies and business leaders acknowledge that geopolitical considerations are central when crafting and executing strategic goals. Unfortunately, however, most organizations are not prepared for the unknown and ever more dangerous terrain of the new national security economy.

This became clear to one of the authors when a technology company he advised considered launching a new business in Europe. Several jurisdictions in Western and Eastern Europe were discussed, including a former Soviet republic in the Baltics. Although the country in question was not Ukraine, and these discussions took place a few months prior to Russia's February 2022 invasion, the likelihood of military action was a realistic possibility given the Kremlin's escalating rhetoric, Russia's 2014 annexation of Crimea, and the deployment of more than 100,000 Russian troops to border areas near Ukraine.

Given Russia's provocations and Moscow's warnings about the consequences of former Soviet republics becoming more aligned with Europe, it was obvious to us at least that Russian military aggression was a possible risk for a new venture in the proposed country. To be clear: we weren't predicting that war was imminent, would occur, or even that the likelihood was high. We were describing a low likelihood, high-impact risk. Nevertheless, the comment caught meeting participants off-guard. The risk was unexpected and a concern that had not been considered. This failure reflected not only a lack of imagination (or the readiness to connect news headlines to business plans) but also the absence of a systematic framework to account for this risk. Worryingly, this experience is not uncommon even among the most sophisticated and experienced business leaders and organizations.

Why? First, many companies typically lack personnel with meaningful knowledge about politics, economics, and other disciplines needed to connect the dots. Second, even when a business has the talent, relevant geopolitical insights and risk factors are typically not part of formal strategy-formulating processes and are often only an ancillary part of risk reviews. Third, geopolitical issues do not always find their way to the relevant decision-maker, and even when they do, it is often because of an urgent situation and not part of a disciplined organizational process.

Upskilling an organization to enhance understanding of new national security economy risks takes far more than a subscription to *The Economist*, perusing newsletters, and attending a conference every now and then. It is not plug and play like a software update. It demands a nuanced understanding of factors that may seem disparate or distant and can only be integrated into an organization's strategic decision-making through genuine knowledge and processes to consider and act on that knowledge.

2. Reconceptualizing assets and their protection

In an introductory economics course, students learn that the traditional factors of production are land, labor, entrepreneurship, and capital. Historically, the value of most companies was derived from physical assets, such as gold, crops, inventory, or real property like factories and land. Businesses had a standard set of tools to protect those assets: security to guard a warehouse, insurance, hedging transactions, or executing robust legal agreements to protect intellectual property, among them. Most of these methods are not new; some have been around for centuries.

Those traditional assets still represent important value. But technology has expanded the definition, composition, and value of assets for many companies. Most fundamentally, the technologies that underpin what many refer to as the Fourth Industrial Revolution rely on data. Accordingly, the value of many organizations is increasingly reflected in intangible rather than tangible assets. A generation ago, you could, in many cases, physically touch a company's most valuable asset like a new factory or the ground of a new drilling site.

Today, data is an additional factor of production. The most valuable companies derive their value from intangible assets: algorithms, customer data, and intellectual property. These intangible assets are often stored far away, in a distant server. Major cloud service providers like Amazon Web Services, Google Cloud, and Microsoft Azure store data across centers that are often located in multiple jurisdictions. The rise of cloud computing will only accelerate this trend.

This transition to digital assists is having a profound effect. According to one World Economic Forum estimate, approximately 60 per cent of global GDP is now digitized in some way, and that number is growing.[24] The importance of those digital assets is clear when considering what drives value for many companies. Aon reports that intangible assets comprised 84 per cent of the value for the S&P 500.[25] This has important implications. First, the largest and most valuable companies in the world derive their value from intangible assets. Second, a large portion of these intangible assets are tied to data, specifically the ability to accumulate, analyze, and act on data. Third, technology companies have been the most successful in capturing the value of intangible assets. In today's competitive global landscape, the battle for supremacy will be driven by technology companies that capture relevant data, use AI to generate meaningful insights from that data, and then swiftly execute superior business decisions.

This dynamic suggests that technology companies will not only be primary actors in the new national security economy, but they will be among those most significantly impacted as well. Despite its many controversies—or perhaps because of them—Facebook remains a representative example of how data leads to dollars. Prior to its May 2012 IPO, Facebook argued in the *New York Times* that this was the key to its future.

> The company has been busily collecting that data for seven years, compiling the information that its more than 800 million users freely share about themselves and their desires. Facebook's value will be determined by whether it can leverage this commodity to attract advertisers, and how deftly the company can handle privacy concerns raised by its users and government regulators worldwide.[26]

Facebook's share price when it listed was $38. Ten years later, Facebook reported that it was approaching 3 billion users and that data-based advertising drives its prodigious revenue. As of this writing, its share price topped $650. The company rebranded itself as "Meta" in 2021 to both prepare for and herald the arrival of the metaverse that it intended to master and dominate.

Yet, despite its success in commoditizing data—or again, perhaps because of it—Meta confronts intensifying regulatory scrutiny and has been cast into the spotlight after its data was misused during the 2016 US presidential elections by Cambridge Analytica,[27] ostensibly at the behest of Russian interlocutors to favor Republican candidates, namely Donald Trump and Ted Cruz, and to support the separation of the United Kingdom from the European Union during the Brexit referendum in 2016.[28]

The experience of Chinese company Didi Chuxing ("Didi") is a cautionary tale showing how technology companies can be whipsawed by technological and political developments. Didi, like Uber, is a ride-hailing company and is the dominant platform in China. In fact, Uber sold its China business to Didi in 2016.

Despite signals that Chinese regulators might object due to data security concerns, Didi conducted its initial public offering on the New York Stock Exchange (NYSE) on 30 June 2021. Those regulators did object, and over the subsequent six months, Didi's app was removed from app stores, and the company was barred from signing up new customers in China.[29] These regulatory actions lead to Didi's share price collapsing by 75 per cent. Ultimately, Didi announced that it would delist from the NYSE and eventually conduct another offering on the Hong Kong Stock Exchange. The Chinese authorities' actions reverberated through capital markets. Regulatory uncertainty effectively stopped Chinese companies from listing on US exchanges, and a spate of activity followed in Hong Kong as the Hong Kong Stock Exchange was expected to be a primary beneficiary of Chinese companies "coming home" either through re-listing or the pursuit of secondary listings in Hong Kong. The Didi saga is a bellwether for other US-listed Chinese companies that negotiate pull and push pressures from political and regulatory forces on both sides of the Pacific.

Increasingly, companies and governments understand that data is the currency of corporate value. As Georgetown University's Lizhi Liu cogently observes, "data has changed the basis of power. A state's strength lies in not only its military or trading power, but also its capacity to collect, refine, and utilize data."[30] As companies become more reliant on data due to advances in technologies like artificial intelligence and become more efficient in unlocking the value of data they collect, the worth of the companies that possess these digital assets will increase. As a result, data, technology fueled by data, and the companies that dominate by effectively leveraging data will have an outsized impact on the world. Consequently, and perhaps unwittingly, they will also become key players in the new national security economy and will need to be better attuned to the geopolitical implications of their assets.

3. Whose data? Protecting customers

In his 2019 letter to JP Morgan's shareholders, CEO Jamie Dimon, wrote: "The threat of cyber security may very well be the biggest threat to the US financial system."[31] Two years later, JP Morgan's International Council, which includes Dimon and luminaries such as former Secretary of State Condoleezza Rice and former Secretary of Defense Robert Gates, went further, calling cyber "the most dangerous weapon in the world—politically, economically and militarily."[32]

According to the Cyentia Institute, over 60 per cent of the Fortune 1000 have had at least one data breach in the last decade.[33] According to some estimates, a cyberattack occurs every thirty-nine seconds.[34] This increase in cyber risk has focused the attention of businesses. Gartner reported that 88 per cent of boards of directors viewed cybersecurity as a business risk and not just a technology risk, meaning that this shift in risk perception signifies that technology risk is not a discrete risk that exists in its own silo but is interconnected with the rest of an organization's core business and its survival.[35]

Even non-technology companies must make cybersecurity a top priority. Most companies are one significant data breach away from unwanted public and regulatory scrutiny, declining value and share prices, and significant litigation. For example, Equifax, a leading American credit bureau, reported a data breach in July 2017 that reportedly exposed the records and confidential information of almost 150 million Americans, just under half the population of the United States.[36] Over the next few months, Equifax's share price fell over 40 per cent,[37] and it later agreed to a $425 million global settlement with the US Federal Trade Commission.[38] It's telling that the Department of Justice issued criminal indictments against Chinese military personnel for allegedly being responsible for the Equifax breach.[39] Businesses may not be alert to the security implications of their work and their assets, but potential adversaries are.

Given the centrality of data for the digital economy, businesses and government are increasingly concerned about unauthorized access, use, and abuse of data by cyber threats. The types of threat actors have multiplied and motivations for cyberattacks vary. As the value of intangible assets for companies has increased, the cyber threat landscape has heightened accordingly and grown more complex. While attacks might have previously focused on corporate espionage (like efforts to steal vaccine data mentioned in Chapter 2), today's motivations also include financial gain, nation-state spying, intentional disruption of systems, or even politically motivated revenge. Bad actors include former employees, competitors, terrorists, nation-states, and even criminal enterprises that provide hacking as a service. Plainly, there are many different sources of threat with discrete and sometimes overlapping motivations.

North Korea is an archetype for a multi-faceted threat actor. Over the last decade, North Korea has launched cyberattacks for every possible reason. Attacking financial institutions as well as hacking cryptocurrency players provides the regime with access to funds to finance its operations. In 2025 alone, North Korea was responsible for at least half of the roughly $2.7 billion in crypto assets stolen.[40] In addition to extortion or theft, North Korea also utilizes cyberattacks to fulfill what the Pyongyang government deems to be political or strategic objectives. This can take the form of hacks to access the personal data of 1,000 North Korean defectors now living in South Korea,[41] or the attempted theft of proprietary vaccine information from pharmaceutical companies during the early days of the COVID-19 pandemic.[42] North Korea reportedly attempts an average of 1.5 million cyberattacks daily against South Korea businesses, governments entities, and individuals.[43] While extreme, this is a difference of degree, not of kind, and businesses must be alert to the growing risk and significance of cyber threats and prepare for them accordingly.

The ways that companies collect, use, and protect data have important business and regulatory implications. The behavior of the largest technology companies illuminates this point. Large technology companies are increasingly moving to a platform-like structure that is sometimes referred to as a "super-app" model. A super-app is an all-in-one application that allows users to do all their business in one place; users don't have to leave the application to conduct a transaction. Super-app business models initially emerged in Asia but have since spread worldwide. Tencent's WeChat started as a chat app but now includes a range of functions including ride-hailing, online shopping, ordering food, and making payments. The super-app is the primary application type used by WeChat and AliPay in China, Kakao in South Korea, Line in Japan, and Grab in Southeast Asia.

In the West, however, companies have been slower to adopt the super-app model, instead favoring to keep the scope of their applications narrower. This resulted in users opening one app to message friends, another to transfer money, and yet another to order a car. Believing that super-apps yield a more nuanced understanding of user behavior because they have access to greater amounts of accurate, detailed data, some Western companies are now trying to transition to the super-app model—Meta is perhaps the best example—but these efforts significantly lag behind their counterparts in Asia.

Additionally, most of the commonly used technology applications are "platform companies." A platform company, or a "multi-sided platform" as economists call them, facilitates direct engagement between market players, often relying on network effects for growth.[44] For example, Meta is a platform company that accumulates a prodigious amount of data through assets like Facebook, Instagram, and WhatsApp that can then be monetized. Meta facilitates engagements amongst users, third-party sites, content developers, and advertisers. Platform companies like Meta and the super-apps described above employ an ecosystem that is "sticky"—users are incentivized through convenience and scale to stay within that ecosystem to fulfill as many of their technology-facilitated needs as possible.

Sharing data within a company group, however, is a potential internal risk as questions about user data and privacy are inevitably triggered. In 2021, Meta caused international furor after announcing that it intended to amend its data-sharing policies to allow more WhatsApp user information to be shared with Facebook, which would better inform Facebook's advertising business, its biggest source of revenue. That triggered a groundswell to delete WhatsApp from phones as users grew concerned that the app would share content and details of texts with Facebook. Users also bristled at being compelled in the updated terms and conditions to accept this change or not be allowed to continue using WhatsApp's services. The pushback highlights the sensitivity that many users have about how data is shared, even internally within the same company.

Accordingly, when assessing risks, the way a company uses data internally can be as important a concern as protecting it from external threats. Controversies around use of personally identifiable information (PII) have and will continue to trigger greater regulatory scrutiny, which creates another category of compliance risks for companies amassing data and sharing it across different areas of their platform. Companies must communicate clear guidelines regarding use of PII. This is a legal requirement in some jurisdictions, but even when not required by law, companies that are proactive about this issue can engender greater customer loyalty, and it will help reduce or avoid excessive regulatory scrutiny. A willingness to lean forward on protecting PII is a critical piece of the regulatory landscape for companies, especially as the value of data increases.

4. Weak link in the chain: Protecting production processes

Supply chain risk can be broken down into a few broad factors: access to materials and inputs needed for manufacturing, the manufacturing and assembly process, which includes workers, and delivery of goods through logistics and transportation networks. When supply chains are global, the risk for each factor increases. A manufacturing process may rely on a key component that is only available from a foreign source, which could be restricted or banned due to regulatory action. Workers may not be available to make or ship goods because they are subject to lockdown measures, a recurrent problem in Asia during the COVID-19 crisis. Extended supply chains are especially threatened by disruption to logistics and transportation if shipping capacity is constrained, air freight is grounded, or ports are operating at limited capacity.

During the halcyon days of globalization, extended supply chains and the need to keep them intact was an almost unstated imperative. Arbitraging labor in the developing world to sell products in the developed world relied on an intricate connection of just-in-time logistics, resources, and transportation. This is no longer the only supply chain model. Jörg Wuttke, chairman of the European Union Chamber of Commerce in China, captured the evolving logic of business in the post-COVID world when he declared that "The globalization of putting everything where production is the most efficient—that is over."[45]

Recent events have exposed flaws in that supply chain model. Natural disasters, like the 2011 Great East Japan Earthquake, the intensifying impact of climate change, and the COVID-19 pandemic, have demonstrated the fragility of globalization's East to West supply chains. Political and business leaders have realized that comparative advantage in supply chains can lead to unintended vulnerabilities that leave countries and companies exposed and unable to access key production inputs or even basic goods. Europe has awakened to the cost of its reliance on Russian energy, which affords Moscow influence over European decision-making. BlackRock's Larry Fink highlighted this point in his annual letter, noting that "while dependence on Russian energy is in the spotlight, companies and governments will also be looking more broadly at their dependencies on other nations."[46] Accordingly, companies and countries are trying to reduce vulnerabilities by becoming more self-sufficient or avoiding concentration risk by sourcing production inputs from multiple locations and rebalancing manufacturing capacity. The result is an unraveling of the connectivity that has underpinned the globalization of the last thirty years.

As a result, on both sides of the Pacific, extended supply chains are increasingly viewed as a risk both in boardrooms and in legislatures. China and the West are both pivoting to greater self-sufficiency in key areas. China's Made in China 2025 plan, discussed earlier, is a prime example of this push.

Consider the story of Mike Bowen. Bowen joined Prestige Ameritech, a US medical mask manufacturer, founded to provide the US medical industry with masks. Its sales pitch was supply chain resilience, except it was difficult to compete with low-cost Chinese competitors that were starting to dominate the personal protective equipment market.[47] Bowen would warn customers that, "The entire US mask supply was under foreign control ... If there's a pandemic, we're going to be in trouble."[48]

That warning proved prescient when COVID-19 hit. Though the pandemic provided a lifeline for American mask manufacturers, it was not enough to sustain many manufacturers once the situation started to normalize. As Bowen wrote in an email to the White

House, "Hospitals drive the mask market … Any plan that allows imported masks to cost less than US made masks will result in a foreign government controlled US mask supply—as currently exists."[49] This sentiment prompted President Biden in his 2022 State of the Union Address to declare his intent to revive manufacturing in the United States and promote American-made goods. This trend—localizing supply chains to minimize their global footprint and potential strategic threat—is a significant risk factor for businesses in the new national security economy.

Traditional analysis focuses on the so-called hardware of supply chains, like manufacturing locations, access to commodities, transportation hubs and spokes, and optimal delivery schedules. Supply chain software can be a problem too. This software includes the protection of intellectual property, specifically the protection of manufacturing processes, control of product molds, and confidentiality of technical specifications.

Technology's impact on global supply chains is a study in progress. Much has been written about regulations, politics, and policy initiatives related to supply chains, and business schools spend countless hours trying to understand how to optimize supply chain operations and logistics. Nonetheless, the advances in technologies described in this book have not been fully reflected in how supply chains might be re-conceptualized. For example, it remains unclear whether advances in 3D printing might reduce manufacturing costs to the point where it is viable to re-locate manufacturing from China back to the United States or for a European company to open its new factory in Germany rather than in Indonesia.[50] Technological advancements will eventually impact manufacturing, which may reshape long, intricate, supply chains. This in turn will affect the interdependence of thick globalism.

What is clear today, however, is that for many industries, adjusting supply chains creates difficult choices. In a recent report, Christos Cabolis, Chief Economist at the IMD World Competitiveness Center, explained that "When European automotive suppliers map their semiconductor supply chains, they discover something unsettling: each optimal configuration for cost efficiency is inverse to its optimal configuration for geopolitical resilience. Every decision that reduces risk tends to increase costs considerably. This is the new reality. Every business decision forces a choice between competing imperatives."[51] These kinds of tradeoffs capture the challenges that businesses grapple with in the new national security economy.

5. Hitting a moving target: The challenge of law and regulations

Much of the government response to the new national security economy is done through de facto or de jure application of regulations or laws. Laws and regulations add an increased layer of complexity for almost all businesses. This nuanced, evolving regulatory and legal environment presents a particularly difficult challenge for most businesses. Historically, industries like health care or banking have been highly regulated across multiple jurisdictions, but other industries are now falling under greater scrutiny because of technology. This becomes more difficult in two key ways in the new national security economy.

First, large technology companies encounter increasing regulation. Given the centrality of technology to the new national security economy, technology companies are most impacted by heightened regulatory scrutiny based on tax, competition, privacy, and national interests. Companies need to recognize that their business description is changing and understand and anticipate the potential regulatory frameworks that they might be required to address.

Ford Motor Company, the iconic company that represents, perhaps more than any other, twentieth-century American manufacturing prowess, has been forced to transition in the twenty-first century. The title of a 2017 *New York Times* article captures Ford's dilemma well: "Can Ford Turn Itself into a Tech Company?"[52] This is an existential question that many so-called traditional companies must now ask. Under pressure to produce electric vehicles as well as develop autonomous driving technologies, Ford has been compelled to embrace a new ethos, one that goes well beyond opening an office in Silicon Valley. As companies like Ford pursue the next iteration of their strategic evolution, they must anticipate the regulatory and legal consequences. These companies are used to the regulatory environment that frames the automobile business, but growing reliance on technology and data and the resulting value that is being created means that they will also find themselves subject to some of the same rules and scrutiny that technology companies bear.

This is already happening at the intersection of finance and technology. Here, the portmanteau that has emerged is fintech, a combination of finance and technology, which represents the use of technology to enhance traditional banking and financial services. A related concept is "techfin," which refers to the entrance of large technology companies into financial services.[53] Prime examples include Apple's eponymous credit card, the Apple Card, and Facebook's ambitious cryptocurrency, Libra, which was proposed and then scrapped.[54] For fintechs and techfins, the convergence of money and data will yield a parallel confluence of regulatory hurdles, which will need to be managed across jurisdictions.

This is especially true for platform internet companies. Large American and Chinese technology companies operate in a complicated and increasingly punitive regulatory environment. Companies like Amazon, Alphabet (Google), and Meta (Facebook) have paid large fines due to violations of privacy and user data. Calls to break them up for anti-competitive behavior are a consistent refrain. For example, Amazon was given an almost €750 million fine in 2021 for noncompliance with Europe's GDPR principles,[55] while X was fined €120 million in 2025 for violations of the Digital Services Act.[56] In 2021, China instituted its own robust legal and regulatory actions against major technology companies in a variety of industries, including transportation (Didi, mentioned earlier), education, and food delivery for issues related to anti-competitive behavior, data security, and public policy concerns related to inequality.[57] An analysis of new Chinese policies and regulations indicates that while some changes are a result of global competition between the West and China, there are also purely (or predominately) domestic political considerations that business leaders need to internalize and acknowledge to understand the model of Chinese state capitalism explored in Chapter 3.

The second driver is the strategic competition between the West and China. Both sides have introduced a raft of policies, processes, and regulations that target the other. For example, Chinese companies accessing US capital markets now face significant opposition in both Beijing and Washington. As highlighted in the example of Didi, a Chinese company that wants to list in the United States receives scrutiny from both Chinese and US regulators. This has resulted in Chinese companies losing access to US capital markets, and many non-Chinese investors are unable to invest in some of the world's most dynamic companies.

In the short term, a conundrum exists for companies. US-listed Chinese companies that abide by US laws may be violating Chinese rules and vice versa. In June 2021, China introduced an anti-foreign sanctions blocking law that along with two earlier laws aims to counter Western sanctions by allowing countersuits against individuals and entities that impose sanctions against China.[58] How the blocking laws will be implemented and enforced is currently unclear, but they show the seriousness with which the two countries use law and regulatory action as a tool of strategic competition. Law professor Ji Li, an expert on US–China business law, captures this dilemma when he writes, "As the US government enacts and enforces more laws to contain China, the Chinese government reacted with legal countermeasures. The proliferation of conflicting laws poses an intractable compliance dilemma for MNCs [multinational corporations] exposed to the world's two largest economies."[59]

Financial markets also overlap with many aspects of the new national security economy and bridge domestic and international politics. For instance, tackling increasing housing prices is part of the longer-term vision of Chinese state capitalism and President Xi's pledge of "common prosperity" is a domestic priority.[60] Although China's household savings rate is among the world's highest, some combination of regulatory constraints on investment options and risk aversion has historically made bank deposits and real estate the preferred investment vehicles for the typical Chinese saver. Despite a weakening property market in recent years, real estate remains the primary source of wealth for Chinese households, though the landscape is evolving due to the growth of wealth management products offered by domestic financial institutions and policy efforts to cultivate the domestic equity market.

The historical preference for real estate, combined with cultural norms emphasizing home ownership, resulted in a rapid increase in property prices, especially in so-called tier 1 cities like Shanghai and Shenzhen. The average home price in Shanghai is approximately $1 million, with average home prices in Shenzhen and Beijing starting to approach that mark. These cities are now among the most expensive for home ownership in the world. As a result, some China investment analysts believe that one long-term objective of the recent crackdown on overseas IPOs is to cultivate the domestic equity market by keeping attractive Chinese companies at home. It is projected that capital will be relocated from real estate investment to domestic equity markets, providing another means of investment for the average Chinese investor, while also creating meaningful alternatives to Western capital markets.

Although diligent companies have been cognizant of and paid close attention to legal and regulatory risk, these risks have assumed a different, more complicated complexion because of technology and strategic competition. Balancing competing regulatory demands will not be easy, and for most global companies this responsibility should not be relegated to outside counsel, even though lawyers have a vital role to play.

Ultimately, businesses may need to sacrifice short-term profit for more freedom over the long term, a lesson that the most influential tech companies are learning. This thinking became apparent in conversations with a senior executive of a large Chinese technology company that was preparing for its eventually successful US listing. The company had to be extremely careful in organizing and operating its business to ensure that it had maximum flexibility in terms of navigating both US and Chinese regulations.

Though originally launched in China, the business limited its reliance on the domestic Chinese market and pursued overseas market expansion instead. This meant that its Chinese users have limited access to its services. Additionally, the company monitors the number of its Chinese users to stay below 1 million, since exceeding this threshold would trigger a cybersecurity review by the Cyberspace Administration of China. And since it only has limited operations in China that are largely focused on R&D, the company does not have to rely on China-based auditors but instead can engage US-based auditors. This allows the company to avoid issues arising from the Holding Foreign Companies

Accountable Act. The company could have easily generated more revenue in the short term by expanding its China-based users, but its executives consciously sacrificed early growth to gain regulatory flexibility, giving them the opportunity to pursue a listing in the United States. This is one example of the strategic thinking that the new national security economy demands from executives.

Also important to note is the decision to base the company's R&D operations in China, another testament to the rise of China's technology prowess. The company's R&D is conducted solely in China and then rolled out to users and markets globally. For this and many other companies, China is a source of innovation and value, and this trend will continue, something that might have been difficult to imagine a generation ago.

The two strands—technological regulation and China–West strategic competition—create uncertainty and represent an example of the challenges created by regulatory change in the new national security economy.

The new normal

Collectively, the five key risks outlined above lead us to a broader problem of amplified execution risk. The risks associated with the new national security economy enhance uncertainty around execution and contribute to increasingly binary outcomes where a project is either a success or failure; there is no middle ground.

Every business manages execution risk when implementing a strategy, rolling out a new product, or completing a deal. None of this is new, and business is full of situations where execution is key, but the new national security economy creates more complexity and uncertainty in the zero-sum dynamic that is enhanced by the risks we identify. Charles "Chip" Kaye, Chairman of Warburg Pincus, one of the world's largest private equity firms, captured this sentiment when he stated that "Pretty much no one has seen [these conditions] during the arc of their investment career."[61] This is unchartered territory.

For many traditional types of business risk, the likelihood of successful execution can be increased by allocating more professionals to the task, speeding up the process to gain a first-mover advantage, or throwing more money at the problem. Any single option, or some combination of the three, can ameliorate traditional execution risk. Some of the risks that are part of the new national security economy, however, cannot be reduced or eliminated by these responses and the stakes are high: the realization of a potential danger can lead to a catastrophic outcome, such as being unable to complete a deal, being barred from a market, or in the worst-case scenario, being shut down.

The case of China General Nuclear Power Corporation (CGN), a state-owned enterprise, is one example of the enhanced execution risk created by the new national security economy. As detailed by University of Southern California law professor Angela Zhang in her work *Chinese Antitrust Exceptionalism*,[62] CGN sought expansion into the European market and created a joint venture arrangement with Électricité de France (EDF) to build and operate three nuclear power plants in the United Kingdom. Given CGN's small economic footprint, this transaction would normally not require a review by the European Commission (EC). But Chinese investment into Western jurisdictions is increasingly radioactive, and the EC, like other governments in the West, created a rationale whereby the joint venture between CGN and EDF would fall under its purview. In the EC's reasoning, all Chinese energy companies are a single meta-entity, linked together by Chinese state control. This agglomeration pushed the revenue figures above the threshold triggering EC review of the deal between EDF and CGN. Though the arrangement was eventually allowed to proceed, the story is a cautionary tale and another example of the risks created by the new national security economy.[63]

This is the new normal for business. The global nature of business in this environment is amplified by strategic competition, technology, and increased connectivity that pervades supply chains, data, and regulatory environments. Specialized knowledge is required to negotiate this terrain, and the premium for such knowledge is higher than ever. The organizations and leaders that succeed will do so because they have acquired and operationalized this new and increasingly valuable knowledge of the new national security economy.

7

FRAMING THE BUSINESS RESPONSE

The Lotte Group, one of South Korea's largest conglomerates, traces its roots to Korean magnate Shin Kyuk-ho, who started the company while living in Japan during the late 1940s. Initially a small business focused on selling chewing gum, it emerged as a sprawling collection of business interests that included food and confectionary, supermarket chain stores, construction, shopping malls, hotels, and several other industries that span the globe.

Ever eager to tap growth opportunities, Lotte was optimistic about entering the Chinese market and planned a long-term strategy that included shopping malls, supermarkets, and large entertainment complexes across the country. It eventually built over 100 of its eponymous supermarkets, Lotte Mart, in China.[1] All seemed to be going well, as Lotte, like other South Korean companies, benefited from the glow created by *Hallyu*, sometimes called the Korean wave.[2] Hallyu refers to the global popularity of South Korean music, media, and culture. Its reach is evident in the success of bands such as BTS, animated productions like *KPop Demon Hunters*, and films like *Parasite*, which won the Academy Award for Best Picture and the prestigious Palme d'Or at the Cannes Film Festival.

At the time, Hallyu was already surging in China, and South Korean firms were taking advantage. South Korean pop music (K-Pop) regularly played in Chinese stores and restaurants, and television shows (K-Dramas) were aired on Chinese television. Movie theaters in Beijing and Shanghai showed South Korean movies, and Chinese consumers eager to own the latest Korean products frequented the growing number of South Korean stores and shopping malls found in many Chinese cities. The tourism industry was another beneficiary of South Korea's coolness as Chinese tourists flocked to South Korea. In 2016, the peak of Chinese tourism to South Korea, nearly one out of every two visitors was Chinese.[3] This fascination helped transform the bilateral relationship: China emerged as South Korea's largest trading partner, accounting for approximately 25 per cent of South Korean trade.[4]

The boom wouldn't last. Those halcyon days ended abruptly when South Korea and the United States announced the deployment of the Terminal High Altitude Area Defense (THAAD) anti-missile system in the summer of 2016.[5] THAAD, a defensive system designed to shoot down missiles, was sent to South Korea to counter advances in North Korea's ballistic missile program. Announced in July 2016 and deployed the following year, THAAD greatly enhanced the ability to monitor and, if necessary, interdict North Korean ballistic missiles that posed an ever-deadlier threat to South Korean and US forces.

China, however, was greatly alarmed by the THAAD deployment. Its chief concern, at least officially, was that the system's powerful radar could be configured to monitor locations as far as 3,000 kilometers away, providing the ability to scan deep into China, potentially neutralizing Beijing's own ballistic missile capabilities.[6] Despite repeated efforts by US officials and experts to allay Chinese concerns, Beijing remained unmoved.

Instead of responding against the US, China focused its anger on South Korea. Though the catalyst was a security issue defined alternatively between North and South Korea (and the US) or between the US and China, China applied pressure on South Korea, and South Korean companies bore the brunt. Shortly after Seoul announced the THAAD deployment, South Korean business interests in China were hammered. Concerts and performances by South Korean music groups in China were restricted and broadcasts of South Korean shows and music were curtailed.[7] South Korean movies were no longer available in cinemas, South Korean products were boycotted, and the once never-ending tour groups that inundated Seoul and other locations in South Korea slowed to a trickle.[8]

Undaunted, South Korea continued plans for the deployment. The government identified a piece of land as an ideal location for the initial THAAD battery: part of a golf course, distant from any urban area. The South Korean government proposed a deal with the owner of the golf course that would swap the piece of land for another owned by the government. Both sides agreed, and the land was exchanged. Lotte, owner of the golf course, had little idea what was in store. A tremendous backlash ensued as Chinese consumers avoided their stores, and ongoing projects were halted or subject to administrative delays.

After investing billions of dollars in the Chinese market, Lotte waited—hoping the storm would pass. It soon learned that the patriotism of Chinese consumers and their willingness to boycott Lotte goods surpassed Lotte's capital and patience, compelling many of its subsidiaries to exit the Chinese market, including its Lotte Mart business.[9] It is estimated that the THAAD debacle cost Lotte some $7 billion.[10]

Until the summer of 2016, the average South Korean business person had no idea what THAAD was or why it was necessary. The idea that consumer product executives might have to prepare for any business risk related to a missile defense system was even more remote. But by the end of the summer, most business leaders in South Korea, and every South Korean business person engaged with the Chinese market, were well aware of THAAD and its impact on business performance.

The Lotte example makes plain the power of the new national security economy, and how strategic and political competition impact businesses in a real and profound way. While it is difficult in hindsight to have predicted Lotte's troubles, this case clearly demonstrates the risks associated with this new business landscape and the creativity needed to address them.

Traditional risk is often easy to track: A leads to B. Causation and result are both clear and generally businesses are able to monitor and, if proactive, even mitigate such risks. In the new national security economy, however, both the cause of risk and its associated results assume a different form. Not only can unexpected risk emerge suddenly (i.e., A manifesting as an unknown risk), but outcomes are also less predictable. Thus, instead of A leading only to B, less apparent but high-stakes C and D outcomes may also emerge. The cause of risk catches organizations and leaders off-guard, while the accompanying effects extend in a more expansive, complicated way. Furthermore, even if A is a known risk to an organization, it can still result in unexpected outcomes due to the factors that underpin the new national security economy.

Lotte was not able to properly assess this risk and outcome calculus. Lotte, a company not involved in the defense industry, was asked to exchange a piece of land for the deployment of defensive missiles focused on deterring North Korean attacks. On its face, this decision does not involve China or the average Chinese consumer that shops at a Lotte supermarket. As details of China's objection to the missile deployment emerged, however, Lotte was dragged into a geopolitical conundrum that included China, North Korea, South Korea, and the United States. A Lotte source shared, "We agonized over whether to accept the government's proposal ... But we were unable to say 'No' because it was related to national security. If we say no to the government, we can't do business in Korea."[11] Meanwhile, Lotte confronted a backlash in China.

In its boardroom and among its senior strategists, Lotte's leadership almost certainly never anticipated it would be forced to exit the Chinese market over a single real estate transaction, especially after committing so much time and capital to expanding there. Unfortunately for Lotte, not only was the risk unclear, but so were its compound effects. This is the reality and risk of the new national security economy for global businesses.

Framing uncertainty

How should businesses and leaders view this new reality? Revisiting a framework developed during a different period of uncertainty might be helpful. Initially drawing on leadership research conducted by Warren Bennis and Burt Nanus,[12] the US Army War College developed the acronym VUCA to describe their view of a post-Cold War world. VUCA stands for volatility, uncertainty, complexity, and ambiguity and was coined to characterize the unknown operating conditions of a world following the fall of the Soviet Union. Over time, VUCA entered business jargon to describe the challenges leaders faced guiding organizations in an increasingly globalized, interconnected, and rapidly changing world.[13]

Though the terms included in VUCA—volatility, uncertainty, complexity, and ambiguity—are used in everyday speech, each term has a specific meaning within the VUCA framework.[14] Volatility relates to an unexpected challenge or an unstable situation that often causes rapid change. The catalyst for the change is usually identifiable, so the reason for the volatility is usually not difficult to understand. For example, the sharp increase in the price for personal protective equipment as well as the difficulty in procuring such equipment at the beginning of the COVID-19 pandemic is an example of volatility, but the reason for such volatility was easy to recognize.

Uncertainty typically represents scenarios where there is insufficient knowledge about a topic or situation. The situation itself may not be particularly complex, but facts are missing. For instance, when Tesla introduced its Model S in 2012, traditional automobile manufacturers were unsure how electric vehicles would impact them.

Complex situations exist when there are multiple strands and variables. The different variables may be simple and easy to digest individually, but the sheer volume of variables creates complexity. Complexity can overwhelm organizations and leaders, causing them to miss important information. Large private organizations and governments collect large amounts of data from various sources, but without systems, processes, and professionals, much of that data will be ignored, unactionable in a timely manner, or misunderstood. This can be critical. A global company may lose a competitive advantage or a government may be exposed to a strategic threat, which could have been avoided through a proper understanding of the relevant data and variables.

Lastly, ambiguity is concerned with understanding causal relationships. When a situation is highly ambiguous, it is difficult to determine the direction of causality and to distinguish between causation and correlation. What causes an outcome? Ambiguity is further exacerbated when we do not know what we do—and don't—know. The classic example of this is former US Secretary of Defense Donald Rumsfeld's response when asked in 2002 about weapons of mass destruction in Iraq: "there are also unknown unknowns—the ones we don't know we don't know. And if one looks through the history of our country and the other free countries, it is the latter category [unknown unknowns] that tends to be the difficult ones."[15] Rumsfeld was widely panned for his response, but he was spot on about the challenges associated with ambiguity and unknown unknowns.

Engaging with organizations over the last few years through leadership training and company talks has made it quite clear that VUCA shapes the reality for today's business leaders. In fact, many have started using the term again. Armed conflict, pandemic, political protest, social unrest, and the US-China trade dispute are just a few factors that are routinely cited as part of a VUCA environment by organizations that we meet with. In addition to those conditions, businesses also still deal with traditional challenges related to competitors, culture, developing new products, and pursuing growth.

Accordingly, business leaders must now deal with an ever more complex decision-making landscape. Previously, the primary focus of business leaders was financial performance, but business purpose has expanded (the focus of the next chapter). Business leaders are required to be alert to issues that may not have existed or weren't on their radar or agendas for most boards even a few years ago. How should organizations and leaders respond?

While there are several areas that organizations and their leaders should focus on to navigate the new national security economy, this chapter will highlight areas of opportunity at the intersection of strategic practice and this new operating environment. Some practices are cultural in nature, while others are technical; collectively, they can improve the ability of firms to compete, collaborate, and execute to achieve strategic objectives as well as avoid strategic pitfalls.

Foundational principles

To recap, the new national security economy is shaped by four concepts: globalization which created deep connectivity within and between societies, as well as new risks created by increased interdependence; geopolitical competition between China and the West; advanced technologies ranging from AI to 5G that are key areas in the struggle for dominance; and finally, emergence of the private sector as a primary actor in this competition.

Building on that foundation, we offer a two-pronged conceptual model for private actors to use as they consider this new operating environment. Though most of our advice is focused on large global companies, it is still valuable for smaller companies and should be adjusted to fit each business's respective industry and global footprint.

First, we propose a five-pillar framework—GRAPL—that businesses and leaders can use to reorient themselves. As these points are explored in-depth in other sections of the book, we will only briefly elaborate on each of the concepts here.

1. Global business is now different

Business has evolved in monumental and profound ways over the last thirty years. From the fall of the Berlin Wall and the rise of dynamic East Asian economies, then super-charged by the emergence of the internet, businesses and business leaders came of age during a period of globalization in which democracy, free markets, and trade were assumed to be both interconnected and net goods. These global linkages have started to dissolve, and there is a growing belief that the world is deglobalizing—a trend that most leaders or organizations are not familiar with nor are they prepared for.

2. Risk in the new national security economy is holistic

As the Lotte example makes clear, risk in the new national security economy is not always linear nor direct and often leads to unexpected outcomes. McDonald's exiting the Russian market following Moscow's invasion of the Ukraine is another case. Unlike many global businesses, McDonald's had a significant footprint in Russia with over 850 locations, which were responsible for approximately 7 per cent of the company's revenue.[16] The status of employees, the impact of sanctions due to McDonald's use of certain Russian banks, financial loss, and impact on its iconic brand by staying in the market created a confluence of unexpected risks following the invasion that were difficult to identify *ex ante* using traditional risk perspectives. A symbol of capitalism in the former communist country, McDonald's exit has had a meaningful impact on McDonald's financially, on its shareholders, and other stakeholders, as well as symbolically.[17]

In the new national security economy, there are novel risks that do not lend themselves to easy compartmentalization and often straddle areas of concern, which can be difficult for organizations to imagine, let alone tackle. These risks are holistic and often compound, creating low-likelihood but high-impact scenarios that traditional risk management models, frameworks, and systems are ill-equipped to detect or address.

3. Accept politics is a key component of business

Politics is an ever more important component of the business environment. Of course, the impact of politics on business is not a new idea. The PEST framework, developed by Harvard Business School professor Francis Aguilar, is regularly taught in business schools to help evaluate the external operating environment by analyzing political, economic, social, and technological factors. Business leaders have often wanted to create a clear delineation between politics and business, but in a landscape where geopolitics and domestic politics draw scrutiny from across the political spectrum and where leaders and organizations must engage regularly on social media, any gulf that might have existed has been bridged. In the new national security economy, "local" politics has been redefined as businesses operate across multiple jurisdictions. Geopolitics cannot be an afterthought when businesses operate internationally.

4. Purpose is paramount

One reason politics plays a growing role in business is because of a reinvigorated debate about the purpose of business. For many of today's leaders and organizations, shareholder profit maximization was sacrosanct, but this assumption has been revisited with the rise of stakeholder capitalism and key roles private enterprises play in the new national security economy. The debate about corporate purpose and the willingness of businesses to address political and social issues reflects a new approach that is still being shaped. In the new national security economy, a focus on purpose raises questions about corporate national identity and the relationship between corporate and national interests, which will be explored further in the next chapter.

5. Learning about the new national security economy is essential

Having conducted extensive corporate training and executive education, it is clear to us that high-performing organizations are committed to enhancing the knowledge and skills of their people. Most of these learning opportunities, however, are overwhelmingly focused on discrete business skills ranging from financial management to decision-making. From hiring to training, organizations now need to account for knowledge about the new national security economy. This is a pressing skill set in the current operating environment.

* * *

The second prong of our framework is captured by the acronym PIPE: people, identity, process, and engagement. Where GRAPL is focused on a strategic, conceptual orientation of an organization and its leaders, PIPE is centered on tactical approaches that businesses can inculcate to enhance standard operating procedures.

1. People power

Meeting with highly competent business leaders over the last few years, most of whom studied traditional business disciplines or engineering and do not have training in international affairs, regional studies, or related disciplines, we have observed that the extent of learning about new national security economy concepts often begins and ends with the news. Those with more than a passing interest in geopolitics will proudly share that they have a subscription to *The Economist*. We're fans of *The Economist* too, but for companies and leaders operating on a global stage, it's not enough. No single publication (not even this book) can be the sum of their understanding of the world and the new national security economy.

The AI era notwithstanding, understanding the new national security economy still requires the right people, with the right knowledge, and most importantly the right insight. Unlike commoditized financial data, which is readily available via a Bloomberg terminal, much of the knowledge relevant for the new national security economy cannot be found with a simple Google search or just using your favorite AI large language model. And like the best investors and business operators, the key is not knowing a particular piece of knowledge but understanding how that knowledge can be connected to other dots. This is the alpha-generating skill that we believe will drive superior decision-making and performance in the new national security economy.

We are not saying an organization needs to hire legions of people with backgrounds in international affairs, but a few select hires can greatly enhance the capabilities of many firms. Possible future issues can be addressed early on by proactive employees who possess key geopolitical, geoeconomic, and other related knowledge. Such early interventions can prevent larger problems from emerging and metastasizing. Additionally, companies should have a rolodex of experts that they can contact for additional insight as needed. This may take the form of consulting firms dedicated to certain areas, expert network providers, but can also include think tank professionals, university faculty, and independent researchers.

The composition of this constellation of expertise will depend on the industry. The key point is that companies need to be prepared to mine all possible sources, both inside and outside their organizations, for insight and understanding. And as a first step, many companies should do a talent inventory to assess what people and skill gaps exist and need to be bridged. Just as a firm assesses talent needs and development, existing processes can be used to address the people aspect of the new national security economy—the first line of defense for most organizations.

2. Identify your identity

Many companies maintain some sort of explicit or implicit national identity. Sometimes this identity is overt in the case of state-owned or controlled enterprises. In Chapter 3, we focused on state-owned enterprises in a Chinese context. China, however, is not the only example. State ownership takes many forms.

During the mid- to latter half of the twentieth century, governments in Eastern and Western Europe employed state ownership in varied ways, and in the extreme, this took the form of nationalization, where the state exercised complete ownership of a particular company or industry. In Asian economies guided by the developmental state model, there were mutually supporting relationships between government and business. In all those cases, the line between the public and private sector is hard to trace and business priorities can be less clear.

The rise of sovereign wealth funds (SWFs) has given governments stakes in both publicly traded companies as well as private enterprises, in many cases in countries that don't operate the fund. Increasingly, the largest equity owners in many markets are SWFs. The size and investments of SWFs will only continue to grow, and they will become increasingly influential. For instance, Singapore's primary sovereign investment vehicles, GIC and Temasek, are active investors in a wide range of alternative investments such as technology startups and real estate.[18] South Korea's National Pension Service, which manages the world's third-largest pension fund, is the biggest investor in the South Korean equity market.[19] Beyond Asia, Norway's SWF, managed by Norges Bank Investment Management, is the largest SWF in the world with over $1.3 trillion of assets, while Saudi Arabia's Public Investment Fund has been an investor in a variety of assets including infrastructure projects and a partner with Softbank in creating the $100 billion Softbank Vision Fund, which is an investor in companies like Uber and ByteDance (the parent company of social media platform TikTok).

The US is also getting into the game. Consider two examples. In August 2025, the US government became a nearly 10 per cent shareholder in Intel, the American semiconductor manufacturer.[20] A few months earlier, it also received a "golden share" in Nippon Steel's takeover of US Steel. State ownership, once anathema to many free-market conservatives, has become part of the playbook for the new national security economy; even in the US, President Trump commented, "I think we should take stakes in companies ... Now, some people would say that doesn't sound very American. Actually, I think it is very American."[21]

State ownership of private assets tweaks business purpose and introduces various degrees of national identity. There is a continuum of how state ownership impacts businesses. Sometimes a state investor is like any other shareholder; in other cases, state capital casts a long political shadow. Consequently, it is important for companies to understand who its investors are, the source of their capital, and the motivations they may have beyond a return on investment.

Ownership, however, is just one way in which identity may be expressed. Even if a company has no shareholders that represent sovereign capital, a firm can still reflect a political identity. Following the landmark 2010 US Supreme Court case, *Citizens United vs. the Federal Election Commission*, US campaign finance rules were altered, and many restrictions on financial contributions by companies and industry groups to candidates were removed.[22] This led to a rise in corporate political spending, often creating its own controversy.[23] A firm's political spending can also be considered a form of expressing a political identity. Whatever form it takes, a company needs to actively control its identity otherwise there is a danger that its narrative will be written by others.

As mentioned in the previous chapter, John Chipman, from the International Institute for Strategic Studies, has urged multinational companies to develop a corporate foreign policy.[24] His prescription rested on the assumption that companies can no longer hide behind the façade of being a neutral actor. While the desire for a company to remain neutral and be perceived as an independent entity is understandable, it is not a viable approach for at least three reasons. First, it does not reflect reality. Companies can be direct actors like the British East India Company, or they can be dragged into the political process as in the Lotte example. Either way, companies take on a national identity. Second, by maintaining the illusion of neutrality, the strategic mindset will remain reactive rather than proactive. Third, stakeholders increasingly call on companies to have interests beyond maximizing profit, which becomes even more pertinent when the stakeholder is a nation.

Consider the following thought experiment. If you were to ask Jay Y. Lee, executive chairman of Samsung Electronics and head of the Samsung conglomerate, to describe the political identity of his business, he would likely say that Samsung is a Korean company that operates globally. If you were to ask Tim Cook a similar question, he would likely say that Apple is a global company that happens to be headquartered in the United States. Economies that developed through high rates of state coordination or intervention like in China, Japan, South Korea, and other examples already include businesses as engines of state growth. Thus, such businesses are often imbued with a social purpose, which shapes their political identity.

Before facing political scandal and impeachment in early 2017, South Korean President Park Geun-hye pardoned 6,527 individuals.[25] The pardons were not unusual: it's common practice for South Korean presidents to pardon prisoners as part of the country's 15 August Independence Day celebrations. Among those pardoned was Chey Tae-won, chairman of SK Group, who was serving a prison sentence related to misuse of corporate funds and embezzlement. SK Group is South Korea's third-largest conglomerate with significant interests in energy, chemical, telecommunications, and semiconductors. As part of Chey's release, the Ministry of Justice issued a statement that Chey and other leaders were released "to give them a chance to develop the country's economy," while President Park commented: "The pardons have included some businessmen, for the tasks we face, economic revival and job creation."[26] Don't be shocked: South Korean presidents often give pardons to business tycoons to stimulate a sluggish economy. Samsung Electronics' Lee was granted a similar pardon in 2022 by then newly elected South Korean President Yoon Suk-yeol under the same rationale—supporting a struggling economy.[27]

At least in certain contexts, businesses have a distinct, observable national interest ranging from state-owned enterprises to those receiving state support like favorable industrial policies or presidential pardons of a firm's top executives. Only after recognizing this political identity can a business craft an approach to navigate the new national security economy, perhaps even a version of its own corporate foreign policy where relevant, which integrates with other aspects of their business strategy. Are they explicitly expected to serve a political interest because of the nature of their home country's economic system or is the expectation implicit, perhaps shaped by their investors? The answers to such questions guide a company's approach and response.

With geopolitics emerging as a more pronounced variable impacting business operations, companies must start considering these questions before a crisis and continually iterate as conditions change. Specifically, what is your organization's identity? Is that the correct identity, one that is consistent with your values, principles, and purpose? If not, what can be done to change the organization's perceived identity?

3. Process

With people and identity in place, the next important element is enhancing processes or creating new ones to generate new perspectives and to incorporate those perspectives into strategic formulation, execution, and decision-making processes. Even little actions such as having an up-to-date political calendar noting election cycles and important legislation for markets where a company is active are important. Organizations need to track the vital political and government data that impacts businesses, and most business people do not. For example, organizational and leadership maps of government regulators and bodies are typically publicly available and can be found with a simple online search. Most business people do not do this. These are the simple elements that can be incorporated into strategic and operational planning as well as short- and long-term risk assessments. In the new national security economy, the more complex the operating environment, the more relevant this information becomes.

Having the right processes allows companies and leaders to ask the right follow-on questions, prepare contingencies, and identify opportunities that might otherwise be missed. Global financial institutions are subject to a complex mosaic of regulatory guidelines, laws, and reporting requirements that change frequently. These institutions must stay abreast of recent updates, and after the 2008 Global Financial Crisis, many banks, asset managers, and other financial services firms have invested heavily in compliance functions—often by hiring more people, which is expensive and not scalable. Some firms use technological solutions to optimize their systems and processes. Sia Partners, a global consultancy headquartered in France, developed an AI-powered tool that screens, flags, and helps disseminate regulatory updates, which clients can deploy to streamline the compliance process and to ensure they are ameliorating the type of preventable risk we discussed earlier.

As part of an effective strategic process, organizations and leaders must also look beyond first-order effects to second- and third-order effects. Our experience is that while most organizations can identify the first-order impact of an action, it is difficult for them to look beyond primary effects and outcomes. They struggle with recognizing secondary or tertiary outcomes. For example, an organization might identify the ways that a new technology will help them reduce costs or streamline a process, as well as the immediate negative externalities it might generate, such as triggering staff cuts. Most business executives end their analysis here. They underestimate advantages that might accrue to their competitors, they often do not account for new types of competition, and are not fluent in the practice of thinking about second- and third-order effects. Simply stated, they need to consider the impact of the initial impact.

We can illustrate this thinking with an example. As autonomous vehicles become more prevalent, it is easy to predict that taxi and truck drivers will be displaced. This first-order effect is apparent. Another first-order effect will likely be some impact on vehicle ownership. The spread of autonomous vehicles could reduce rates of vehicle ownership in urban areas as ride-sharing platforms relying on autonomous vehicles become more convenient, leading to a possible decline in vehicle ownership. With fewer vehicles, a second-order effect may be a reduced need for parking in new apartment buildings, allowing space that would have been allocated to parking to be used for additional units. This could then contribute to a possible third-order effect of lowering home prices as supply increases. Without the discipline of a process or framework, it would be difficult for most people to make a possible connection between autonomous vehicles and home prices.

In many ways, we are witnessing a high-stakes version of this play out now. The US has instituted a range of policies to control China's access to and development of AI-related technologies, particularly around semiconductors. Ostensibly, the first-order effect of these policies by both the Biden and Trump administrations was to weaken China's AI industry and for the US to extend its lead in this critical technology. It seems that the second-order and beyond implications of these policy decisions were not fully considered. For instance, the speed at which China has strengthened its domestic AI infrastructure to reduce reliance on overseas suppliers and inputs was likely underestimated. This reflects a broader trend across the Chinese economy. As China's self-sufficiency increases, America's leverage over China decreases. Additionally, China's retaliatory restrictions on American access to rare earth minerals appeared to catch Washington off-guard, another unanticipated higher-order effect of American trade policies. Overall, the US does remain ahead in AI for now, but NVIDIA CEO Jenson Huang's regular observation that "China is nanoseconds behind America in AI" is telling. Some policies may be producing unintended outcomes that can only be fully considered by rigorously examining both the immediate moment and what may lie ahead.

Predictably, looking beyond first-order effects requires making assumptions, which may not come to fruition. Nevertheless, organizations and leaders need to grow more comfortable with and fluent in such analysis to understand and appreciate the impact of change. The goal is not to predict the future but to imagine possible futures, and plan or mitigate accordingly. Such thinking will be an increasingly integral part of business planning in the new national security economy.

Regardless of which systems or tools that are utilized, it shouldn't just be a spreadsheet somewhere on a company's network. Besides being accurate and timely, the information must be available and presented in such a way as to facilitate analysis that contributes to effective decision-making. Many firms do this by constructing dashboards that make it easy for relevant individuals, leaders, and decision-makers to see essential metrics and data in an easy to digest way. Fit-for-purpose technology solutions to assist in monitoring and disseminating information are often key components of robust processes.

4. Engagement: Shaping the future

Once a company has aligned its people, identity, and processes, it is poised for robust strategic engagement. The nature of this strategic engagement will depend on industry, jurisdiction, and context. Depending on the situation, companies may need to engage in lobbying and coordination with industry groups, and otherwise shape values and standards in the marketplace.

Recently, large technology companies have faced intense scrutiny not just in the United States, but globally, over issues such as data privacy, data use in algorithms, cybersecurity, tax, and anti-trust behavior. Much of the tech industry's efforts in engaging regulators and governments has often been reactive while under investigation or inquiry. At this point, remedial actions that companies take mostly address damage control and mitigation. Frankly, many of the issues that large technology companies face today were predictable. A more strategic perspective that incorporated elements of the principles outlined in this book would have prompted leaders of these companies to address these issues much more aggressively early on. This trend is not unique to the United States—similar regulatory actions are occurring in China, South Korea, and other countries.

A strategic perspective that incorporates the new national security economy allows companies to move from reacting to anticipating. In the technology context, companies could have considered forms of self-regulating or responsible innovation rather than engaging in questionable behavior that would eventually, if not invariably, lead to scrutiny. Admittedly, this is difficult in the "growth at all costs" climate that consumes many technology firms. Nevertheless, there is an increasing appetite for technology firms to embrace responsible innovation by considering the impact that their work has and will have on society. Even Facebook's parent company, Meta, considered by many to be the poster child for technology gone wrong, has introduced principles around responsible innovation.[28]

Of all the people the authors have met or interviewed working in this space, Jon Zieger, a co-founder and the first executive director of Responsible Innovation Labs ("RIL"), captures best the promise of responsible innovation. Having assembled a collection of renowned business leaders, leading venture capital firms, and well-known technology companies all committed to responsible innovation, RIL is advancing a proactive, forward-looking approach that departs from business as usual. RIL's charter states:

> As an industry, we cannot manage our products and businesses solely aimed at growth and profits in the near term, as critical as they are. Nor can we prioritize speed and scalability over all other considerations. Instead, we must integrate and prioritize suitable constraints, transparency, and frameworks that support sustainable growth *and* good. That is the essence of responsible innovation, and it requires a new focus in the way the innovation community approaches and values its work.[29]

This type of engagement exemplifies strategic vision with the potential to shape the future. And since technological advancement will always outpace the ability of the law to regulate, companies should work with other market participants to create standards. There is precedent for this in financial markets and in telecommunications. In 2021, several large US companies through the influential Business Roundtable proposed a number of policies to guide the development and use of AI.[30] The impact of this initiative remains unclear, but such forward-looking, proactive collaboration is necessary and represents evidence of the strategic engagement that is needed in the new national security economy.

Bringing it all together

The strategic orientation provided by GRAPL and the tactical steps embedded in PIPE position a company to sidestep the negative impacts of the new national security economy as well as to benefit from identifying and executing on opportunities created by this new paradigm. Leaders and organizations that fail to do so will encounter additional issues and miss chances to deliver superior performance in a VUCA operating environment.

In March 2021, YJ Kim, CEO of Magnachip Semiconductor Corporation, a South Korean technology company focused on providing circuits for displays and solutions to power semiconductors, delivered a momentous win for his shareholders by signing an agreement that would sell Magnachip to entities managed by Wise Road Capital, a Chinese private equity firm. Wise Road Capital's offer was an all-cash deal that would pay $29 a share, a more than 50 per cent premium to Magnachip's stock price at the time.[31] Though listed on the New York Stock Exchange (NYSE), Magnachip has no meaningful presence in the United States. Core functions like management, research, and sales are either conducted in South Korea or in other foreign jurisdictions.[32] Given this lack of US exposure, the parties did not report the transaction to CFIUS, the Committee on Foreign Investment in the United States, feeling that it wasn't necessary. Martin Yang, an equity research analyst at Oppenheimer & Co. commented once the deal was announced, "given MX's [Magnachip's NYSE ticker symbol] minimal exposure to the US market, CFIUS will not pose threats to the deal's closing."[33]

This optimism proved misplaced. CFIUS contacted the parties and informed them that they would be conducting a review based on national security risks with the expectation that its recommendation would be to prevent the transaction.[34] Consequently, the parties announced the transaction's termination in December 2021.[35]

A prime example of a missed chance, such outcomes are the reality of the new national security economy. Though there might not have been any difference in the end, we believe that by incorporating elements of our two-pronged framework, the parties in the transaction would have been more aware and better prepared by considering mitigating steps and preparing possible contingencies that would have given the deal more viability, increasing the odds of success.

Geopolitical risk creates costs. Professors Vivek Astvansh, Wesley Deng, and Adnan Habib examined 4,600 American companies over three decades of data and discovered that the negative impact of geopolitical risks can be persistent and can set a company's innovation plans back three to five years. Unlike share price, innovation and supply chains don't immediately bounce back when a crisis ends. It can take time for these key business components to recover and reset, which represents a tremendous cost and potential risk for companies, industries, and nations that are in strategic competition:

> our findings illustrate how individual business and global politics are tightly intertwined. … rising geopolitical risk is likely to have a substantial stifling effect on innovation across the board. And the only way to address this underlying issue is for political and business leaders (alongside other key players, such as lawmakers and media plat-

> forms) to work together to reduce global tensions and build a more peaceful—and innovative—future.[36]

Business leaders must develop a "beyond compliance mentality." Of course, businesses need to follow the laws and regulations of relevant jurisdictions, but this is a minimum standard. Business leaders can no longer think of such issues as "mere" compliance matters that do not add value. This approach is outdated and will harm the long-term prospects of the business. Unfortunately, most businesses are ill-equipped to assess and analyze risks associated with the new national security economy because they lack the right people, right processes, and a proper understanding of the role of businesses in such a world. The failure to adjust will have substantial consequences.

8

BUSINESS PURPOSE IN THE NEW NATIONAL SECURITY ECONOMY

While the heavily armed masked men that infiltrated Crimea in February 2014 bore no markings nor identification, there was little doubt that they were Russian military forces.[1] There was an outcry from American and European political leaders as the so-called "little green men" occupied key infrastructure and secured government buildings in a bid to take control of the strategic region. Business leaders, however, remained largely quiet. Christophe de Margerie, former CEO of French energy company Total (now Total Energies), when asked about a response to Russia's invasion, offered a typical response, "We are involved in business; we are not involved in politics."[2] Business's response was business as usual.

Eight years later, when Russian soldiers crossed the border into Ukraine, the response was significantly different as business leaders decried the aggression. Shortly after Russia's invasion, Yale School of Management professor Jeffrey Sonnenfeld, an expert in corporate governance and CEO leadership, started compiling a list of largely Western companies that had either suspended or withdrawn operations from Russia. Sonnenfeld's original list contained a few dozen companies, but it swelled, and the number of companies that eventually curtailed their Russian operations expanded to over 1,000.[3] During the intervening period between Russia's absorption of Crimea in 2014 and its 2022 invasion, something had changed. Expectations of business had transitioned from a focus on maximizing profitability to a broader agenda focused on stakeholders.

Though it's difficult to identity a single tipping point, social movements like Black Lives Matter, which arose in response to the killing of often unarmed, young black men by US law enforcement; the response to sexual harassment of women that culminated in the #MeToo movement; and heightened awareness of climate change were calls to action that stakeholders issued to business. When President Trump took office in 2017, his divisive policies and rhetoric prompted employees and consumers to lean on companies to take a stand on a range of issues from immigration to promoting diversity. Cumulatively, these events shifted thinking.

Social media amplified the pressure driving this shift. Political and business leaders could directly reach a global audience in 140-character messages (now 280 characters). Leaders using platforms like Twitter (now X) and Facebook, however, were not engaging in one-way communication. These social media channels allowed for anyone with a message or criticism to reach out to leaders and organizations as well. CEOs of companies were now directly addressing customer complaints, offering condolences on the most recent tragedy, and talking about new products all via social media.

This reciprocal communication dynamic also created an expectation that organizations and leaders needed to respond to a broader range of comments, questions, and grievances than in the past. Before social media, a shareholder often waited until an annual general meeting to question a company's position or hope for media coverage. Today's heightened expectations require organizations and leaders to have a voice and a position on myriad issues beyond narrowly defined business performance.

Companies are now players in much more pronounced ways in political and social causes. This phenomenon is not limited to domestic politics. The role of companies as critical players in the new national security economy is also connected to this broader sense of business purpose. And with a shift from shareholders to stakeholders, for some companies, their most important stakeholder could be a nation; consequently, advancing a national interest becomes an important company objective. This insight is a key aspect of the interplay between business purpose, companies, and the new national security economy.

Ukraine revisited

In the days and weeks that followed the 24 February 2022 invasion, humanitarian groups, nonprofits, and concerned citizens rallied around the world to aid the Ukrainian people. This aid took the form of in-kind contributions of needed supplies, monetary contributions, and even cryptocurrency transfers directly to the Ukrainian government.

Like many of you, our social media feeds and email inbox were filled with posts and messages containing information about how to help. Calls for help and the response weren't new. There have been similar campaigns following earthquakes, floods, public health crises, and in response to other armed conflicts. Shortly after Russia's invasion, one email appeared that outlined ways to assist Ukraine. This email, however, was not from the Red Cross or a former colleague with familial ties to Ukraine. Instead, it was from Readdle, a technology company that develops a suite of widely used productivity apps like PDF Expert, Scanner Pro, and Spark, an email manager. Readdle's apps have been downloaded over 200 million times, cementing its status as a fixture in Apple's App Store.

Despite using Readdle's products for nearly a decade, we'd never considered Readdle's identity. After all, Readdle's previous communications had always focused on its products. It turns out that Readdle's CEO, Igor Zhadanov, and his cofounders are Ukrainian and started their business in Odesa, a port city on the Black Sea where the company still maintains operations.

Readdle's email stated it was a Ukrainian company and invited users to "stand with us. Please support Ukraine by lobbying your governments, protesting peacefully, and donating money to support the people of Ukraine." The email included links for contributions to either humanitarian aid or defense support. Readdle, a company that is part of Ukraine's burgeoning technology scene, created its own foreign policy in response to the war. Readdle would soon be joined by hundreds of other companies, including many of the world's largest businesses, representing some of the most globally iconic brands.

Business purpose debate

Business purpose is too expansive a topic to fully cover here. To frame business purpose in the context of the new national security economy, however, a brief background is necessary.

In the American legal tradition, the business purpose debate stretches back to at least 1919, with many corporate law courses referencing the 1919 Michigan Supreme Court case, *Dodge v. Ford Motor Company*.[4] During the 1930s, the debate moved from the courts to academia as Columbia Law School professor Adolf Berle, and Merrick Dodd, a law professor at Harvard Law School, exchanged views in the *Harvard Law Review*. Berle, along with Gardiner Means, wrote *The Modern Corporation and Private Property*,[5] a classic in the field of corporate governance. Initially, Berle's stance was characterized as pro-shareholder, while Dodd adopted a pro-stakeholder position that embraced a broader corporate responsibility that encompassed not only shareholders but all those who had a "stake" in the continuing functioning of the organization, a list that would include employees and society writ large. Berle later clarified his stance, aligning it more closely with Dodd's.

Following this period, the question of business purpose was relatively dormant through World War II and the Cold War but was revived by the Chicago School of Economics and specifically the work of Milton Friedman, who received the Nobel Memorial Prize in Economic Sciences in 1976. In 1970, Friedman wrote an article for *The New York Times Magazine* titled "A Friedman Doctrine: The Social Responsibility of Business Is to Increase Its Profits." He rejected an expansive view of corporate purpose and made his point with a simple declaration that has become imprinted on the minds of business students and leaders. Friedman wrote:

> there is one and only one social responsibility of business—to use its resources and engage in activities designed to increase its profits so long as it stays within the rules of the game, which is to say, engages in open and free competition without deception or fraud.[6]

Acting on Friedman's dictum, a generation of business leaders adopted the shareholder profit maximization mantra. It quickly became the dominant view and remained so until a shift began in the second decade of the twenty-first century.

Though discussion of the purpose of business existed at the fringes of academic debate and among a minority of business organizations, it was thrust back into the spotlight when BlackRock CEO Larry Fink issued his 2018 Letter to CEOs titled "A Sense of Purpose." In it he wrote:

> We also see many governments failing to prepare for the future, on issues ranging from retirement and infrastructure to automation and worker retraining. As a result, society increasingly is turning to the private sector and asking that companies respond to broader societal challenges. Indeed, the public expectations of your company have never been greater. Society is demanding that companies, both public and private, serve a social purpose. To prosper over time, every company must not only deliver financial performance, but also show how it makes a positive contribution to society. Companies must benefit all of their stakeholders, including shareholders, employees, customers, and the communities in which they operate.[7]

Fink's words carry weight. BlackRock, the world's largest asset manager, has more than $10 trillion in assets under management and is a sizeable shareholder in many of the world's largest companies through its various investment products. Fink's letter triggered a wave of discussions in boardrooms and classrooms around the world. Global consulting firms like McKinsey & Co. started issuing reports about the importance of business purpose and accounting for stakeholders, not just shareholders. Investment banks like Goldman Sachs found themselves fielding questions about corporate social responsibility initiatives at campus recruitment events. Shareholder value maximization was no longer the only game in town, and Friedman's admonition about the social purpose of business being to increase profits was no longer sacrosanct.

In 2019, the Business Roundtable, an organization comprising CEOs from prominent American companies, including Apple's Tim Cook, J.P. Morgan's Jamie Dimon, and Julie Sweet from Accenture, issued a document called the Statement on the Purpose of a Corporation,[8] which was signed by 181 CEOs. This document was significant because over the last forty years the Business Roundtable had supported a model of corporate governance referred to as "shareholder primacy." Shareholder primacy echoes Friedman in its focus on prioritizing the economic benefit of shareholders with other stakeholders generally not considered. With 2019's Statement on the Purpose of a Corporation, the Business Roundtable and the CEOs of leading US companies instead offered support to a business purpose that has become known as stakeholder capitalism.

Stakeholder capitalism accounts for the interests of customers, employees, suppliers, and communities as well as shareholders. Shareholders are undoubtedly key stakeholders, but stakeholder capitalism focuses on producing long-term value for shareholders, and not focusing on maximizing short-term profits, which historically came at the expense of other stakeholders. Accordingly, an important piece of logic underpinning stakeholder capitalism is that by focusing on long-term shareholder interests, shareholders should better align with other stakeholders. Its proponents argued that this alignment amongst stakeholders, with a focus on the long term, should also lead to better firm performance.

Advocates argue that companies with a purpose that goes beyond merely generating a profit are better able to attract talent, create better organizational culture, engage customers more meaningfully, and have strong financial performance in the long run. Critics counter that the rhetoric of stakeholder capitalism sounds good, but in practice, it is just window dressing for many companies. For example, Harvard Law School's Lucian Bebchuk and Roberto Tallarita examined the signatories of the 2019 Business Roundtable statement. Among the companies that responded to their queries, they discovered that approximately 98 per cent of CEOs had not received board approval before signing.[9] Given that, skepticism still exists about how serious companies are about making changes to their business purpose.

A consistent theme throughout *Tech-Gemony* is the role of the state in the private sector (the golden share mentioned in the previous chapter is one example). Nowhere is this more pronounced than when the state becomes a shareholder and national purpose shapes business purpose. The state combines two stakeholder roles, as both shareholder and government. Goldman Sachs Global Institute leaders Jared Cohen and George Lee capture the importance of this trend linking capital and statecraft, and its accompanying tension, in a *Foreign Policy* article:

> Call it the rise of instrumental capital: the use of state-directed funds to pursue the dual mandate of generating financial returns and projecting state power. This capital is patient, long-term, and aligned with the domestic and international agendas of leaders. How countries invest is, increasingly, how they compete. In this new paradigm, governments are more than regulators of markets; they are now among the most consequential asset owners and allocators of capital in the global economy.[10]

This shift to stakeholder capitalism assumes special meaning in a world defined by a globalized, connected operating environment. Businesses operate in and across several communities and must account for their diverse and periodically competing interests. As businesses consider the local, regional, national, and—for global companies—the international communities in which they operate, political questions inevitably arise. Recognizing the growing influence of companies, specifically technology companies, Denmark appointed a tech ambassador for its "TechPlomacy" initiative.[11] Other European countries followed Denmark's lead. The new role was motivated by the realization that certain technology companies impacted Denmark more than many traditional nation-states. Business has emerged as a reluctant actor on the political stage.

The politics of business purpose

On 27 January 2017, a week into his first term, President Trump signed Executive Order 13769, Protecting the Nation from Foreign Terrorist Entry Into the United States.[12] Though eventually revoked, this controversial decision essentially paused the admission of refugees into the United States but also prevented citizens from seven majority-Muslim countries from entering the country as well, even if they had valid visas issued by the State Department. The countries covered by the Executive Order were Iran, Iraq, Libya, Somalia,

Sudan, Syria, and Yemen. The Trump administration justified the order by asserting national security, specifically the prevention of terrorists from entering the United States. The order quickly became known as the "Muslim travel ban" and was widely condemned as discriminatory. Crowds assembled at airports in New York City, Los Angeles, and other locations to protest the ban, while lawyers mobilized to represent affected travelers by seeking temporary relief from aspects of the executive order.[13]

Several CEOs quickly spoke out against the travel ban. Within this group, leaders of technology companies were particularly vocal in their opposition, citing its discriminatory nature and the impact it would have on their ability to recruit and retain talent. This issue hit especially close to home, since the technology industry is populated by immigrants; Meta's Mark Zuckerberg, Apple's Tim Cook, Alphabet's Sundar Pichai, and Microsoft's Satya Nadella were among the most prominent CEOs who spoke out against a rule that was cloaked in the rhetoric of national security but felt discriminatory in nature. Pichai, an immigrant himself, commented via his Twitter account following the ban: "Immigration has contributed immensely to America's economic success, making it a global leader in tech, and also Google the company it is today. Disappointed by today's proclamation—we'll continue to stand with immigrants and work to expand opportunity for all."[14] The response to the immigration ban is one example of the increasing activism by CEOs in debates that are not confined to business.

Participating in such debates can mean more than issuing a tweet or a press release. Lyft, the ride-hailing company, donated $1 million to the American Civil Liberties Union to assist efforts in representing individuals impacted by the Trump travel ban.[15] Lyft is particularly interesting because in 2017, Lyft was still a private company that was not profitable and would not conduct its IPO until approximately two years later in March 2019. Nevertheless, Lyft's leadership felt it was important to vocally and financially support those who challenged the order. Lyft's behavior garnered positive media attention compared to its competitor Uber, which many felt was late in opposing the ban, leading to a #DeleteUber campaign.[16] Politics combined with social media has become a new dimension of the business landscape.

In recent years, there are many examples of companies and their leaders engaging in political activism, even when disconnected from issues directly related to their business. An often-overlooked outcome of the adoption of stakeholder capitalism is the increasing role of business in politics. At a broader level, businesses and business leaders are increasingly expected by employees, customers, and even shareholders to have opinions and policies on a broad range of issues that a generation ago a CEO would not have been expected to engage with.

The rise of what professors Aaron Chatterji and Michael Toffel dubbed CEO activism captures the increasing participation of business and business leaders in politics. They describe CEO activism as CEOs "taking public stands on political and social issues unrelated to their companies' bottom lines."[17] These are often politically sensitive topics related to race, gender, sexual orientation, and a host of other issues that CEOs historically might not have engaged. But times have changed and with instant access to stakeholders through various social media channels, CEOs are increasingly lending their voices to causes they feel are important to them, their customers, and their companies.

Jeff Immelt, GE's former CEO, has been outspoken in his support for such activism as a statement of purpose and authenticity. He argues that "it's insincere to not stand up for those things that you believe in ... We're also stewards of our companies; we're representative of the people that work with us. And I think we're cowards if we don't take a position occasionally on those things that are really consistent with what our mission is and where our people stand."[18]

The transition from the Biden administration to the second Trump term has caused some to question such proclamations by business leaders and the overall viability of stakeholder capitalism. If evidence of stakeholder capitalism is narrowly interpreted as support solely for corporate social responsibility and environmental, social, and governance initiatives, then that version is under retreat during President Trump's second act. Casting a broader, objective lens that is disconnected from ideology, however, spotlights the reality that interests beyond profitability remain salient, just manifested differently. Politics matter—it's only that the pendulum has swung to the other end of the spectrum. And in geopolitics, nations as dual-hatted stakeholders advancing a national interest as part of their ownership reinforce the shift to a stakeholder focus, reifying prioritization of values like national interest beyond pure profit maximization.

In the new national security economy, companies may even use the national interest as a key value as part of their business purpose. In M&A, potential acquirers are sometimes categorized as financial buyers or strategic buyers. Financial buyers typically prioritize the best return possible, while strategic buyers are focused on how the company they are acquiring might fit into long-term plans and consequently might be willing to pay more than a financial buyer. Law professors Jeffrey Gordon and Curtis Milhaupt, of Columbia Law School and Stanford Law School respectively, extend this idea by studying cross-border M&A pursued by Chinese companies through the lens of a "national strategic buyer," introduced in Chapter 3.[19] A national strategic buyer is not solely motivated by commercial considerations but by broader national interests. Of course, this logic extends beyond China. Nation-states are the ultimate strategic buyers. Thus, the politics of business purpose are not only encompassed by domestic political considerations but by geopolitics as well.

The new national security economy puts a premium on this sense of purpose. The emerging risk profile will create situations that are new and novel for most CEOs; they will be forced to address issues that their predecessors never had to face. Even without a playbook, business purpose properly defined and reflected in an organization's culture and actions will help guide them through the uncertainty.

An end to globalization?

Four years after penning his polemic arguing that businesses have to reconsider their purpose, BlackRock's CEO Larry Fink circulated another annual letter that was equally provocative. In his 2022 letter, briefly referenced in Chapter 6, he stated that:

> the Russian invasion of Ukraine has put an end to the globalization we have experienced over the last three decades ...
>
> The invasion has catalyzed nations and governments to come together to sever financial and business ties with Russia. United in their stead-

> fast commitment to support the Ukrainian people, they launched an "economic war" against Russia ...
>
> Capital markets, financial institutions and companies have gone even further beyond government-imposed sanctions. As I wrote in my letter to CEOs earlier this year, access to capital markets is a privilege, not a right. And following Russia's invasion, we saw how the private sector quickly terminated longstanding business and investment relationships.[20]

The implications of Fink's argument are sweeping. Some will debate the accuracy of his claim that globalization as we know it is coming to an end. We believe that the new national security economy is driving a Balkanization of the global economy; this is not the end of globalization but it will certainly emerge with different contours over time. His observations about businesses' reactions to the invasion of Ukraine, however, do not require time to assess. Though the Russian economy is small relative to some of its European neighbors and is dwarfed by that of the US and China, it is still a source of significant revenue for some companies. When these companies cut ties with Russia, it was usually before any legal obligation to do so; instead, they worried about the reputational cost associated with ignoring such aggression and conducting business as usual. A previous generation of leaders would have echoed Total's former CEO in separating politics and business and pursued profit instead of purpose.

The intersection of the new national security economy and the renewed focus on business purpose, particularly stakeholder capitalism, bridges the traditional narrative of a division between politics and business. The robust international linkages between business and geopolitics, even in a world where elements of globalization are perceived to be in decline, have created a landscape that organizations and their leaders must account for as part of their daily operating environment.

In this dynamic decision-making milieu, one element in the psychology of business decision-making must be highlighted. Social network analysis illuminates the dense networks of key business nodes like London, New York City, and the Bay Area. These networks are created by common employers, similar educational backgrounds, and shared social and professional affiliations. None of this should be a surprise—business leaders and board members of influential companies always have significant overlap, creating thick relationships where information and views are commonly communicated and shared.

As the original cohort of companies announced their intention to curtail or withdraw business activities in Russia, a form of social proof emerged. Social proof is a psychological concept in which individuals copy the behavior of others to reflect a normative expectation for a particular situation. An action is taken because others are doing it. In areas with dense networks, the influence of social proof can be particularly strong. As a company's choices became recorded and publicized on Professor Sonnenfeld's list and then reported by media outlets, companies that were ostensibly purpose driven had to react as peer organizations made choices consistent with their purpose and demands by stakeholders. For example, many business organizations that initially separated from the Russian market were professional service organizations, such as consulting and law firms. Once a large global law firm suspended its Russian operations, it became difficult for its competitors to continue operating there. The pressure intensified when the law firm's clients also removed themselves from the Russian market. This interaction among dense networks and the impact of social proof is another expression of the new national security economy.

The power and hope of purpose

One of us (Lee) was on the board of directors of a company that was one of the most active technology investors in Myanmar. The firm had made investments across several high-growth sectors, including fintech, logistics, and key consumer areas. The future seemed exceptionally bright until the COVID-19 pandemic, which was followed by political unrest and civil war. The situation was bleak, with reports of employee deaths from disease or violence. The firm's leadership worked relentlessly to protect people and assets, preserve employment where possible, and provide needed assistance when able. Earning an MBA or reading a business magazine does not prepare leaders for these excruciating moments.

The role and impact of business in society has expanded. Since 2001, Edelman, the global public relations and media firm, has conducted an annual survey on trust, dubbed the Trust Barometer, which evaluates the level of trust respondents feel about business, government, media, and NGOs.[21] The annual surveys are typically based on responses from over 30,000 people in about thirty countries. In 2021, the Trust Barometer indicated a significant shift as respondents reported trusting business more than government, media, and NGOs. And in the surveys since then, from 2021 to 2025, business has remained more trusted than government, media, and NGOs.[22] Intriguingly, this occurs as businesses expand their sense of purpose and, during the pandemic and now war, have had to shoulder responsibilities beyond financial performance. In 2022, Richard Edelman, CEO of the eponymous firm, commented:

> Business is the most trusted institution in the world. It has been for two years. It is important to note, though, that it is "my employer" that is much more trusted than business in general ... Geopolitics is the new test of trust.[23]

Trust and purpose are linked. The debate over business purpose and the battle between shareholder value maximization and stakeholder capitalism have tilted in favor of stakeholders. Part of this evolution reflects the growing influence of geopolitics on business decision-making—and the influence of businesses leaders on geopolitics. This transition also impacts companies they lead by shaping organizational culture and purpose.

Organizational success requires that businesses and business leaders have a clear sense of purpose, an understanding of stakeholders' needs and interests, and most importantly, the courage to make choices consistent with that sense of purpose.

PART IV

A PERILOUS MOMENT

9

A PERILOUS MOMENT

Shift happens. Since the end of World War II, the locus of international power and wealth has been the trans-Atlantic community. Asia accounted for less than one-fifth of global wealth in the 1950s.[1] The world has essentially been run by Western governments: they established the primary institutions of the international order, such as the United Nations and its associated bodies, the World Bank, and the International Monetary Fund, and their operating systems—the values, norms, rules, and principles that guided them. The West's victory in the Cold War ostensibly signaled the end of ideological division and competition and the ascension of a single set of values and accepted principles—democracy and market-oriented economics. While there has been an even more acrimonious debate about the nature of that order, and whether it was in fact global, its aspirations to universalism have not been questioned.

This ambition has created the central paradox of today's world. It is founded on the principles of individual equality and openness, which some states exploit by accepting some, but not all, of those principles yet demand their inclusion as members nevertheless. As a result, illiberal states are part of liberal institutions. (Of course, at times the United States too has been reluctant to be bound by international institutions that it helped create.) Their inclusion rests on another foundational principle of the global order, one that well predates the pre- and post-Cold War versions: the sovereignty of states, which protects dissenters from being forced to accept the entire menu of liberal norms.

Membership had its benefits. Efforts to ensure that all nations shared in the promise afforded by that order proved so successful that they planted the seeds of its demise. Asia was especially good at seizing those opportunities and flourished, subjecting the old order to increasing stress as a result. As the new millennium began, Asia accounted for 32 per cent of global GDP (measured by purchasing power parity). That share increased to 46.38 per cent in 2025 and is anticipated to exceed half of global wealth (52 per cent) by 2040. In almost every category, Asia's share of global cross-border flows—trade, capital, people, knowledge, transport, culture, and resources—is climbing.[2] Leavening is to be expected.[3]

Understandably, this growth prompted a desire in rising states to have more say in the rules of the system and to rewrite them to better reflect their histories, cultures, and preferences. That they profited from existing rules does not mean that they accept them as is or don't seek their amendment or revision. The success of the rest and the resulting rebalancing of wealth and opportunity have prompted a similar reconsideration of those rules among countries that wrote them. The result is a mix of governments, with distinctive political economic models, jostling for position. "This heterogeneity is not so much a rush to excellence as the projection of the domestic characteristics of the major powers onto the international arena."[4]

The greatest challenge comes from illiberal governments that want to preserve their access to the global commons while maintaining political economic systems that contravene other elements of the global order. They support some pillars of the existing order, in particular the absolute sovereignty and autonomy of states (at least their state), while rejecting others, such as the human rights covenants that are included in institutional charters. Some profess great respect for the equality of all states and global democracy but are quick to insist that some states are more equal than others. Examples include then Chinese Foreign Minister Yang Jiechi's insistence at a 2010 meeting of Asia-Pacific nations that "China is a big country and there are small countries, and that's a fact," and Russian President Vladimir Putin's belief that nations on Russia's periphery do not have the right to make decisions that compromise Moscow's security. The 2022 invasion of Ukraine is the sharpest expression of that mindset.

Chinese claims go further. The Chinese government believes that its extraordinary success in creating domestic prosperity offers an alternative model for other developing states and would be supported by those nations if its domestic practices were not subject to international criticism. It is precisely China's success and the prospect of a rewriting of the rules that triggers anxiety in the West and underlies much of the debate over the new national security economy.

That doesn't mean that China wants to take over the world. Rather, the Chinese leadership wants safe space to advance and protect its national interests. As one recent study concluded, "China seeks regional preeminence and global prominence to cement its global presence and influence across all dimensions of power."[5] Since hegemony and conquest are not the goal, then the result is likely to be a pluralistic world order. Contrary to warnings that this is a formula for conflict as different governments and ideologies tussle for position, others, including Chinese President Xi Jinping, see hope in "a shared future for mankind."[6] Timofei Bordachev, program director of Russia's Valdai Club, welcomes diversity: "amid the new framework and conditions, it may become more likely that no significant state will be so dissatisfied with its position that it will push for revolutionary change."[7] A lot rides on "may," and there are signs of enough dissatisfaction that destabilizing, if not revolutionary, change is possible.

Supercharging this process is the belief that economic interdependence is not working and may even be a threat to national security. China has long leaned toward autarky, especially in technology.[8] This inclination has been consistently reinforced, and enshrined as national policy with adoption of the Dual Circulation Strategy in 2020, which seeks to build domestic demand while reinforcing ties to the outside world as a way of reinforcing other countries' dependence on China. The foundation of this policy is Xi's oft-stated dictum that "core technologies must be in our own hands," the product of a worldview in which world powers inevitably block the rise of new or potential challengers.[9] Barry Naughton, a professor of economics at University of California San Diego, concludes that Beijing "hope(s) to foster the continuing presence of GPNs (global production networks) in China—perhaps held there by the attractions of the Chinese market—as a continuing source of technology transfer, while also aggressively squeezing out the foreign links in the GPNs."[10] Xi validated that analysis when he told the Central Committee for Financial and Economic Affairs that China would become a "gravitational force" in the economy, attracting resources from across the world to become an indispensable link in global supply chains.[11]

Meanwhile, Chinese government agencies are scrubbing the national IT infrastructure of foreign parts and components[12] and the Communist Party has passed a historic resolution—only the third in its 100-year history—that identifies the internet "as the main arena, battleground, and frontline of the ideological struggle" and demands that the party retain control of that domain.[13] This dogged insistence reinforces claims that the Beijing government "no longer maintains ambiguity about the nature of the Chinese system and whether it will converge with OECD norms: it emphasizes the differentness and says it won't."[14]

The US appears determined to reinforce those divisions. Washington has banned Chinese companies from participating in national telecommunications networks (and pressed allies and partners to do the same). It has denied many of those same companies access to key components, semiconductors in particular, that are critical to their ability to stay on the frontier of technology (and pressed allies and partners to do the same). Chinese researchers are being excluded from research projects in US companies and universities (and it has pressed allies and partners to do the same). It has denied US financiers investment opportunities in promising Chinese startups and denied Chinese funders investment opportunities in US companies. As detailed in Chapter 1, China is now retaliating in tit-for-tat fashion.

There is a spirited debate over the consequences of this split, which is described at various times as deglobalization, regionalization, diversification, reshoring, or decoupling. Executives are divided and the anecdata is mixed.[15] Foreign businesses in China complain about increasingly difficult, if not hazardous, operating conditions; Chinese businesses in the West issue the same charges. Between 2016 and 2020, overall direct investment between the US and China shrank from $62 billion to $16 billion, a 75 per cent fall, and the tech sector plummeted 96 per cent over the same time.[16] The stock of US investment in China has remained steady at $127 billion in 2023, but annual investments are falling. And Chinese direct investment in the US plummeted $44 billion in 2023, down 16 per cent since 2019.[17] Yet, despite a 30 per cent drop in US investment in China in 2020, overall inbound direct investment in the mainland from all countries increased about 7 per cent, allowing China to surpass the US as the world's leading recipient of FDI.[18]

Macro considerations counsel pessimism, nevertheless. The mindsets in major capitals, driven by growing concern about vulnerabilities, the desire to spur domestic production, and acceptance of a world of long-term rivalry and competition, will make separation a reality. This is evident in government programs to develop competing standards, promote national champions, and to wall off data and inhibit its flow. All reinforce technological bifurcation. The US Chamber of Commerce worries that "one cannot expect a return to engagement as we knew it. Decoupling is likely to continue in one form or another, even if it does evolve in a more measured, targeted way."[19] The IMF sees signs of geopolitical fragmentation.[20]

While no government will take the blame (or the credit) for instigating this downward spiral—each claims that it is only responding to actions by the other—the result is unmistakable: a slippery slope, in which:

> the technologies that are defining the future, and which are increasingly integrated into every sector of the economy, are being divided between two of the world's largest economies, each of which has a growing firewall separating itself from the other. The US is moving toward a world in which Chinese technology should be purged from supply chains servicing Americans, while China is creating state-sponsored national champions to dominate a self-reliant ecosystem of indigenous technology integrated across its entire market.[21]

The epistemology of national security

There is much that can and should be done to address the challenges of the new national security economy. The starting point is adopting a new mindset, a new approach to national security. Recognition of the role of economic security is part of this new thinking, but only a part. Our world is punctuated by new threat vectors, new uncertainties and new instabilities. At the societal level, there are vulnerabilities created by the thickening weave of connections in daily life, the digital technologies that enable them, and the permeable information space of democratic societies.

The threat horizon extends in other directions as well. Governments and businesses must pay more attention to supply chain resilience and stability, identifying vulnerabilities and taking steps to ensure supplies of critical materials and components. They must protect critical national infrastructures (however they may be defined), more closely scrutinize foreign investments, and control access to critical and emerging technologies.

All of this takes place as the world moves toward multipolarity. A global order that was created by nations with a shared political and culture heritage must adjust as newcomers demand their say. Governance is becoming more complex as more actors (not just states; businesses, civil society groups, and even rogue elements want a say) are involved and interests shift depending on the issue. One expression of this multipolarity is intensifying competition between the West and China. This is not just a rivalry between Washington and Beijing, even if they are the chief protagonists. The West is rightly considered an agglomeration of countries, and it is not monolithic in its views of China. Tensions within the West are also mounting. The second Trump National Security Strategy is read by some as a declaration of war against Europe.

This struggle is multidimensional, and while the military plays a role, Bill Burns, former director of the CIA, is right to note that technology is "the main arena for competition and rivalry with China."[22] Its unique form of capitalism, its hybrid companies, and the expanding reach of the Chinese Communist Party into virtually all spheres of economic activity challenge foundational tenets of international economic governance. Alarms are ringing, because, as one US intelligence official explained, "there's a lot at stake with a lot of these technologies. If we lose supremacy in these areas ... we could be eclipsed as an international superpower."[23]

Ultimately, threat horizons extend literally in all directions: not just a 180-degree perspective or even a 360-degree outlook but one that takes in 720 degrees. One analysis concludes that relations among states will be marked by "securitization"—the identification of ever more spheres of activity as matters of "security" rather than ordinary politics—along with Balkanization (fragmentation) and weaponization (the use of dependencies to create leverage over other states).[24]

Foundational assumptions must go

The notion of a single global order is quaint and outdated. In a world marked by increasing equality and greater diversity, consensus will be harder to find; the result is what we call "thin globalism." There will be acceptance of a very few universal principles: state sovereignty is likely to be one of them. Even if there is agreement on an economic paradigm—capitalism—its practice will be variegated. Economist Branko Milanovic espies two distinct forms of capitalism: a liberal, meritocratic variant in the West and the "state-led political or authoritarian" version that China practices.[25] The gulf between the two is wide and growing, as Chapter 3 makes plain.

Economic integration will continue, but interdependence will be attenuated. Advanced technologies and their accoutrements will be segregated. This new regionalism will be different from that of previous eras. Previously, geography was dispositive. Now, however, digital connectivity helps undermine the tyranny of distance and allows the formation of blocs based on different organizing principles, such as values.[26] Talk of a G10[27] and a T12,[28] as well as the increasing geopolitical alignment of trade partners, are acknowledgements of this new reality.[29]

Jarring as it will seem, the West must accept that its views and values will no longer be dispositive. This will have profound implications for human rights as debates about its meaning and content will intensify. This will pose real challenges for business as activities that span the emerging blocs will be subject to differing rules and regulations. It isn't clear how governments or companies will reconcile their conflicting mandates.

The idea that China can't compete must also be discarded. Shenzhen and Beijing are two of the world's top five science and technology clusters, and China is now more innovative than Japan on many metrics.[30] The country has access to resources and is an effective competitor across a range of industries. It will challenge Western leadership in some of the most important new sectors, if it isn't already.[31] And it can do so without theft or other illegal activities. In other words, defensive measures alone are not the answer.

It's also time to scrap the mythology that dominates discussion of the American entrepreneur. Since the end of the Cold War, and the extraordinary success of America's digital businesses, there has been a rewriting of history. The prevailing narrative is of solitary individuals toiling away in a garage, animated by deep and unshakable faith in an idea. Grit and determination pay off with a world-changing—and very profitable—product. It's a nice tale—and revisionism of the highest order. Individual effort is important, but those technologies emerged from a soil fertilized by decades of US government money, usually, but not exclusively, from the Department of Defense. As we note in Chapter 3, Silicon Valley wouldn't exist without federal defense contracts that provided vast sums of money and a guaranteed customer long before technologies were commercially viable.

An agenda for success

As the West contemplates this emerging world, it needs a new strategy. An effort is unfolding but it is haphazard, impulsive, and uncoordinated. An ad hoc approach was understandable a few years ago (given the novelty of the situation), but it's a dangerous indulgence today. Continued delay, a lack of vision, and an absence of urgency will have geopolitical consequences.

That strategy needs to be articulated and presented. It isn't enough to have a "guiding philosophy" or "an approach." Governments must create documents to explain the whys and the whats of policy, to provide expectations for action and baselines for evaluation. National security strategies need to be more inclusive, encompassing not only traditional threats but those of the digital economy, adopting that 720-degree perspective. There should be more attention to and clarity regarding the economic dimensions of national security. A good idea is the creation of a technology policy officer close to the chief executive (president or prime minister) to direct and coordinate policy and ensure a whole-of-society approach. That person should be working closely with an economic security minister.

The United States must begin with an industrial policy. Debate on this subject has been reinvigorated, but it continues to be dominated by myths and untruths about the past. The US Innovation and Competition Act marks a shift in policy and reflects a rethinking of assumptions about the relationship between the public and private sectors. The bill has its problems—like all omnibus legislation—but it is a good start. The Biden administration's National Security Strategy acknowledged and embraced this reality. It endorsed "a modern industrial and innovation strategy" and endorsed strategic public investment, "the backbone of a strong industrial and innovation base in the twenty-first century global economy."[32] Industrial policy will be part of a broader economic security strategy that meets the needs of a larger proportion of the population of developed countries. In recent years, Western economic diplomacy—and certainly that of the US—has sought to advance the interests of a small group of constituents, those representing capital, over that of labor. That priority has contributed to domestic inequality and undermined support for free trade.

Other Western governments have been slow to recognize the nature of the China challenge, but they are catching up. Unfortunately, they have displayed no greater capacity for speed, focus, or strategic thinking. To their advantage, they don't have the same baggage when it comes to acknowledging the role and value of industrial strategy, but they have related problems, which will be taken up below.

Each country's strategy will reflect its particular circumstances, and they will have diverse policies and priorities. Nevertheless, there will be common features. Every effective strategy will rest on three pillars. The first consists of defensive measures, and there is an expansive list to consider. Greater attention to cybersecurity is a necessity, with emphasis on protection of critical infrastructure. Control of access to technologies will be a central feature of any strategy, for which trade and investment measures and research guidelines will be prominent. Central to defense will be resilience, which would ensure that the effects of attacks are reduced. The sad truth is that there are just too many possible vectors of attack. Perfect defense in cyberspace is essentially impossible.[33] A focus on resilience rather than absolute protection is akin to what nuclear strategists call deterrence by denial. It reduces the incentive to attack by minimizing the ensuing damage. And, if an adversary is not deterred, the impact of an attack is still diminished.

The second pillar consists of the offensive measures that are needed to spur innovation, research, and development. This includes money, of course, and more of it must come from the public sector. Private sector funding is necessary, but not sufficient. It has different objectives and different timelines. A more aggressive government program will require new tolerances for failure, but it will also demand ancillary changes, especially the creation of a cadre of bureaucrats that can oversee those efforts to ensure that there is no waste and that publics get the requisite bang for their bucks. In addition, governments need to do more to protect intellectual property while simultaneously—and somewhat paradoxically—ensuring that the fruit of that work is better distributed so that others can build upon it. There needs to be more effort to ensure that there is a skilled and expanding cadre of workers. That means supporting STEM learning early in the education process and keeping doors open to talented immigrants.

The third pillar is multilateral coordination. Offensive and defensive measures will be implemented with varying degrees of creativity and intensity, but those efforts will invariably come up short in the absence of cooperation and coordination among allies and like-minded partners in almost every field of endeavor. They should be working together to promote innovation. Onshoring and promoting indigenous production are tempting; "friend-shoring," or the routing of supply chains through allies and partners, makes more sense. Moreover, it makes no sense for one country to restrict access to valued technology if it is legally available by other means, from other vendors or research initiatives. To ensure those work arounds do not exist, there must be information sharing on a variety of levels. Most basically, there must be agreement on the identification of priority technologies to be protected. There is now talk of a "fifth export control regime" that would focus on the emerging technologies critical to success in the geopolitical competition of the twenty-first century.[34]

There also needs to be a shared understanding of the threat—the bad actors, the predatory practices. The guiding principle should be "collective resilience."[35] While this seems obvious, it will not be easy. There is an ingrained reluctance, if not resistance, to intelligence sharing, born of justifiable concern about the protection of that information. Sometimes there is hesitation about accepting such intelligence as a result of the obligations created by its receipt. Finally, the private sector's involvement adds another layer of complications. Businesses often hesitate to share critical information because it is proprietary (in the case of technologies or commercial relationships) or because disclosure (for example, cyberattacks) could create reputational risks.

Cooperation is needed in multilateral forums to ensure that diplomatic initiatives reflect the values of the like-minded countries. China's Belt and Road Initiative is a world-historical effort that seeks to transform the way that the world works, reshaping and redirecting international connectivity in both form and substance. Governments that are concerned about the proliferation of Chinese technology or the influence bestowed by Beijing's largesse must respond in kind. They must create options, offering alternatives that address the social needs that the BRI fills. The continuing metamorphosis of the West's signature program—from the Blue Dot Network to the Build Back Better World to Partnership for Global Infrastructure Investment (BDN to B3W to PGII)—is a sign that those governments understand the challenge and proof that they haven't been able to rise to the moment. A smart approach is required. In some cases that will mean the provision of tangible deliverables, but the West must accept that it cannot match China dollar for dollar. Instead, there will be a premium on creative efforts to provide services that blunt the more pernicious effects of strategic or values-blind lending. Moreover, as important as the technology provided are the standards that guide its use. Like-minded governments should work together in international venues to ensure that those standards promote their values and the use of their technologies.

Finally, they must be adept and aggressive in crafting a narrative about what they are and are doing and why to win over publics at home and in other countries. For the first time in a generation, there is a real competition to shape the information space and produce a compelling narrative. The West cannot assume that its story is better or more compelling. While Western governments must accept that worldwide human rights are neither defined nor prioritized as they are in their countries, they should presume that their position on those issues is a source of comparative advantage in the competition with illiberal regimes. They should draw sharp contrasts between their policies and priorities and the results they create, essentially weaponizing human rights. In addition, those governments should coordinate their actions to punish scofflaws to maximize the impact of sanctions and demonstrate that misbehaviors have consequences. A premium on cooperation will also raise the bar for action and ensure that this tool is not used indiscriminately or to excess.

As governments rethink industrial policy, they must be judicious, avoiding the excesses of the past. Governments should support industries, along with research and development. That does not mean picking corporate winners and losers. They will need to develop or acquire the talent to assess projects to ensure that decision-making remains independent and avoids capture. They will have to accept that there will be losses; not every project will be a winner. Governments will have to be equally alert to the temptations of protectionism and mercantilism; the line between "good" industrial policy and "bad' is often in the eye of the beholder (or the beneficiary) but can also be bright and clear. Transparency is one powerful tool to avoid excesses.

In the United States, the debate over economic security has implications for the practice of its democracy. National security authorities are vested in the president; constitutional scholars warn that declaring something "a national security concern" is a blank check for the executive, one that in the economic sphere would clash with the Constitution's grant to Congress of "exclusive and plenary" authority over regulating commerce with foreign nations. Critics charge that the Trump administration's definition of national security was designed "to justify expansive executive authority in collision with longstanding foreign policy goals or the administration of the regulatory state."[36] The growth of the new national security economy facilitates the creep of executive authority and will create greater imbalance among the three branches of government.

A capacious definition of national security facilitates capriciousness in US foreign policy. While the president was assumed to require extensive power in this field because of a need to act with urgency in a national emergency, that authority can and often is abused. A threat to economic security can be easy to manufacture—jobs losses, for example, are a constant—and provide easy justification for protective or retaliatory measures. The second Trump administration is an ongoing exercise in the extension of presidential authority by the declaration of national security emergencies. Guardrails are absolutely necessary.

A new world for business

One may disagree with his politics, but Peter Thiel recognized the defining features and dangers of the twenty-first-century economy very early. In a 2002 speech, while CEO of PayPal, Thiel highlighted two major trends. "First, the globalization of the economy. The economy is growing interconnected ... Second, the quest for security. In this globalized, decentralized world, violence and terrorism are widespread and hard to contain."[37] Speaking less than a year after the September 11 attacks, a focus on terrorism reflected the moment. But Thiel's broader concern was security and the vulnerabilities and dangers that stem from connectivity and the integration it facilitates. Acting on this thinking, Thiel cofounded Palantir the following year. Now counting US military, intelligence, and other government agencies and departments among its most significant clients, Palantir exemplifies the new national security economy, prominently stating on its home page that: "Our software powers real-time, AI-driven decisions in critical government and commercial enterprises in the West, from the factory floors to the front lines."

Unfortunately, the path for most organizations and leaders in the new national security economy will likely not be as clear as Palantir's. Given that, superior strategy and tactical thinking will both be necessary and at a premium. Readers can use the GRAPL and PIPE frameworks from Chapter 7 as a structured starting point to engage on salient issues. Building on those frameworks, additional guidance is offered below.

First, and most important, be proactive. Organizations and their leaders must develop the agility that comes from a commitment to constantly iterate as risks emerge. This demands new learning, processes, insights, and improved decision-making. This may seem daunting—a journey without a final destination—but all great companies adapt to market dynamics and the evolution of the wider world in which they operate. Companies must commit to a continual process of refinement as they determine what works, enhance capabilities, and build on that foundation to navigate the new national security economy. Unfortunately, heightened geopolitical concerns, a sluggish global economy, and a worldwide pandemic have prompted many executives to adopt a passive, defensive posture as they try to wait out the storm. That won't work. Taking calculated risks and playing offense, even amid turbulence, is critical.

This will be aided by adoption of a broader perspective, one that transcends narrowly defined shareholder-based decision-making. The singular focus on revenue or profit is outdated. It promotes tunnel vision, elevates the possibility of reputational damage, and obscures the strategic perspective necessary to consider alternative courses of action. This isn't new: perhaps the most famous such incident occurred in the 1980s, when a subsidiary of Toshiba sold machine tools to the Soviet Union that allowed it to build quieter submarines. That indifference to broader national security concerns resulted in the spectacle of members of the US Congress smashing Toshiba radios with sledgehammers on the Capitol steps.[38]

Second, an easy way to acquire that perspective and enhance expertise is by selecting the right board members. Ideally a board comprises directors who can contribute a diverse range of skills and experience to support a company and advise its leadership. Companies should ensure at least one board member has knowledge and skills related to understanding and communicating relevant geopolitical issues. Many boards of listed companies have tried to enhance expertise in cybersecurity or issues related to inclusion and diversity. Expertise in the new national security economy should also be sought when considering new board members.

Third, companies should use scenario planning to anticipate new national security economy risks. These exercises should bring together employees from different functions and geographies and explore their unique perspectives. These exercises can commence by identifying existing geopolitical hot spots or those that are on the horizon, then asking a simple question: how will that impact us? These are not—cannot be—one and done. The world changes, demanding regular assessment and revision. There is a burgeoning industry that offers help and training. Think tanks, law firms, and consultancies are all getting into the game. Some governments are as well. Finland's National Defense Courses helps inform leaders from business and civil society about crises ranging from conflict to environmental disasters and how to address them.[39] The National Defense Courses aim "to give participants a total overview of Finland's foreign, security and defense policy," and they "are meant to improve cooperation between different sectors of society and facilitate networking of people working in the various fields of comprehensive security."[40]

These types of exercises offer considerable benefits. They force companies to contemplate issues, impacts, and responses; good leadership will develop contingency plans. They allow individuals within an organization to build relationships and networks, which facilitates the sharing of information. Often companies have indicators of potential crises, but those warning signals are not shared with decision-makers. Regularly conducting exercises helps build a culture where national security concerns assume weight and substance and acting upon them becomes a feature of business operation and practice. Incorporating even some of our suggestions will increase business resilience, helping not just to cope with but also to grow in the face of new challenges.

Fourth, companies must rethink reputational risk. This problem assumes new significance in the new national security economy. Globalization has extended and obscured risk as supply chains thread their way around the world, often without recognition of the optics or dangers they pose. Critical supplies and materials can be sourced from volatile and unstable areas or those with human rights abuses, or poor environmental or labor standards. Customers and clients may use products or services in ways that violate the norms, principles, and values of sellers' countries, customers, and stakeholders. Nokia announced it would withdraw from Russia after the invasion. Yet shortly after, *The New York Times* published a story describing how Nokia had sold "equipment and software connecting the [Russian] government's most powerful tool for digital surveillance to the nation's largest telecommunications network."[41] Records showed that Nokia knew it was enabling a Russian surveillance system, but it had become a top supplier of equipment and services, deals that yielded hundreds of millions of dollars in annual revenue, even as Putin became more belligerent abroad and more authoritarian at home. These decisions were legal and made sense from a shareholder-focused, profit-oriented perspective. Assessed against the backdrop of all of Nokia's stakeholders, its organizational purpose, national identity, and the broader geopolitical landscape, it was a failure.

Executives must recognize too that incentives and motivations are changing. While hackers might have once threatened to disclose the fruits of their labor to extort money, in an era of state-based technological competition there is every incentive to damage a company. Reputational harm is the goal, not just a source of leverage.

Ultimately, businesses must recognize that incidents that create reputational damage are inevitable: the issue is not if, but when. The old playbook that counseled caution and playing defense no longer applies. Tony Langham, cofounder of Lanson, a British public relations firm, warns that "It is sometimes risky as a company or a leader to play it safe. That can now sometimes be more risky than taking a position."[42]

Fifth, security concerns should be embedded from the beginning in a company's strategic plan. Often, a plan is constructed, and mitigation is considered later. Consider a technology company developing a new product that potentially raises privacy issues. Frequently, the engineer considered her job to be making the coolest and most effective product possible, while negative externalities created by the product would be managed by regulators, lawyers, and lawmakers. Acknowledging that responsible innovation is necessary, technology companies are starting to accept that a different approach is required, especially in the age of AI. Responsible organizations must understand that geopolitical and national security concerns are not afterthoughts or add-ons but integral components of a company's strategic outlook.

Building resilient societies

The penetration of digital technologies deeper into our daily lives means that ordinary citizens have a vital role in promoting and protecting national security. Estimates of the number of internet-connected devices per household in North America and Europe range from a dozen to fifty as 5G technologies and the Internet of Things proliferate.[43] Plainly, the public must become aware of the vulnerabilities created by connectivity and the need to act accordingly. Citizens must become savvier consumers, demanding more from the companies that provide goods and services as well as becoming more competent and reliable practitioners of cybersecurity, practices that include everything from regularly updating software to resetting installed passwords to being alert to cyber scams.

Greater competence in handling technology must be complemented by greater sophistication when navigating the modern information environment. This includes digital literacy training and basic civics education. Consumers must "learn to discern"[44] and acquire good habits for engaging with information. While this assignment is likely to get easier as societies become better habituated to new forms of information engagement and consumption and become "digital by default," there will always be a need to teach the proper skills to assess and evaluate information. It isn't enough to be able to tell "true" from "false" or "real" from "fake" news: rather, citizens need to better understand the information ecosystem and figure out how to navigate it without getting distracted, misdirected, or misled.

The adjustment of skills and expectations in a connected world demands a whole-of-society approach. Government will typically lead, since it has the resources to educate and inform. But in a world in which a growing number of targets—infrastructure, assets in strategic sectors, or critical nodes in supply chains—are privately held, business too will be in front. The result is a conceptual shift, "from an underpinning belief in deterrence as a passive, dormant posture to which governments are solely responsible, to an active and dynamic state of mind that reaches across society, where everyone plays a part."[45] This isn't unprecedented. During the Cold War, some Western countries practiced "Total Defense," which mobilized all elements of society—businesses and general population—to defend the homeland and ensure continuity of daily life in the event of war or some other national crisis. The same thinking animates public awareness campaigns for preparation and survival in earthquake-prone cities, such as San Francisco or Tokyo.

Success in those efforts demands trust in government, however, which can be difficult since many misinformation campaigns target that very commodity. The fate of the US Department of Homeland Security's Disinformation Governance Board, proposed and scuttled in the spring of 2022—the result of a "deliberate disinformation campaign"—is testimony to the challenges any such effort will face.[46] The board was set up to "coordinate countering misinformation related to homeland security" by studying best practices to combat disinformation and to help counter viral lies and propaganda that could threaten domestic security. A coordinating body within the Department of Homeland Security with no authority to make operational decisions, it would have no power or ability to declare what is true or false, or compel action against certain types of speech.[47] Critics nevertheless unleashed "a textbook disinformation campaign" against the board and its proposed executive director, Nina Jankowicz, a leading researcher on disinformation. They charged it with being an Orwellian "Ministry of Truth" that would police speech and subjected her to vicious personal attacks and smears. Within three weeks, Jankowicz had resigned, the Department of Homeland Security had suspended the initiative, and supporters conceded that there was little hope of it being reinstated. The fate of the board is a grim indication of the difficulties that any such effort will encounter and, simultaneously, a reminder of the supreme importance of its success.

A big ask

Commonsensical though they sound, implementing these recommendations will be difficult. There are obstacles aplenty. Some are bureaucratic, some are political; the scale of the ambition is another inhibiting factor. And, of course, habits and culture (business and national) will also discourage innovation and adaptation. In some cases, particularly those for business executives, our suggestions challenge foundational beliefs.

Members of the public will have to become more discerning consumers, less passive, more focused on security and more demanding of companies that provide digital goods and services. This is difficult when those tools and technologies are specialized and advanced and typically presented as a black box. After all, technology manufacturers have promoted a plug and play approach that minimizes the consumer's roles and responsibilities. Citizens don't have to become technicians, but they must be alert to the implications of misuse or poor technology hygiene. They must recognize that they are part of a digital ecosystem in which the weakest link imperils the whole.

This responsibility is greater still when citizens navigate the information space. The proliferation—democratization?—of channels has resulted in a flood of information, and we need more than ever the ability to tell fact from fiction, to judge probabilities and certainties, and to distinguish data from distractions, the wheat from the chaff. Critical reading and thinking skills are no longer a luxury but a prerequisite of good citizenship. That process will encourage an ecumenical approach to information gathering, breaking silos and exposing consumers to a genuine marketplace of ideas. Upon reflection, this may be our most fanciful suggestion.

Business obstacles are no less entrenched. There is, for a start, the continuing allure of the Chinese market. Surveys of foreign businesses show ongoing commitment to China, with as many as 72 per cent prepared to expand their presence there even as geopolitical tensions rise. That is understandable given that, as one report noted, China is the "main engine" of global profit growth for many foreign companies; 58 per cent of respondents in one poll considered overall return on investment in China to be higher than other places.[48] One estimate of the cost of decoupling puts it in excess of $3.5 trillion over five years,[49] 60 per cent of which stems from losing access to China's market.[50] Jim McGregor, chairman of greater China for APCO Worldwide, warned that the sheer size of that market could encourage companies to "decouple from America." "It's not defending what is good for America. You can say that's terrible, but that's the way our system works."[51]

John Neuffer, president of the Semiconductor Industry Association, explained that "Revenue from that big market fuels our big research investments, which allows us to innovate and drive America's economic growth and national security."[52] The stakes couldn't be higher. According to a Boston Consulting Group analysis of the semiconductor market, "broad unilateral restrictions on Chinese access to US technology could significantly deepen and accelerate the [global market] share erosion for US companies," resulting in drops in revenue that lead to severe cuts in R&D and capital expenditures, and the loss of 15,000 to 40,000 highly skilled direct jobs in the US semiconductor industry. The study authors warn that South Korea would likely overtake the US as world semiconductor leader in a few years, with China taking leadership in the long term.[53] While this analysis focused on denial of access, a deliberate withdrawal could have the same effect.

When the market isn't the draw, the siren song of Chinese investment can do the trick. Many Silicon Valley executives and entrepreneurs are convinced that tighter scrutiny of Chinese capital will deprive them of much-needed funds or undermine development of new technologies. As one expert explained, "The US government cannot summarily assert that AI is inherently dual-use and place all AI technology under export controls; doing so could undermine the very ecosystem of capital, data, and know-how that fuels AI innovation."[54] Still, anecdotal evidence suggests those fears are exaggerated and that if those companies lose access to Chinese funds, other sources will fill the gap.[55]

The second problem for business is the required overhaul of basic tenets. Our suggestions are revolutionary. While we believe that they will pay off, they require executives to take risks. Most of their jobs aren't structured to encourage that behavior: performance-based remuneration may be reduced when exercising "patriotic restraint." The most fundamental challenge is accepting that transactions will not necessarily be guided by calculations of profit and loss. This not only undermines the most basic principle of business education but erodes a pillar of economic logic as well. Moreover, this is not an obvious or ubiquitous phenomenon: not all transactions occur under these circumstances. Executives must not only be prepared to reject basic principles of their education and experience, but they must also acquire the skills to understand when they should do so—and when traditional policies, practices, and principles apply. They must learn to discriminate among opportunities. This will necessitate a new relationship with governments since they are likely to have information those executives need to make correct decisions.

Governments must adapt as well. They must figure out how to acquire and then share information that will be of value to executives in this new business environment. They must develop ways to promote a reciprocal exchange of information to facilitate this process, which raises a host of issues about security, privacy, intelligence collection, and protection.

Another obstacle that governments will face is the inherent difficulty in forging and maintaining a unified position. A consensus seems to have emerged among developed nations that China is a competitor and "systemic rival" that has its own views of global order. But the contours of a broad agreement do not guarantee collective action on specific items and issues: the devil is invariably in the details. There are legitimate differences in interpretation of actions and policies, different definitions of the interests at stake or, to be crass, the possibility of benefiting from another country's misfortune (as happened when wine exporters swooped into China to take the share of that market that Australia lost during its trade fight with Beijing).[56]

Cohesion is also threatened by disputes and conflicts within the West. Sadly, these are all too common, and there is the very real danger of another "America First" president turning the spotlight on Europe's trade practices. A united front, the prerequisite of any campaign to deter misbehavior by revisionist powers, is difficult if not impossible if (or when) the West fights among itself. The debate over "strategy autonomy" is an attempt to give structure and content to the European desire and ability "to maintain a close partnership with the United States, but they also need to protect themselves when US policies run counter to their interests and values."[57] There is a steady drumbeat of concern in the capitals of US allies and partners about the reach of US restrictions that threaten even companies from "friendly" countries. Governments and businesses in Europe and Australia are already nervous about a hardening US approach that could limit their autonomy when dealing with China.[58]

Of course, success can be dangerous. Recognition that the real twenty-first-century geostrategic competition is economic and that technology is central to its outcome has pushed governments into a zero-sum logic that has risks of its own. The US–China trade war was one expression of the intensifying competition. As noted in earlier chapters, Beijing has created a raft of measures to check or balance US policies. Karen Sutter, an analyst at the US Congressional Research Service, argues that these new measures "attempt to create parity with the United States by mirroring certain US authorities and practices" in these fields.[59] Determined not to blink, the two governments are increasing pressure, even as costs grow and general welfare diminishes. There is concern in each capital about the domestic political cost of appearing to falter in this struggle and a fear that rational behavior will be perceived as weakness by the other side.

All this results in what Australian scholars Darren Lim and Victor Ferguson call a "technology security dilemma."[60] Both the US and China (and others) believe that "the national security of a leading power will be eroded if it falls behind in technological innovation vis-à-vis a strategic rival, or otherwise loses competitiveness or market share in high-tech industries." Washington alleges that China is stealing technology, protecting its domestic market and promoting national champions to eclipse the US as the leading technology power, but the countermeasures it has adopted are denounced by China as "part of a broader containment strategy designed to slow or stop China's economic development and rightful rise as a science and technology leader."[61] With nearly 1,000 Chinese companies penalized in one way or another by the US for national security or human rights reasons,[62] Chinese allegations make some sense. One Chinese scholar from the China Institutes of Contemporary International Relations, a think tank affiliated with the country's intelligence agencies, summarized the situation simply: "the United States regards China's development as a threat, while China believes that it must accelerate its development to cope with potential US suppression measures."[63]

There is no finer distillation of the security dilemma. Steps taken to protect one side's legitimate national security trigger a downward spiral "fueled by mutual insecurity." This logic drove the first half of the Cold War, dominating policy in Washington and Moscow until the dangers became inescapable. Sobered by near brushes with catastrophe, the US and the Soviet Union recognized the need to halt the steady march toward confrontation and destruction and agreed on confidence building measures that could reassure each side about the other's nonaggressive intentions, which eventually yielded arms control agreements.

There aren't any such efforts today, in either the military or the economic space. Worse, business, one of the traditional shock absorbers in the relationship, is now a source of tension, and significant parts of the foreign business community in China are fanning the flames (not without justification). There may be room for discussion of restrictions on the military applications of some technologies—such as the use of artificial intelligence in warfare—but multilateral conversations on these issues have been slow and begrudging. Some Chinese observers warn that talks may even exacerbate the dilemma as multilateral coalitions are more proof of the West's malign intent, and the forums where compromises are supposed to be hammered out instead reinforce suspicion. The Biden administration's commitment to work with like-minded governments and its pursuit of enhanced cooperation to help establish:

> a regional technology alliance countering China ... will substantively strengthen Chinese determination to fight for supremacy in the latest innovations. ... Unified actions by these allies, either on sanctions or on exclusive standards-settings, will drive China to isolate and centralize its effort—which will in turn give the allies more reason to isolate China even more.[64]

Connectivity has created new weapons of mass disruption (and destruction), and rules and norms must be established to ensure that they are not used indiscriminately, if at all. Even without an eventual confrontation, the failure of China and the US and other like-minded countries to reach a modus vivendi will limit, or perhaps worse, deprive the world of opportunities to work together to tackle global problems. Climate change, global health, eliminating food insecurity, promoting equitable, sustainable development—all demand cooperation. During the Cold War, the US and the Soviet Union figured out ways to bound and contain their competition: the West and its rivals must do the same in this perilous moment.

NOTES

INTRODUCTION

1. Cited in John Carlin and Garret Graf, *Dawn of the Code War: America's Battle Against Russia, China and the Rising Global Cyber Threat*, Hachette Book Group, 2018, pp. 41–2.
2. Harlan Grant Cohen, "Nations and markets," *Journal of International Economic Law*, 2020, 23(4), p. 802.
3. Shaun Breslin, "Do we need a new vocabulary for talking about European strategy?" in Michiel Foulon and Jack Thompson, eds, *The Future of European Strategy in a Changing Geopolitical Environment: Challenges and Prospects*, The Hague Center for International Security, December 2021, pp. 7–9.
4. US Department of the Treasury, "Remarks by Secretary of the Treasury Janet L. Yellen on the Way Forward for the Global Economy," 13 April 2022, at https://home.treasury.gov/news/press-releases/jy0714.
5. The White House, The 2017 National Security Strategy of the United States, 2017, p. 17.
6. European Commission, "Joint Communication to the European Parliament, the European Council and the Council on European Economic Security Strategy," Brussels, 20 June 2023 at IMMC.JOIN%282023%2920%20final.ENG.xhtml.1_EN_ACT_part1_v8.docx.
7. G7 summit declaration 2023.
8. Harlan Grant Cohen, "Nations and markets," *Journal of International Economic Law*, 23(4), 2020, p. 794.
9. Natsuya Yuzue and Takashi Sekiyama, "Defining economic security through literature review," *Frontiers of Political Science*, 7, 2025, 1501986.
10. Felix Mallin and James Sidaway, "Five theses on geoeconomics," *EPA: Economy and Space*, 57(1), 2025, p. 150.
11. Ruben Gonzalez-Vicente and Han Cheng, "China, geoeconomics and the 'new' state capitalism," *Environment and Planning A: Economy and Space*, 57(1), 2025, p. 129.

12. Brendan Thomas-Noone, "What the Cold War can teach Washington about Chinese tech tensions," Brookings Institution, TechStream, 12 January 2021 at https://www.brookings.edu/techstream/what-the-cold-war-can-teach-washington-about-chinese-tech-tensions/.
13. They might want to check out *Kill Chain* by Christian Brose.
14. Vannevar Bush, *Science, the Endless Frontier*, 1945, p. 7.
15. Camille Gijs, "Trump: The EU is 'nastier than China,'" Politico.eu, 12 May 2025, at https://www.politico.eu/article/trump-eu-nastier-than-china/.
16. Bill Priestap and Holden Triplett, "Beyond economic espionage," *Lawfare*, 3 March 2021 at https://www.lawfareblog.com/beyond-economic-espionage.
17. Sam Sutton, "Of corn and crypto," Politico newsletter, 24 May 2022, at https://www.politico.com/newsletters/digital-future-daily/2022/05/24/of-corn-and-crypto-00034823.
18. Farhad Manjoo, "The Ukrainian cyberwar that wasn't," *The New York Times*, 11 March 2022 at https://www.nytimes.com/2022/03/11/opinion/russia-ukraine-cyberattacks.html.
19. UK Government, "Estimating the impact of sanctions on Russia's war efforts," 13 June 2025, https://www.gov.uk/government/publications/estimating-the-impact-of-sanctions-on-russias-war-efforts.
20. Ben Werschkul, "This economic weapon could cut Russia off from the 21st-century economy," yahoo!Finance, 20 February 2022 at https://finance.yahoo.com/news/this-economic-weapon-could-cut-russia-off-from-the-21st-century-economy-124135370.html.
21. The White House, "Remarks by President Biden on Russia's unprovoked and unjustified attack on Ukraine," 24 February 2022, at https://www.whitehouse.gov/briefing-room/speeches-remarks/2022/02/24/remarks-by-president-biden-on-russias-unprovoked-and-unjustified-attack-on-ukraine. The effectiveness of those efforts should be questioned, however. Russia has staved off economic collapse and continues to fight. The IMF estimated that the country's economy grew 4.3 per cent in 2024 because of massive war spending. IMF, Russian Federation country data, at https://www.imf.org/en/Countries/RUS.
22. Karen DeYoung and Michael Birnbaum, "U.S., allies plan for long-term isolation of Russia," *The Washington Post*, 16 April 2022 at https://www.washingtonpost.com/national-security/2022/04/16/us-nato-isolate-russia/?variant=95d42e19c24b03e7.
23. Cisssy Zhou, "China scrambles for cover from West's financial weapons," *Nikkei Asia*, 13 April 2022 at https://asia.nikkei.com/content/9bafa81f4679968125375fa4b29a666f.
24. Karen DeYoung and Michael Birnbaum, "U.S., allies plan for long-term isolation of Russia," *The Washington Post*, 16 April 2022 at https://www.washingtonpost.com/national-security/2022/04/16/us-nato-isolate-russia/?variant=95d42e19c24b03e7.
25. President of Russia, "Joint Statement of the Russian Federation and the People's

Republic of China on the International Relations Entering a New Era and the Global Sustainable Development," 4 February 2022 at http://en.kremlin.ru/supplement/5770.

26. Xinhua News Agency, "Ukraine issue reveals U.S. pursuit of hegemony, bullying: FM spokesperson," 13 April 2022 at http://en.people.cn/n3/2022/0413/c90000–10083291.html.

1. RETHINKING NATIONAL SECURITY IN A NEW TYPE OF COMPETITION

1. Swapna Venugopal Ramaswamy, "Trump says 'We're in one now' about a trade war with China," *USA Today*, 16 October 2025 at https://ca.news.yahoo.com/trump-says-were-one-now-231635779.html.
2. Ibid.
3. Ayeshea Perera, "Why the US needs China's rare earths," BBC, 16 October 2025 at https://www.bbc.com/news/articles/c1drqeev36qo (the name is a bit of a misnomer: these minerals are abundant in nature, but they are very hard to find in pure form and are very hazardous and dirty to extract).
4. Osmond Chia, "China has found Trump's pain point - rare earths," BBC, 17 October 2025 at https://www.bbc.com/news/articles/ckg1jr18z4ko.
5. Geopolitics Substack, "Expert suggests China could differentiate the U.S. from other countries in rare earth controls to avoid being accused of 'strangling the whole world,'" 17 October 2025 at https://www.geopolitechs.org/p/is-chinas-new-rare-earth-control.
6. Ibid.
7. We're talking about strategic trade controls.
8. Volker Bruhl, "The economic rise of China—an analysis of China's growth drivers," *International Economics and Political Science*, 2025, p. 16.
9. Henry Farrell and Abraham Newman, "Weaponized interdependence," *International Security*, 44(1), 2019, pp. 42–79.
10. Geopolitics Substack, "China's new restriction on rare earths: What they cover and why they were made?" 11 October 2025, at https://www.geopolitechs.org/p/chinas-new-restriction-on-rare-earths.
11. Xinhua, 29 May 2018 at http://www.xinhuanet.com/english/2018–05/29/c_137213175.htm.
12. Julian Baird Gewirtz, "China's long march to technological supremacy," *Foreign Affairs*, 27 August 2019 at https://www.foreignaffairs.com/articles/china/2019–08–27/chinas-long-march-technological-supremacy.
13. Rush Doshi, Prepared Statement before the US Senate Committee on Commerce, Science, and Transportation, Subcommittee on Security. For the Hearing "The China Challenge: Realignment of U.S. Economic Policies to Build Resiliency and Competitiveness," The United States, China, and the Contest for the Fourth Industrial Revolution, 30 July 2020 at https://www.commerce.senate.gov/services/files/6880BBA6–2AF0-4A43-8D32-6774E069B53E. Doshi and his col-

leagues also argue that China bifurcates discourse, applauding collaboration and interdependence to foreign audiences while warning against it to domestic ones. See Rush Doshi, Emily de la Bruyere, Nathan Picarsic, and John Ferguson, "China as a cyber great power: Beijing's two voices in telecommunications," Brookings Institution, April 2021 at https://www.brookings.edu/wp-content/uploads/2021/04/FP_20210405_china_cyber_power.pdf.

14. Rush Doshi, Emily de la Bruyere, Nathan Picarsic, and John Ferguson, "China as a cyber great power: Beijing's two voices in telecommunications," Brookings Institution, April 2021 at https://www.brookings.edu/wp-content/uploads/2021/04/FP_20210405_china_cyber_power.pdf.
15. Office of the Director of National Intelligence, "Global Trends 2040: A More Contested World," March 2021, p. 56, 58.
16. National Commission on Innovation and Competitiveness Frontiers, Council on Competitiveness, "The 2025 State of Competitiveness: A New Age of Disruption and Discontinuity," April 2025, p. 2.
17. John Allen and Giampiero Massolo, "Introduction," in *The Global Race for Technological Superiority*, edited by Fabio Rugge, ISPI-Brookings, 2019, p. 7.
18. European Political Strategy Center, Rethinking Strategic Autonomy in the Digital Age, 2019.
19. Ministry of Economy, Trade and Industry, "New Industrial Structure Vision," nd, https://www.meti.go.jp/english/publications/pdf/vision_171222.pdf.
20. Klaus Schwab, The Fourth Industrial Revolution, The World Economic Forum, 2016, p. 12, https://www.weforum.org/stories/2016/01/the-fourth-industrial-revolution-what-it-means-and-how-to-respond/.
21. See, for example, "Technology Trends Survey: Future Emerging Technology Trends," Version 3, HQ Supreme Allied Commander Transformation, Defense Planning Policy and Analysis Branch, February 2015, at https://www.act.nato.int/images/stories/events/2012/fc_ipr/technology_trend_survey_v3.pdf.
22. Office of the Director of National Intelligence, "Global Trends 2040: A More Contested World," March 2021, p. 60.
23. Eric Schmidt, "I used to run Google. Silicon Valley could lose to China," *The New York Times*, February 2020, at https://www.nytimes.com/2020/02/27/opinion/eric-schmidt-ai-china.html.
24. PWC, "AI adoption could boost global GDP by an additional 15 percentage points by 2035, as global economy is reshaped: PWC research," 29 April 2025, at https://www.pwc.com/gx/en/news-room/press-releases/2025/ai-adoption-could-boost-global-gdp-by-an-additional-15-percentage.html.
25. Bughin et al., "Notes from the AI frontier: Modeling the impact of AI on the world economy," McKinsey & Company, September 2018, at https://www.mckinsey.com/featured-insights/artificial-intelligence/notes-from-the-ai-frontier-modeling-the-impact-of-ai-on-the-world-economy.
26. McKinsey & Company, "The economic potential of generative AI: The next productivity frontier," 24 June 2023, at https://www.mckinsey.com/capabilities/

tech-and-ai/our-insights/the-economic-potential-of-generative-AI-the-next-productivity-frontier#business-value.

27. Henry Farrell, "The twilight of tech unilateralism," Programmablemutter.com, 19 August 2025, at https://www.programmablemutter.com/p/the-twilight-of-tech-unilateralism.
28. Radina Gigova, "Who Vladimir Putin thinks will rule the world," CNN.com, 2 September 2017, at https://edition.cnn.com/2017/09/01/world/putin-artificial-intelligence-will-rule-world/index.html.
29. Gregory Mandel, "Regulating emerging technologies," Temple University Beasley School of Law, Legal Studies Research Paper Series, Research paper No 2009–18, 8 April 2009, p. 2.
30. McKinsey Global Institute, "The bio revolution," May 2020, p. vi.
31. Ibid., p. 1.
32. The National Academies of Sciences, Engineering, and Medicine, "Safeguarding the bioeconomy," NAS Press, 2020; McKinsey Global Institute, "The bio revolution," May 2020, p. 8.
33. Biotechnology Innovation Organization International Convention, "Investment, Innovation and Job Creation in a Growing US Bioscience Industry 2018," 5 June 2018 at https://www.pharmtech.com/view/report-shows-bioscience-industry-has-2-trillion-economic-impact-0.
34. McKinsey Global Institute, "Disruptive technologies: Advances that will transform life, business and the global economy," May 2013 at https://www.mckinsey.com/~/media/McKinsey/Business%20Functions/McKinsey%20Digital/Our%20Insights/Disruptive%20technologies/MGI_Disruptive_technologies_Executive_summary_May2013.pdf.
35. Jamie Metzl, *Hacking Darwin*, 2019, p. 11.
36. "Nanotechnology—global market trajectory and analysis," July 2020 at https://www.researchandmarkets.com/reports/338364/nanotechnology_global_market_trajectory_and#relb0-5130907.
37. Christine Kobza, "Could 3-D printing change global trade? The potential effects of 3-D printing on global supply chains," US International Trade Commission, Executive Briefings on Trade, December 2019.
38. McKinsey Global Institute, "Disruptive technologies: Advances that will transform life, business and the global economy," May 2013, https://www.mckinsey.com/~/media/McKinsey/Business%20Functions/McKinsey%20Digital/Our%20Insights/Disruptive%20technologies/MGI_Disruptive_technologies_Executive_summary_May2013.pdf.
39. Statista.com, "Global robotics market revenue 2018–2025," 10 March 2021, at https://www.statista.com/statistics/760190/worldwide-robotics-market-revenue/.
40. McKinsey Global Institute, "Jobs lost, jobs gained: Workforce transitions in a time of automation," January 2017.
41. Ian Stewart, Debapratim De, and Alex Cole, "Technology and people: The great

job-creating machine," Deloitte Perspectives, August 2015, at https://www2.deloitte.com/uk/en/pages/finance/articles/technology-and-people.html.

42. Cited in Brad Glosserman, "All hail the rise of humanoid robots," *The Japan Times*, 22 April 2025, at https://www.japantimes.co.jp/commentary/2025/04/22/world/rise-of-humanoid-robots/.
43. GMSA, "The Mobile Economy 2025," 2025, p. 5 at https://www.gsma.com/solutions-and-impact/connectivity-for-good/mobile-economy/.
44. Defense Innovation Board, "The 5G Ecosystem: Risks and Opportunities for the DoD," April 2019, p. 7.
45. Bernard Marr, "How much data do we create every day? The mind blowing stats everyone should read," *Forbes*, 21 May 2018, at https://www.forbes.com/sites/bernardmarr/2018/05/21/how-much-data-do-we-create-every-day-the-mind-blowing-stats-everyone-should-read/?sh=5744cf0560ba.
46. Marija Naumovska, "How much data is created every day: 40 key figures and insights," 11 February 2026 at https://www.designrush.com/agency/big-data-analytics-companies/trends/how-much-data-is-created-every-day.
47. David McCormick, Charles Luftig, and James Cunningham, "Economic might, national security and the future of American statecraft," *Texas National Security Review*, May 2020, https://tnsr.org/2020/05/economic-might-national-security-future-american-statecraft/.
48. Brad Glosserman, "All hail the rise of humanoid robots," *The Japan Times*, 22 April 2025, at https://www.japantimes.co.jp/commentary/2025/04/22/world/rise-of-humanoid-robots/.
49. McKinsey Global Institute, "Disruptive technologies: Advances that will transform life, business and the global economy," May 2013, p. 11, at https://www.mckinsey.com/~/media/McKinsey/Business%20Functions/McKinsey%20Digital/Our%20Insights/Disruptive%20technologies/MGI_Disruptive_technologies_Executive_summary_May2013.pdf.
50. Eric Schmidt, "Emerging technologies and defense," testimony before the Senate Committee on Armed Services, 23 February 2021, at https://www.armed-services.senate.gov/imo/media/doc/Schmidt_02–23–21.pdf.
51. Dylan Patel et al., "America is missing the new labor economy—Robotics, Part 1," Semianalysis, 10 March 2025, at https://newsletter.semianalysis.com/p/america-is-missing-the-new-labor-economy-robotics-part-1.
52. Brad Glosserman, "All hail the rise of humanoid robots," *The Japan Times*, 22 April 2025, at https://www.japantimes.co.jp/commentary/2025/04/22/world/rise-of-humanoid-robots/.
53. Dylan Patel et al., "America is missing the new labor economy—Robotics, Part 1," Semianalysis, 10 March 2025, at https://newsletter.semianalysis.com/p/america-is-missing-the-new-labor-economy-robotics-part-1.
54. This list is from Robert Atkinson, "The case for a national industrial strategy to counter China's technological rise," The International Technology and Innovation Foundation, 13 April 2020, at https://itif.org/publications/2020/04/13/case-national-industrial-strategy-counter-chinas-technological-rise.

55. Organization for Economic Cooperation and Development, "Innovation and growth: Rationale for an innovation strategy," 2007, p. 3 at https://www.oecd.org/sti/inno/39374789.pdf.
56. Robert Manning, "Technonationalism vs. the Fourth Industrial Revolution," *Global Asia*, 14(1), March 2019 at http://www.globalasia.org/v14no1/cover/techno-nationalism-vs-the-fourth-industrial-revolution_robert-a-manning.
57. The list of works on this topic is long—but it is not the focus of our study. Our concern is the more important role that these technologies play in determining national competitiveness and international leadership. Those who want to dig in can see Zachary Davis et al., eds, *Strategic Latency Unleashed: The Role of Technology in a Revisionist Global Order and the Implications for Special Operations Forces*, Center for Global Security Research, Lawrence Livermore National Laboratory, January 2021, at https://cgsr.llnl.gov/content/assets/docs/StratLatUnONLINE.pdf.
58. James Manyika and William H. McRaven, chairs, "Innovation and national security: Keeping our edge," Council on Foreign Relations, September 2019, p. 4 at https://www.cfr.org/report/keeping-our-edge/pdf/TFR_Innovation_Strategy.pdf.
59. Linda Weiss, *America Inc.?: Innovation and Enterprise in the National Security State*, Cornell University Press, 2014, p. 2.
60. Export Control Reform Act of 2018, Public Law 115–232--Title XVII, Subtitle B, section 1752, (3), p. 10 at https://www.bis.doc.gov/index.php/documents/regulations-docs/2263-legal-authority-for-the-export-administration-regulations-1/file.
61. National Commission on Innovation and Competitiveness Frontiers, Council on Competitiveness, "The 2025 State of Competitiveness: A New Age of Disruption and Discontinuity," 2025, p. 1.
62. Frank Tang, "China unveils 'strategic emerging industries' plan in fresh push to get away from US technologies," *South China Morning Post*, 24 September 2020, at https://www.scmp.com/economy/china-economy/article/3102911/china-unveils-strategic-emerging-industries-plan-fresh-push.
63. Office of the United States Trade Representative, "Findings of the investigation into China's acts, policies and practices related to technology transfer, intellectual property and innovation under section 301 of the Trade Act of 1974," (USTR Report) 22 March 2018, p. 15.
64. Tai Ming Cheung, Thomas Mahnken, Deborah Seligsohn, Kevin Pollpeter, Eric Anderson, and Fan Yang, "Planning for innovation, understanding China's plans for technological, energy, industrial, and defense development," US–China Economic and Security Review Commission, 28 July 2016, p. 33, https://www.uscc.gov/research/planning-innovation-understanding-chinas-plans-technological-energy-industrial-and-defense.
65. Jerry Coughter, "Recent research: ASPI Report says U.S. cedes lead in critical technologies research," SSTI.org, 12 June 2025, at https://ssti.org/blog/recent-research-aspi-report-says-us-cedes-lead-critical-technologies-research.
66. Special Competitive Studies Project, "Welcome to the arena: Who's ahead, who's behind and where we are headed in the U.S.-China technology competition," January 2025, p. 4.

67. Renli Wu, Christopher Esposito, and James Evans, "Shifting power asymmetries in scientific terms reveal China's rising leadership in global science," *Proceedings of the National Academies of Science*, 122(44), 2025, e2414893122.
68. Niccolo Pisani et al., "China's rise as global technology powerhouse: A trajectory of international collaboration and specialization in high-impact research," *Research Policy*, 54(8), 2025.
69. Ashley Tellis, "Overview: The return of US-China strategic competition," in Strategic Asia 2020, National Bureau of Asian Research, 2020, p. 30.
70. White House Office of Trade and Manufacturing Policy, "How China's economic aggression threatens the technologies and the intellectual property of the United States and the world," June 2018, p. 20, https://trumpwhitehouse.archives.gov/wp-content/uploads/2018/06/FINAL-China-Technology-Report-6.18.18-PDF.pdf.
71. Office of the Director of National Intelligence, "2025 annual threat assessment of the national intelligence community," March 2025, p. 13.
72. Reuters, "China theft of technology is biggest law enforcement threat, FBI says," 6 February 2020.
73. Christopher Ford, "U.S. national security export controls and Huawei: The strategic context in three framings," US State Dept. Bureau of International Security and Nonproliferation, Arms Control and International Security Papers, 1(8), 22 May 2020, p. 4.
74. Dion Rabouin, "US trade deficit grows to largest since July 2008," Axios.com, 4 September 2020 at https://www.axios.com/us-trade-deficit-china-canada-mexico-aeaf4aac-e8d3–4310-b149–6c06fae5b18f.html.
75. Finbar Bermingham and Frank Tang, "China's exports boomed in August, as trade surplus with the US widened by 27 per cent," *South China Morning Post*, 7 September 2020, at https://www.scmp.com/economy/china-economy/article/3100473/chinas-exports-boomed-august-imports-continued-slump-showing.
76. Chad Bowen, "China bought none of the extra $200 billion of US exports in Trump's trade deal," Peterson Institute for International Economics, 19 July 2022, at https://www.piie.com/blogs/realtime-economics/2022/china-bought-none-extra-200-billion-us-exports-trumps-trade-deal.
77. US Department of State, Bureau of International Security and Nonproliferation, Fact Sheet, "Military-Civil Fusion and the People's Republic of China," 28 May 2020, at https://www.state.gov/remarks-and-releases-bureau-of-international-security-and-nonproliferation/mcf-and-the-prc/.
78. Christopher Ford, "U.S. National Security Export Controls and Huawei: The Strategic Context in Three Framings," US State Dept. Bureau of International Security and Nonproliferation, Arms Control and International Security Papers, 1(8), 22 May 2020, p. 5.
79. Ibid., p. 6. Emphasis in the original throughout.
80. Robert Williams, "The innovation-security conundrum in US-China relations," *Lawfare*, 24 July 2018 at https://www.lawfareblog.com/innovation-security-conundrum-us-china-relations.

81. Eurasia Group, "The Geopolitics of Semiconductors," September 2020, https://www.eurasiagroup.net/live-post/geopolitics-semiconductors.
82. Editorial Board, "The US is seeking to constrain China's rise," *Financial Times*, 21 May 2019.
83. Darren Lim, "The U.S., China and technology war," *Global Asia*, 14(1), March 2019, https://globalasia.org/v14no1/cover/the-us-china-and-technology-war_darren-lim.
84. Of course, there remains a range of views regarding how intense that rivalry is, the prospects for engagement and the best ways to deal with Beijing given this competition.
85. "A New Great Game—China, the U.S. and Technology: A report by the China Senior Analyst Group," S&P Global, 1 October 2018, at https://www.spglobal.com/en/research-insights/featured/a-new-great-game-china-the-u-s-and-technology.
86. Danielle Cheslow and Giselle Ruhiyyih Ewing, "Orban, Trump enter 'golden age'—at Ukraine's expense," *Politico*, 7 November 2025, at https://www.politico.com/newsletters/national-security-daily/2025/11/07/orban-trump-enter-golden-age-at-ukraines-expense-00642850.
87. The White House, "Fact Sheet: President Donald J. Trump strikes deal on economic and trade relations with China," 1 November 2025, at https://www.whitehouse.gov/fact-sheets/2025/11/fact-sheet-president-donald-j-trump-strikes-deal-on-economic-and-trade-relations-with-china/.
88. Rush Doshi, Emily de la Bruyere, Nathan Picarsic, and John Ferguson "China as a cyber great power: Beijing's two voices in telecommunications," Brookings Institution, April 2021 at https://www.brookings.edu/wp-content/uploads/2021/04/FP_20210405_china_cyber_power.pdf.
89. Julian Baird Gewirtz, "China's long march to technological supremacy," *Foreign Affairs*, 27 August 2019, https://www.foreignaffairs.com/articles/china/2019–08–27/chinas-long-march-technological-supremacy.
90. Ibid.
91. Xi Jinping, "Secure a decisive victory in building a moderately prosperous society in all respects and strive for the great success of socialism with Chinese characteristics for a new era," Beijing, 18 October 2017, https://www.chinadaily.com.cn/china/19thcpcnationalcongress/2017-11/04/content_34115212.htm.
92. Elizabeth Chen, "Semiconductor scandal: A concerning backdrop to Xi's pursuit of 'core technologies,'" China Leadership Monitor, Jamestown.org, 26 March 2021, at https://jamestown.org/program/semiconductor-scandal-a-concerning-backdrop-to-xis-pursuit-of-core-technologies/.
93. Chinese Communist Party, "Proposal of the Central Committee of the Chinese Communist Party on drawing up the 14th Five-Year Plan for National Economic and Social Development and Long-Range Objectives for 2030," at https://cset.georgetown.edu/publication/proposal-of-the-central-committee-of-the-chinese-communist-party-on-drawing-up-the-14th-five-year-plan-for-national-economic-and-social-development-and-long-range-objectives-for-2030/.

94. Xi Jinping, "Explanation of the recommendations of the Central Committee of the Communist Party of China for Formulating the 15th Five-Year Plan for National Economic and Social Development," 28 October 2025, at https://english.news.cn/20251028/efbfd0c774fd4b1c8daeb741c0351431/c.html?utm_source=substack&utm_medium=email.
95. Main development goals for economic and social development during the "Fifteenth Five-Year Plan, Period (Studying and Implementing the Spirit of the Fourth Plenary Session for the 20th CPC Central Committee 十五五"时期经济社会发展的主要目标 (学习贯彻党的二十届四中全会精神)," *People's Daily*, 4 November 2025, at https://sinocism.notion.site/Main-Development-Goals-for-Economic-and-Social-Development-during-the-Fifteenth-Five-Year-Plan-Per-2a184ece41d780cf8e33d85763c3bfeb.
96. Adam Tooze, "Slouching towards (Red-Green) utopia. Voicing the muted politics of China's renewable energy revolution," Chartbook, 2 November 2025, at https://adamtooze.substack.com/p/chartbook-414-slouching-towards-red?utm_campaign=email-half-post&r=ht&utm_source=substack&utm_medium=email.
97. Jue Wang, "The Implication of Global Technological Innovation on US-China Strategic Competition," in Marianne Schneider-Petsinger et al., "US-China strategic competition: The quest for global leadership," Chatham House, November 2019, p. 26 at https://www.chathamhouse.org/sites/default/files/publications/research/CHHJ7480-US-China-Competition-RP-WEB.pdf.
98. Yu Jie, "A power shift in international technology standard setting," in ibid., p. 29, https://www.chathamhouse.org/sites/default/files/publications/research/CHHJ7480-US-China-Competition-RP-WEB.pdf, citing Xi Jinping, "The rejuvenation of the Chinese nation is a dream shared by all Chinese', 6 June 2014, *Xi Jinping: The Governance of China*, p. 70; E. Kania, *The Right to Speak: Discourse and the Chinese Power*, Centre for Advanced China Research, 2018, https://www.ccpwatch.org/single-post/2018/11/27/The-Right-to-Speak-Discourse-and-Chinese-Power.
99. Alex Capri, "Techno-nationalism: What is it and how it will change global commerce," *Forbes*, 20 December 2019, at https://www.forbes.com/sites/alexcapri/2019/12/20/techno-nationalism-what-is-it-and-how-will-it-change-global-commerce/?sh=7403d4a7710f.
100. Elsa Kania, "Strategic innovation and great power competition," TheStrategyBridge.org, 31 January 2018, at https://thestrategybridge.org/the-bridge/2018/1/31/strategic-innovation-and-great-power-competition.
101. Article 9 reads: "Aspiring sincerely to an international peace based on justice and order, the Japanese people forever renounce war as a sovereign right of the nation and the threat or use of force as means of settling international disputes. In order to accomplish the aim of the preceding paragraph, land, sea, and air forces, as well as other war potential, will never be maintained. The right of belligerency of the state will not be recognized."
102. ISEAS Yusof Ishak Institute, "Seventh State of Southeast Asia Survey," April 2025.

103. Office of the Prime Minister, National Security Strategy, 17 December 2013, p. 2.
104. Shiro Armstong, "Economic security in Japan: Evolution, context and emerging questions," RIETI Discussion Paper series, 24-E-083, December 2024, p. 3.
105. Kazunobu Hayakawa, "The ripple effect beyond borders: Indirect effects of US export controls on Japanese Firms," VOXEU, 1 August 2025, at https://cepr.org/voxeu/columns/ripple-beyond-borders-indirect-effects-us-export-controls-japanese-firms and Kazunobu Hayakawa et al., "The impact of the strengthening of export controls on Japanese exports of dual-use goods," *International Economics*, 174, August 2023, at https://www.sciencedirect.com/science/article/abs/pii/S2110701723000239.
106. Nemoto Ryo, "Countering China is top priority for Japan PM contender Kishida," *Nikkei Asia*, 4 September 2021, at https://asia.nikkei.com/Editor-s-Picks/Interview/Countering-China-is-top-priority-for-Japan-PM-contender-Kishida.
107. Former National Security Advisor Kitamura Shigeru in, "Magazine articles of the month: What economic security policies should Japan implement?" Foreign Press Center of Japan, 31 August 2021 at https://fpcj.jp/en/j_views-en/magazine_articles-en/p=90454/.
108. Shimizu Masato, "Kishida's 'new capitalism' raises economic reform setback fears," *Nikkei Asia*, 5 October 2021, at https://asia.nikkei.com/Politics/Inside-Japanese-politics/Kishida-s-new-capitalism-raises-economic-reform-setback-fears.
109. In November 2019, the Foreign Exchange and Foreign Trade Act was modified to lower the threshold for scrutiny of foreign direct investment in Japan from 10 per cent to 1 per cent. In April 2020, an Economic Section was created in the National Security Secretariat, part of the Cabinet bureaucracy. Other ministries followed suit and established independent sections to handle economic security. The Basic Policy on Economic and Fiscal Management and Reform 2021 explained that Japan's economic security depends on "expanding and deepening cooperation with like-minded countries while ensuring Japan's autonomy and gaining a competitive advantage under an international order based on basic values and rules." This will be accomplished by "identifying key technologies, enhancing efforts to preserve and develop them, and developing the resilience of core industries." More concretely, this has meant policies to reshore or lure facilities for key elements of global supply chains (semiconductors are one focus); the creation of funds to develop new technologies; and new or tighter controls on technology to prevent their loss or acquisition by hostile governments. The post of Minister for Economic Security was created in October 2021. In May 2022, the Economic Security Promotion Act, legislation that codified many of the objectives of the 2021 Basic Policy, came into law. There have been additional steps since then, including passage of a security clearance law.

2. A WORLD OF INSECURITY

1. Marko Milanovic and Michael N. Schmitt, "Cyber attacks and cyber (mis)information

operations during a pandemic," *Journal of National Security Law & Policy*, 11, 27 May 2020, pp. 248–9.

2. Interpol, "Cybercrime; COVID-19 impact," Lyon, August 2020, pp. 6–7.
3. Takami Kosuke, "As Japan expands telework, hackers follow and crack into networks," *Nikkei Asia*, 25 August 2020, https://asia.nikkei.com/Spotlight/Century-of-Data/As-Japan-expands-telework-hackers-follow-and-crack-into-networks.
4. David Keohane and Peggy Hollinger, "Pandemic brought surge in French cyber attacks, warns Thales CEO," *Financial Times*, 6 April 2021, at https://www.ft.com/content/70e1c40d-acc8-4e8e-8ce3-3e0d9901358a.
5. Lily Hay Newman, "Google sees state-sponsored hackers ramping up coronavirus attacks," *Wired*, 22 April 2020, at https://www.wired.com/story/google-state-sponsored-hackers-coronavirus-phishing-malware/.
6. Hoxhunt, "Pfishing trends report: Updated to 2025," at https://hoxhunt.com/guide/phishing-trends-report.
7. Interpol, "Cybercrime; COVID-19 impact," Lyon, August 2020, pp. 6–7.
8. Marcus Willett, "Lessons of the SolarWinds hack," *Survival*, 63(2), April–May 2021, p. 15.
9. Tom Burt, "Cyberattacks targeting health care must stop," Microsoft Blog, 13 November 2020, at https://blogs.microsoft.com/on-the-issues/2020/11/13/health-care-cyberattacks-covid-19-paris-peace-forum/.
10. Marko Milanovic and Michael N. Schmitt, "Cyber attacks and cyber (mis)information operations during a pandemic," *Journal of National Security Law & Policy*, 11, 27 May 2020, p. 248.
11. Amy Ertan and Tarah Wheeler, "The vaccine supply chain is now the most valuable cyber target in the world," Slate.com, 18 March 2021, at https://slate.com/technology/2021/03/world-health-organization-vaccine-cybersecurity-censure.html.
12. Amy Ertan and Tarah Wheeler, "The vaccine supply chain is now the most valuable cyber target in the world," Slate.com, 18 March 2021, at https://slate.com/technology/2021/03/world-health-organization-vaccine-cybersecurity-censure.html; "Coronavirus: North Korea and Russia hackers targeting vaccines," BBC, 13 November 2020, at https://www.bbc.com/news/technology-54936886.
13. Tom Burt, "Cyberattacks targeting health care must stop," Microsoft Blog, 13 November 2020, at https://blogs.microsoft.com/on-the-issues/2020/11/13/health-care-cyberattacks-covid-19-paris-peace-forum/.
14. Marko Milanovic and Michael N. Schmitt, "Cyber attacks and cyber (mis)information operations during a pandemic," *Journal of National Security Law & Policy*, 11, 27 May 2020, p. 249 and Interpol, "Cybercrime; COVID-19 impact," Lyon, August 2020, p. 6.
15. Marko Milanovic and Michael N. Schmitt, "Cyber attacks and cyber (mis)information operations during a pandemic," *Journal of National Security Law & Policy*, 11, 27 May 2020, p. 266.
16. Ibid.

17. The Hidden Costs of Cybercrime, Zhanna Malekos Smith, and Eugenia Lostri, CSIS-McAfee Report, December 2020, p. 3.
18. This is a reference to the iconic US TV crime family the Sopranos, one of the best shows ever made. North Korea is often called the "Soprano state" because of its illicit and illegal money-making activities. To be clear, while North Korea engages in considerable cybercrime and hacking and other such activities, we aren't blaming it for all this crime.
19. Steve Morgan, "Cybercrime facts and statistics," Cybersecurity Ventures, 21 January 2021, p. 1 at https://1c7fab3im83f5gqiow2qqs2k-wpengine.netdna-ssl.com/wp-content/uploads/2021/01/Cyberwarfare-2021-Report.pdf.
20. Robert Mueller, "Remarks to the RSA Cybersecurity Conference," 1 March 2012 at https://archives.fbi.gov/archives/news/speeches/combating-threats-in-the-cyber-world-outsmarting-terrorists-hackers-and-spies.
21. "2021 Cyber Security Statistics: The Ultimate List of Stats, Data & Trends," PurpleSec, nd, at https://purplesec.us/resources/cyber-security-statistics/. The 36 billion figure is from the 2020 report.
22. Steve Morgan, "Cybercrime facts and statistics," Cybersecurity Ventures, 21 January 2021, p. 1 at https://1c7fab3im83f5gqiow2qqs2k-wpengine.netdna-ssl.com/wp-content/uploads/2021/01/Cyberwarfare-2021-Report.pdf.
23. Ransomware Task Force, "Combatting Ransomware," Institute for Security and Technology, April 2021, p. 15 at https://securityandtechnology.org/wp-content/uploads/2021/04/IST-Ransomware-Task-Force-Report.pdf.
24. Ibid., p. 5.
25. Crimes Against Children Research Center, "Online victimization: A report on the nation's youth," June 2000, at http://www.unh.edu/ccrc/pdf/Victimization_Online_Survey.pdf.
26. Verizon, 2020 Data Breach Investigations Report, p. 79.
27. ENISA Threat Landscape Overview, Cyber Espionage January 2019–April 2020, p. 3.
28. The quote and all depressing details are from Bruce Schneier, "The US has suffered a massive cyberbreach. It's hard to overstate how bad it is," *The Guardian*, 23 December 2020 at https://www.theguardian.com/commentisfree/2020/dec/23/cyber-attack-us-security-protocols.
29. Ibid.
30. Cited in Craig Timberg and Ellen Nakashima, "Russian hack was 'classic espionage' with stealthy, targeted tactics," *The Washington Post*, 15 December 2020, at https://www.washingtonpost.com/technology/2020/12/14/russia-hack-us-government/.
31. Bruce Schneier, "The US has suffered a massive cyberbreach. It's hard to overstate how bad it is," *The Guardian*, 23 December 2020 at https://www.theguardian.com/commentisfree/2020/dec/23/cyber-attack-us-security-protocols.
32. Zach Dorfman, "In cyber espionage, U.S. is both hunted and hunter," Axios, 27 January 2021, at https://www.axios.com/american-cyber-warfare-solar-winds-d50815d6-2e03-4e3c-83ab-9d2f5e20d6f5.html.

33. Staff of House Committee on Oversight and Government Reform, 114th Congress, The OPM Data Breach: How the Government Jeopardized our National Security for More than a Generation, 5, 2016.
34. Dan Verton, *Impact of OPM Breach Could Last More Than 40 Years*, FEDSCOOP, 10 July 2015, http://fedscoop.com/opm-losses-a-40-year-problem-for-intelligence-community.
35. Julianne Pepitone, "China is 'leading suspect' in OPM hacks, says intelligence chief James Clapper," NBC News, 26 June 2015 at https://www.nbcnews.com/tech/security/clapper-china-leading-suspect-opm-hack-n381881.
36. Erica Lonergan and Michael Poznansky, "A tale of two typhoons: Properly diagnosing Chinese cyber threats," War on the Rocks, 25 February 2025, at https://warontherocks.com/2025/02/a-tale-of-two-typhoons-properly-diagnosing-chinese-cyber-threats/.
37. Bruce Schneier, "The US has suffered a massive cyberbreach. It's hard to overstate how bad it is," *The Guardian*, 23 December 2020 at https://www.theguardian.com/commentisfree/2020/dec/23/cyber-attack-us-security-protocols.
38. White House, "Press Conference by President Obama after G20 Summit," 5 September 2016 at https://obamawhitehouse.archives.gov/the-press-office/2016/09/05/press-conference-president-obama-after-g20-summit.
39. Shane Harris, *@War: The Rise of the Military-Internet Complex*, Eamon Dolan, 2014, pp. 83–102.
40. Ibid., p. 86.
41. Ibid., p. 94.
42. Kenneth Geers et al., "World War C: Understanding nation-state motives behind today's advanced cyber attacks," FireEye Inc, 2014, p. 3.
43. Steven Levy, "Secret of Googlenomics: Data-fueled recipe brews profitability," Wired, 22 May 2009, https://www.wired.com/2009/05/nep-googlenomics/.
44. Kaveh Waddell, "Report: The U.S. is unprepared for the AI future," Axios, 11 July 2018, https://www.axios.com/the-us-isnt-ready-for-the-ai-future-96649a76–1027–43ba-ae27-e24cb57dd194. html.
45. Christopher Dearing, "Personal information as an attack vector: Why privacy should be an operational dimension of U.S. national security," *Journal of National Security Law and Policy*, 10, 2019, p. 370, citing Paul Ohm, "Broken promises of privacy: Responding to the surprising failure of anonymization," *UCLA Law Review*, 57, 2010, pp. 1701, 1706.
46. Starwood guest reservation database security incident: Marriott International, 30 November 2018, https://perma.cc/2242-X7JU.
47. Josh Fruhlinger, "Marriott data breach FAQ: How did it happen and what was the impact," CSO Online, 12 February 2020, at https://www.csoonline.com/article/3441220/marriott-data-breach-faq-how-did-it-happen-and-what-was-the-impact.html.
48. Greg Roumeliotis and Matthew Miller, "China's Anbang abandons $14 billion bid to buy Starwood Hotels," Reuters, 1 April 2016, at https://www.reuters.com/article/us-starwood-hotels-m-a-anbang-idUSKCN0WX2PE.

49. "CFIUS concerns could stall Anbang's $12.8B Starwood Play," Law360.com, 14 March 2016 at https://www.law360.com/articles/771122/cfius-concerns-could-stall-anbang-s-12-8b-starwood-play; Steven Davidoff Solomon, "If Starwood makes a deal with Anbang, can it be done?" *The New York Times*, 30 March 2016 at https://www.nytimes.com/2016/03/31/business/dealbook/if-starwood-makes-a-deal-with-anbang-can-it-be-done.html.
50. Cory Bennett and Bryan Bender, "How China acquires 'the crown jewels' of U.S. technology," *Politico*, 22 May 2018 at https://www.politico.com/story/2018/05/22/china-us-tech-companies-cfius-572413.
51. Amy Borrett, "Is Chinese investment in Europe's technology industry a threat?" Techmonitor, 14 April 2021 at https://techmonitor.ai/policy/geopolitics/is-chinese-investment-europe-technology-industry-threat
52. Christopher Wray, "The threat posed by the Chinese government and the Chinese Communist Party to the economic and national security of the United States," remarks as delivered, 7 July 2020 at https://www.fbi.gov/news/speeches/the-threat-posed-by-the-chinese-government-and-the-chinese-communist-party-to-the-economic-and-national-security-of-the-united-state.
53. Daniel Hurst, Lily Kuo and Charlotte Graham-McLay, "Zhenhua Data leak: Personal details of millions around world gathered by China tech company," *The Guardian*, 14 September 2020, at https://www.theguardian.com/world/2020/sep/14/zhenhua-data-full-list-leak-database-personal-details-millions-china-tech-company.
54. Christopher Wray, "The threat posed by the Chinese government and the Chinese Communist Party to the economic and national security of the United States," remarks as delivered, 7 July 2020 at https://www.fbi.gov/news/speeches/the-threat-posed-by-the-chinese-government-and-the-chinese-communist-party-to-the-economic-and-national-security-of-the-united-states.
55. In some cases, like the Microsoft Exchange attack, criminals rush in once the vulnerability is exposed to profit on their own.
56. Mara Hvistendahl, *The Scientist and the Spy: A True Story of China, The FBI, and Industrial Espionage*, Penguin, 2020.
57. Catherine Lotrionte, "Countering state-sponsored cyber economic espionage under international law," *North Carolina Journal of International Law and Commercial Regulation*, XL, 2015, p. 464.
58. David Sanger and Tim Weiner, "Emerging role for the CIA: Economic spy," *The New York Times*, 15 October 1995 at https://www.nytimes.com/1995/10/15/world/emerging-role-for-the-cia-economic-spy.html.
59. Bill Warren, "NSA spying and TransAtlantic trade talks," Medium.com, 26 April 2016 at https://medium.com/economic-policy/nsa-spying-transatlantic-trade-talks-4324f715768f.
60. Ellen Nakashima, "Confidential report lists US weapons system designs compromised by Chinese cyberspies," *The Washington Post*, 27 May 2013 at https://www.washingtonpost.com/world/national-security/confidential-report-lists-us-weapons-system-designs-compromised-by-chinese-cyberspies/2013/05/27/a42c3e1c-c2dd-11e2-8c3b-0b5e9247e8ca_story.html.

61. National Counterintelligence and Security Center, "Foreign economic espionage in cyberspace 2018," p. 1., at https://www.dni.gov/files/NCSC/documents/news/20180724-economic-espionage-pub.pdf.
62. Glenn Hastedt, *Espionage: A Reference Handbook*, 2003, cited in Lotrionte, "Countering state-sponsored cyber economic espionage under international law," *North Carolina Journal of International Law and Commercial Regulation*, XL, 2015, fn. 88.
63. Philip Ewing, "Gates: French cyber spies target the U.S.," *Politico*, 22 May 2014 at https://www.politico.com/story/2014/05/france-intellectual-property-theft-107020.
64. Ibid.
65. Tim Johnson, "We knew the US and Russia were hacking powers, but Ethiopia and Pakistan?" *McClatchy*, 25 April 2017 at https://www.mcclatchydc.com/news/nation-world/national/national-security/article146508409.html.
66. Center for Responsible Enterprise and Trade (CREATe.org) and PwC, *Economic Impact of Trade Secret Theft: A Framework for Companies to Safeguard Trade Secrets and Mitigate Potential Threats*, 2014, cited in ECIPE, p. 13.
67. Dennis Blair and Jon Huntsman, Jr., "The IP Commission report: The report of the Commission on the theft of American intellectual property," 11, 2013, p. 15.
68. Press Release, Dept of Justice, "U.S. charges five Chinese military hackers for cyber espionage against U.S. corporations and a labor organization for commercial advantage," 19 May 2014.
69. European Union Intellectual Property Office, "Impact of intellectual property rights intensive industries in the European Union," September 2019, at https://euipo.europa.eu/tunnel-web/secure/webdav/guest/document_library/observatory/documents/IPContributionStudy/IPR-intensive_industries_and_economicin_EU/WEB_IPR_intensive_Report_2019.pdf.
70. Tilburg University, "Trade secret protection in the U.S. and EU," 2016 at: http://arno.uvt.nl/show.cgi?fid=141634. As always, however, experts caution that the number is likely much higher given the problems with the reports the analysis is based upon. See, for example, Scott Ikeda, "85% of cyber espionage is state-affiliated, only 4% tied to organized crime," *CPO Magazine*, 27 November 2020, at https://www.cpomagazine.com/cyber-security/85-of-cyber-espionage-is-state-affiliated-only-4-tied-to-organized-crime/; and Eileen Decker, "Full count?: Crime rate swings, cybercrime misses and why we don't really know the score," *Journal of National Security Law and Policy*, 10, 2020, p. 584.
71. Hosuk Lee-Makiyama, "Stealing thunder: Cloud, IoT and 5G will change the strategic paradigm for protecting European commercial interests. Will cyber espionage be allowed to hold Europe back in the global race for industrial competitiveness?" ECIPE Occasional Paper, No. 2/18, European Centre for International Political Economy (ECIPE), Brussels, 2018, p. 9.
72. Brian Humphreys, "Critical infrastructure; emerging trends and policy considerations for Congress," CRS R45809, 8 July 2019, p. 1. Other definitions are more

detailed. For example, the US government defines critical infrastructure as "sectors whose assets, systems, and networks, whether physical or virtual, are considered so vital to the United States that their incapacitation or destruction would have a debilitating effect on security, national economic security, national public health or safety, or any combination thereof," https://www.cisa.gov/critical-infrastructure-sectors.

73. Humphreys, "Critical infrastructure; Emerging trends and policy considerations for Congress," pp. 2–3.
74. Ibid.
75. An excellent summary of the differences can be found in Lucia Retter et al., "Relationships between the economy and national security," Rand, Europe, 2020, table 3.1.
76. Office of the Director of National Intelligence, The National Counterintelligence Strategy of the US, 2020–2022, 7 January 2020, p. 6.
77. Office of the Director of National Intelligence, "Annual threat assessment of the U.S. Intelligence Community," March 2025, p. 4.
78. Ibid., p. 11.
79. Robert Windrem, "Secret NSA map shows China attacks on U.S. targets," NBC News, 30 July 2015, http://www.nbcnews.com/news/us-news/exclusive-secret-nsa-map-shows-china-cyber-attacks-us-targets-n401211.
80. Daniel Coats, "Statement for the record: Worldwide threat assessment of the US Intelligence Committee," Senate Select Committee for Intelligence, 29 January 2019, p. 6.
81. Office of the Director of National Intelligence, Annual Threat Assessment of the US Intelligence Community, 9 April 2021, pp. 8–15.
82. Understanding the role of connected devices in recent cyber attacks (preliminary transcript): Hearing before H. Comm. on Energy and Commerce, 114th Cong. 27 (2016) (testimony of Bruce Schneier), http://docs.house.gov/meetings/IF/IF17/20161116/105418/HHRG-114-IF17-Transcript-20161116.pdf.
83. Shannon Williams, "Nokia: Cyber attacks on internet-connected devices on the rise," 28 October 2020 at https://securitybrief.co.nz/story/nokia-cyber-attacks-on-internet-connected-devices-on-the-rise.
84. Brandon Vigliarolo, "IBM finds vulnerability in IoT chips present in billions of devices," Tech Republic, 19 August 2020 at https://www.techrepublic.com/article/ibm-finds-vulnerability-in-iot-chips-present-in-billions-of-devices/.
85. Alfred Ng, "IoT attacks are getting worse and no one's listening," CNet, 15 March 2018 at https://www.cnet.com/home/smart-home/iot-attacks-hacker-kaspersky-are-getting-worse-and-no-one-is-listening/?ftag=CMG-01–10aaa1b.
86. Jordan Riddles, "Sonicwall 2024 mid-yearcyber threat report," 23 July 2024 at https://www.sonicwall.com/blog/sonicwall-2024-mid-year-cyber-threat-report-iot-madness-powershell-problems-and-more.
87. Alex Drozhzhin, "Black hat USA 2015: The full story of how that Jeep was hacked," Kaspersky Daily, 6 August 2015 at https://www.kaspersky.com/blog/blackhat-jeep-cherokee-hack-explained/9493/.

88. Kate Kochetkova, "Shock at the wheel: Your Jeep can be hacked while down the road," Kaspersky Daily, 23 July 2015 at https://www.kaspersky.com/blog/remote-car-hack/9395/.
89. Arie Egozi, "Israelis may ban high tech cars from bases: 'Perfect espionage vector,'" *Breaking Defense*, April 2021, at https://breakingdefense.com/2021/04/israelis-may-ban-teslas-other-high-tech-cars-from-military-bases-the-perfect-espionage-vector.
90. Ibid.
91. Staff, "Tesla tells China car cameras not activated outside North America," Reuters, 7 April 2021 at https://www.reuters.com/article/us-tesla-china-idUSKBN2BU12T.
92. Arie Egozi, "Israelis may ban high tech cars from bases: 'Perfect espionage vector,'" *Breaking Defense*, April 2021, at https://breakingdefense.com/2021/04/israelis-may-ban-teslas-other-high-tech-cars-from-military-bases-the-perfect-espionage-vector.
93. It's sometimes called a "value chain" or "third party" attack.
94. Andy Greenberg, "The untold story of NotPetya, the most devastating cyberattack in history," *Wired*, 22 August 2018 at https://www.wired.com/story/notpetya-cyberattack-ukraine-russia-code-crashed-the-world/.
95. Donna Ghelfi, World Intellectual Property Organization, "The 'outsourcing offshore' conundrum: An intellectual property perspective," 2005, p. 8 at http://www.wipo.int/export/sites/www/sme/en/documents/pdf/outsourcing.pdf.
96. Andrew F. Popper, "More than the sum of all parts: Taking on IP and IT theft through a global partnership," *Northwestern Journal of Technology and Intellectual Property*, 253, 2014. Executive Office of the President, Administrative Strategy on Mitigating the Theft of US Trade Secrets, 7 (2013), at http://www.whitehouse.gov//sites/default/files/omb/IPEC/admin_strategy_ on_mitigating_the_theft_of_u.s._trade_secrets.pdf.
97. The White House, "'Fact sheet' Securing America's critical supply chains," 24 February 2021 at https://www.whitehouse.gov/briefing-room/statements-releases/2021/02/24/fact-sheet-securing-americas-critical-supply-chains/.
98. The White House, "Remarks by President Biden at signing of an executive order on supply chains," 24 February 2021 at https://www.whitehouse.gov/briefing-room/speeches-remarks/2021/02/24/remarks-by-president-biden-at-signing-of-an-executive-order-on-supply-chains/.
99. Michael Peel, Jim Brunsden, and Sebastian Payne, "Von der Leyen signals possible emergency Covid vaccine controls," *Financial Times*, 17 March 2021 at https://www.ft.com/content/3f2380d1-d4e1–45f2–9c3d-eea486697dd6.
100. Australian Government Productivity Commission, "Vulnerable supply chains: Interim report," March 2021, pp. 4–5.
101. World Bank, "Resilient industries in Japan: Lessons learned in Japan on Enhancing Competitive Industries in the Face of Disasters cause by Natural Hazards, World Bank, p. 1, http://documents1.worldbank.org/curated/en/75729160404156

7018/pdf/Resilient-Industries-in-Japan-Lessons-Learned-in-Japan-on-Enhancing-Competitive-Industries-in-the-Face-of-Disasters-Caused-by-Natural-Hazards.pdf.

102. Chang-ran Kim and Isabel Reynolds, "Supply chain disruptions face more delays in Japan," Reuters, 23 March 2011 at https://www.reuters.com/article/japan-supplychain-idUSL3E7EN0ND20110323.

103. Japan External Trade Organization, "Government led promotion of fundamental reform manufacturing industry and strengthening of international competitiveness," nd at https://www.jetro.go.jp/en/invest/attractive_sectors/manufacturing/government_initiatives.html.

104. Xi Jinping, "Certain major issues for our national medium- to long-term economic and social development strategy [国家中长期经济社会发展战略若干重大问题], *Qiushi*, 1 November 2020, https://web.archive.org/web/20201111020608/http://www.qstheory.cn/dukan/qs/2020–10/31/c_1126680390.htm, translated by Georgetown Center for Security and Emerging Technology, 10 November 2020, p. 3, at https://cset.georgetown.edu/wp-content/uploads/t0235_Qiushi_Xi_economy_EN-1.pdf.

105. European Union Chamber of Commerce in China and Merics (Mercator Institute for China Studies), "Decoupling: Severed ties and patchwork globalisation," 14 January 2021, p. 3.

106. Ibid., pp. 4–5.

107. Michael R. Wessel, "Prepared testimony US-China relations: Improving US competitiveness through trade," Senate Committee on Finance, 22 April 2021.

108. This section draws on Jeffrey Wilson, "Whatever happened to the rare earths weapon? Critical materials and international security in Asia," *Asian Security*, 18(3), 2018, 358–73 and Congressional Research Service, "An overview of rare earth elements and related issues for Congress," R46618, 24 November 2020.

109. CRS, "An overview of rare earth elements and related issues for Congress," p. 1, 3.

110. Cited in Wilson, "Whatever happened to the rare earths weapon? Critical materials and international security in Asia," *Asian Security*, 18(3), 2018, p. 365.

111. Matt Pottinger, "Is Trump getting played by Xi?" *The Free Press*, 29 October 2025 at https://www.fdd.org/analysis/2025/10/29/is-trump-getting-played-by-xi/.

112. Jeffrey Wilson, "Strategies for security critical supply chain materials," Perth USAsia Centre, April 2020, p. 4. The US government defines a "critical material" as (i) a "non-fuel mineral or mineral material essential to the economic and national security of the United States, (ii) the supply chain of which is vulnerable to disruption, and (iii) that serves an essential function in the manufacturing of a product, the absence of which would have significant consequences for our national security." Executive Order 13817, "A federal strategy to ensure secure and reliable supplies," 82, *Federal Register* 60835 (2017). The full list of critical materials in 2020 included aluminum (bauxite), antimony, arsenic, barite, beryllium, bismuth, cesium, chromium, cobalt, fluorspar, gallium, germanium, graphite (natural), hafnium, helium, indium, lithium, magnesium, manganese, niobium, platinum group metals, potash, the rare earth elements group, rhenium, rubidium,

scandium, strontium, tantalum, tellurium, tin, titanium, tungsten, uranium, vanadium, and zirconium.

113. Wilson, "Strategies for security critical supply chain materials," Perth USAsia Centre, April 2020, p. 4.
114. Global Commission on the Geopolitics of Energy Transformation and IRENA, "A new world: The geopolitics of the energy transformation," IRENA, Abu Dhabi, 2019, p. 40.
115. Brad Glosserman, "Critical raw materials and the challenge of climate crisis in the new millennium, *The Asia Pacific Journal: Japan Focus*, 19(3), 1 October 2021, at https://apjjf.org/2021/19/Glosserman.html.
116. Neil Hume, "Europe faces critical shortage of metals needed for green energy," *Financial Times*, 26 April 2022, at https://www.ft.com/content/72400b96–4c67-4a7f-9a98-53371f5ab421.
117. Jeffrey Wilson, "Strategies for security critical supply chain materials," Perth USAsia Centre, April 2020, pp. 6–8.
118. Ibid., p. 6.
119. Ibid., p. 8.
120. Cited in Keith Bradsher, "Amid tension, China blocks vital exports to Japan," *The New York Times*, 22 September 2010 at https://www.nytimes.com/2010/09/23/business/global/23rare.html.
121. Fergus Hanson, Emilia Currey, and Tracy Beattie, "The Chinese Communist Party's coercive diplomacy," Australian Strategic Policy Institute, 2020, https://www.aspi.org.au/report/chinese-communist-partys-coercive-diplomacy.
122. Ivar Kalstad, "Too big to fault? Effects of the 2010 Nobel Peace Prize on Norwegian exports to China and foreign policy," CMI Working Paper, Bergen Norway, May 2016, at https://www.cmi.no/publications/file/5805-too-big-to-fault.pdf.
123. Hanson et al., "The Chinese Communist Party's coercive diplomacy," Australian Strategic Policy Institute. 2020, p. 19 at https://www.aspi.org.au/report/chinese-communist-partys-coercive-diplomacy.
124. Ibid., p. 21.
125. Nick Carey and Christina Amann, "They can shut us down in two months: Carmakers race China's rare-earths curbs," *The Japan Times*, 21 October 2025.
126. Gabriel Collins, "Russia's use of the 'energy weapon' in Europe," Baker Institute for Public Policy Issue Brief, 18 July 2017, p. 3.
127. Ibid., p. 7.
128. Donald Tusk, "A united Europe can end Russia's energy stranglehold, *Financial Times*, 21 April 2014 at https://www.ft.com/content/91508464-c661–11e3-ba0e-00144feabdc0.
129. Yes, we know that the decisions weren't a ban per se and instead took the ROK off a "white list," which as a result demanded approval of exports and that most shipments were eventually okayed. Still, it was a shock to the ROK—as intended. For more, see Ulrike Schaede, *The Business Reinvention of Japan: How to Make Sense of the New Japan and Why It Matters*, Stanford Business Books, 2020, pp. 69–70.

130. Mitsuru Obe and Kim Jaewon, "Inside the lose-lose trade fight between Japan and South Korea," *Nikkei Asia*, 31 July 2019 at https://asia.nikkei.com/Spotlight/The-Big-Story/Inside-the-lose-lose-trade-fight-between-Japan-and-South-Korea.
131. Schaede, p. 70. Schaede masterfully argues that Japanese businesses have gravitated to positions that dominate key nodes in the high-tech production chain. This private sector strategy creates interdependence that can be weaponized.
132. Peter Harrell and Elizabeth Rosenberg, "Economic dominance, financial technology, and the future of U.S. economic coercion," Center for a New American Security, 29 April 2019, p. 7.
133. Ibid., p. 7.
134. Ibid., p. 9. Sharp-eyed readers might want to spend some time thinking about which of these factors are enduring and which might be available to other countries. For an insider's perspective on how the US has engaged in economic warfare, Edward Fishman, *Chokepoints: American Power in the Age of Economic Warfare*, Portfolio, 2025.
135. Edward Knudsen, "The weaponisation of the US financial system: How can Europe respond?" Policy Paper, Hertie School, Jacques Delors Center, 4 June 2020, pp. 8–9.
136. "Tried to practice" is an important distinction. The vast majority of those efforts have failed. See, for example, Menenvis Cilizoglu and Bryan Early, "Researching modern economic sanctions," *Oxford Research Encyclopedia of International Studies*, 2021, at https://oxfordre.com/internationalstudies/view/10.1093/acrefore/9780190846626.001.0001/acrefore-9780190846626-e-599.
137. The seminal work is Henry Farrell and Abraham L. Newman, "Weaponized interdependence: How global economic networks shape state coercion," *International Security*, 44(1), Summer 2019, pp. 42–79; Daniel W. Drezner, Henry Farrell, and Abraham Newman, eds, *The Uses and Abuses of Weaponized Interdependence*, Brookings Institution Press, 2021.
138. Daniel Drezner, "The uses and abuses of weaponized interdependence in 2021," *The Washington Post*, 2 March 2021.
139. Cited in Peter Singer and Emerson Brooking, *LikeWar: The Weaponization of Social Media*, Mariner Books, 2018, p. 106.
140. Cited in *LikeWar*, p. 148.
141. Daniel Coats, "Statement for the record: Worldwide threat assessment of the US Intelligence Committee," Senate Select Committee for Intelligence, 29 January 2019, p. 5.
142. World Economic Forum, Global Risks Report 2021, 16th edition, insight report, p. 16, http://www3.weforum.org/docs/WEF_The_Global_Risks_Report_2021.pdf.
143. World Economic Forum, Global Risks Report 2021, p. 33.
144. Helen Warrell, "Russia and China waging 'political war', says UK military chief," *Financial Times*, 1 October 2020, at https://www.ft.com/content/8103dcc2-fbb7-4d5d-b5f9-ee9f0db1f1ed.

145. Laura Rosenberger and Lindsay Gorman, "How democracies can win the information contest," *The Washington Quarterly*, 43(2), Summer 2020, p. 75.
146. Laura Rosenberger and Lindsay Gorman, "Foreign interference is a strategy not a tactic," *Lawfare*, 26 October 2020 at https://www.lawfareblog.com/foreign-interference-strategy-not-tactic.
147. Ibid.

3. CHINA REIMAGINES CAPITALISM

1. This story is told by Paul Mozur, "Germany withdraws approval for Chinese takeover of Aixtron," *The New York Times*, 24 Oct 2016, at https://www.nytimes.com/2016/10/25/business/dealbook/germany-china-technology-takeover.html; and Paul Mozur and Jack Ewing, "Rush of Chinese investment in Europe's high-tech firms is raising eyebrows," *The New York Times*, 16 September 2016, at https://www.nytimes.com/2016/09/17/business/dealbook/china-germany-takeover-merger-technology.html.
2. Agence-France-Press, "Chinese firm scraps Aixtron deal after US block," 8 December 2016 at https://www.industryweek.com/the-economy/public-policy/article/21991440/chinese-firm-scraps-aixtron-deal-after-us-block.
3. Mozur and Ewing, "Rush of Chinese investment in Europe's high-tech firms is raising eyebrows," *The New York Times*, 16 September 2016, at https://www.nytimes.com/2016/09/17/business/dealbook/china-germany-takeover-merger-technology.html.
4. Ibid.
5. This and other details are from Emily Feng, "How China acquired mastery of vital microchip technology," *Financial Times*, 29 January 2019 at https://www.ft.com/content/7cfb2f82–1ecc-11e9-b126–46fc3ad87c6\.
6. Silex officials deny the validity of those fears. Feng, "How China acquired mastery of vital microchip technology," *Financial Times*, 29 January 2019 at https://www.ft.com/content/7cfb2f82–1ecc-11e9-b126–46fc3ad87c65\.
7. Cory Bennett and Bryan Bender, "How China acquires 'the crown jewels' of U.S. technology," *Politico*, 22 May 2018 at https://www.politico.com/story/2018/05/22/china-us-tech-companies-cfius-572413.
8. The official was brought into the case by an ATop creditor who worried it would not be repaid. See Bennett and Bender, "How China acquires 'the crown jewels' of U.S. technology," *Politico*, 22 May 2018 at https://www.politico.com/story/2018/05/22/china-us-tech-companies-cfius-572413.
9. Camille Stewart, "Full court press: Preventing foreign adversaries from exfiltrating national security technologies through bankruptcy proceedings," *Georgetown Journal of National Security Law and Policy*, 10, 2019, p. 281.
10. Reuters, "Dutch government took control of Nexperia over fears that it was being gutted—sources," 28 October 2025, at https://www.reuters.com/sustainability/boards-policy-regulation/dutch-government-took-control-nexperia-over-fears-it-was-being-gutted-sources-2025-10-27/.

11. Toby Sterling, "In rare move, Dutch government takes control of China-owned chipmaker Nexperia," Reuters, 114 October 2025 at https://www.reuters.com/world/china/dutch-government-intervenes-chinese-owned-computer-chip-firm-nexperia-2025-10-12/.
12. Victor Menaldo and Nicolas Wittstock, "A tale of TikTok and Tariffs: a historical and analytic perspective on Technology Transfer to China," August 2020, pp. 8–10, at https://cpb-us-e1.wpmucdn.com/sites.uw.edu/dist/0/7926/files/2020/08/Menaldo-Wittstock-Political-Economy-Forum-Version.pdf.
13. Cited in Mariana Mazzucato, *The Entrepreneurial State: Debunking Public vs. Private Sector Myths*, Anthem Press, 2015, p. 36.
14. Kasahara Shigehisa, "The Asian developmental state and the flying geese paradigm," UNCTAD Discussion Papers 213, United Nations Conference on Trade and Development, Geneva, 2013, p. 17.
15. Kenji Kushida, "Inside the castle gates: How foreign companies navigate Japan's policymaking processes," PhD dissertation, Fall 2010, p. 42 at https://escholarship.org/content/qt3q9796r6/qt3q9796r6_noSplash_071f67e76d5338651ac2bb54ea306e5e.pdf.
16. Erik Reinert, *How Rich Countries Got Rich and Why Poor Countries Stay Poor*, Constable, 2007, p. 23.
17. Linda Weiss, *America Inc? Innovation and Enterprise in the National Security State*, Cornell University Press, 2014, p. 20.
18. Eric Schmidt, "Eric Schmidt: I used to run Google. Silicon Valley could lose to China," *The New York Times*, 27 February 2020, at https://www.nytimes.com/2020/02/27/opinion/eric-schmidt-ai-china.html.
19. Seth Center and Emma Bates, "Tech-politik: Historical perspectives on innovation, technology and strategic competition," CSIS Policy Briefs, December 2019, p. 5, at https://csis-website-prod.s3.amazonaws.com/s3fs-public/publication/191219_Tech-Politik.pdf?IQ4IECqSCT0kGewclKhNiZ4JZ.knm.zq.
20. Henrique Choer Moraes and Mikaek Wigell, "The emergence of strategic capitalism: Geoeconomics, corporate statecraft and the repurposing of the global economy," Finnish Institute of International Affairs, FIIA Working Paper, 117, September 2020, p. 5.
21. Wayne Sandholtz et al., *The Highest Stakes: The Economic Foundations of the Next Security System: A BRIE Project*, Oxford University Press, 1992, p. 200.
22. Olga Petricevic and David Teece, "The structural reshaping of globalization: Implications for strategic sectors, profiting from innovation, and the multinational enterprise," *Journal of International Business Studies*, 50, 2019, p. 1488.
23. Lucia Retter et al., "Relationships between the economy and national security: Analysis and considerations for economic security policy in the Netherlands," Rand, Europe, 2020, p. 60.
24. Wayne Morrison, "China's economic rise: History, trends, challenges and implications for the United States," Congressional Research Service, RL33535, 25 June 2019, p. 1.

25. Jacque de Lisle and Avery Goldstein, "Introduction," in *To Get Rich is Glorious: Chinese Economic Reform and Opening at Forty*, Brookings Institution, 2019, p 15. For additional insights on China's rise during this period, Keyu Jin, *The New China Playbook: Beyond Socialism and Capitalism*, Viking, 2023.
26. The World Bank, *China: 2030*, 27 February 2012, p. 114.
27. David J. Teece, "Fundamental issues in strategy: Time to reassess?" *Strategic Management Review*, 1(1), 2020, p. 116, http://dx.doi.org/10.1561/111.00000005.
28. Several supporting policies for implementing the "National Medium- and Long-Term Science and Technology Development Plan Outline (2006–2020)," State Council, Guo Fa [2006] No. 6, issued 7 February 2006, § 29, in USTR p. 12.
29. European Union Chamber of Commerce in China and the Mercator Institute of China Studies, "Decoupling: severed ties and patchwork globalisation," 2021, p. 3.
30. David J. Teece, "Fundamental issues in strategy: Time to reassess?" *Strategic Management Review*, 1(1), 2020, p. 119, http://dx.doi.org/10.1561/111.00000005.
31. Alan Sykes, "The law and economics of 'forced' technology transfer (FTT) and its implications for trade and investment policy (and the U.S.–China Trade War)," *Stanford Law and Economics*, Olin Working Paper No. 544, 17 January 2020, p. 7, at https://ssrn.com/abstract=3523284 or http://dx.doi.org/10.2139/ssrn.3523284.
32. Ibid., p. 8.
33. Office of the United States Trade Representative, "Findings of the investigation into China's acts, policies and practices related to technology transfer, intellectual property and innovation under section 301 of the Trade Act of 1974," (USTR Report) 22 March 2018, p. 22, emphasis in original.
34. Alan Sykes, "The law and economics of 'forced' technology transfer (FTT) and its implications for trade and investment policy (and the U.S.–China Trade War)," *Stanford Law and Economics*, Olin Working Paper No. 544, 17 January 2020, p. 7, at https://ssrn.com/abstract=3523284 or http://dx.doi.org/10.2139/ssrn.3523284.
35. Office of the United States Trade Representative, "Findings of the investigation into China's acts, policies and practices related to technology transfer, intellectual property and innovation under section 301 of the Trade Act of 1974," (USTR Report) 22 March 2018, p. 19.
36. Ibid., p. 22.
37. Ibid., p. 36.
38. Ibid., p. 37.
39. Ibid., p. 43.
40. Ibid., p. 38.
41. Alan Sykes, "The law and economics of 'forced' technology transfer (FTT) and its implications for trade and investment policy (and the U.S.–China Trade War)," *Stanford Law and Economics*, Olin Working Paper No. 544, 17 January 2020, p. 7, at https://ssrn.com/abstract=3523284 or http://dx.doi.org/10.2139/ssrn.3523284, p. 3.

42. Nina Xiang, "China never stole its way to the top, and Biden should say so," *Nikkei Asia*, 10 November 2020, at https://asia.nikkei.com/Opinion/China-never-stole-its-way-to-the-top-and-Biden-should-say-so, and Nicholas Lardy, "China: Forced technology transfer and theft?" Peterson Institute of International Economics, China Economic Watch blog, 20 April 2018, (based on IMF data).
43. A. Andrenelli, J. Gourdon and E. Moïsé, "International technology transfer policies," *OECD Trade Policy Papers*, No. 222, 24 January 2019, p. 6, OECD Publishing, Paris. http://dx.doi.org/10.1787/7103eabf-en.
44. Office of the United States Trade Representative, "Findings of the investigation into China's acts, policies and practices related to technology transfer, intellectual property and innovation under section 301 of the Trade Act of 1974," (USTR Report) 22 March 2018, p. 48.
45. Sean O'Connor, "How Chinese companies facilitate technology transfer from the United States," U.S.-China Economic and Security Review Commission, 6 May 2019, p. 7.
46. Mark Wu, "The 'China Inc.' Challenge to global trade governance," *Harvard International Law Journal*, 57(2), 2016, pp. 267–8.
47. Office of the United States Trade Representative, "Findings of the investigation into China's acts, policies and practices related to technology transfer, intellectual property and innovation under section 301 of the Trade Act of 1974," (USTR Report) 22 March 2018, p. 81.
48. "Xi calls for furthering SOE reform," Xinhua, 18 October 2017, at http://www.xinhuanet.com/english/2017–10/18/c_136688444.htm.
49. "19th CPC National Congress: SCIO briefing on reform of SOEs," 1 October 2017, Embassy of China in Australia, at http://au.china-embassy.org/eng/zggk/t1502282.htm.
50. Office of the United States Trade Representative, "Findings of the investigation into China's acts, policies and practices related to technology transfer, intellectual property and innovation under section 301 of the Trade Act of 1974," (USTR Report) 22 March 2018, p. 84.
51. Wayne Morrison, "China's economic rise: History, trends, challenges and implications for the United States," Congressional Research Service, RL33535, 25 June 2019, p. 25.
52. Jason Arterburn, "The party state in China's military industrial complex: Implications for U.S. national security," Testimony before the U.S.-China Economic and Security Review Commission Hearing on "U.S. Investment in China's capital markets and military-industrial complex," 19 March 2021, p. 3.
53. Jude Blanchette, "Confronting the challenge of Chinese state capitalism," CSIS, 22 January 2021, at https://www.csis.org/analysis/confronting-challenge-chinese-state-capitalism.
54. European Commission, Commission staff working document on the significant distortions in the economy of the People's Republic of China for the purposes of trade defence investigations, SWD(2017)483 final/2, 20 December 2012, pp. 108–9.

55. Office of the United States Trade Representative, "Findings of the investigation into China's acts, policies and practices related to technology transfer, intellectual property and innovation under section 301 of the Trade Act of 1974," (USTR Report) 22 March 2018, p. 85.
56. Asia Society Policy Institute and the Rhodium Group, China Dashboard: Winter 2021 Update, pp. 2–3, at https://aspi-gist-cdn.sfo2.digitaloceanspaces.com/chinadashboard/websites/5ff763b6d9ac920410bd7c65/pages/5ff763b6d9ac920410bd7c67/pdf/The_China_Dashboard--Winter_2021--Full_Report.pdf.
57. Office of the United States Trade Representative, "Findings of the investigation into China's acts, policies and practices related to technology transfer, intellectual property and innovation under section 301 of the Trade Act of 1974," (USTR Report) 22 March 2018, p. 103.
58. Ibid., p. 88.
59. Ibid., p. 89.
60. Curtis J. Milhaupt and Wentong Zheng, "Beyond ownership: State capitalism and the Chinese firm. *The Georgetown Law Journal*, 103(665), 2015, pp. 665–722.
61. US–China Economic and Security Review Commission, "Chapter 2 Section 3: The Chinese Government's evolving control of the nonstate sector," in *2021 Annual Report to Congress*, November 2021, pp. 228–9 at https://www.uscc.gov/sites/default/files/2021-11/Chapter_2_Section_3--Chinese_Government's_Evolving_Control_of_the_Nonstate%20Sector.pdf.
62. Jason Arterburn, "The party state in China's military industrial complex: Implications for U.S. national security," Testimony before the U.S.–China Economic and Security Review Commission Hearing on "U.S. Investment in China's Capital Markets and Military-Industrial Complex," 19 March 2021, p. 3.
63. Jude Blanchette, "From 'China Inc.' to 'CCP Inc.': A new paradigm for Chinese state capitalism," Hinrich Foundation, February 2021, p. 2.
64. Ibid., p. 3.
65. FT View, "China blurs lines between private and state business," *Financial Times*, 9 October 2020 at https://www.ft.com/content/c70ff5bb-dff9–4ad1–9574–75e58c3c87c1.
66. Cited in Jude Blanchette, "From 'China Inc.' to 'CCP Inc.': A new paradigm for Chinese state capitalism," Hinrich Foundation, February 2021, p. 3.
67. Office of the United States Trade Representative, "Findings of the investigation into China's acts, policies and practices related to technology transfer, intellectual property and innovation under section 301 of the Trade Act of 1974," (USTR Report) 22 March 2018, p. 147.
68. Emily Feng, "Chinese tech groups display closer ties with Communist party, *Financial Times*, 11 October 2017, at https://www.ft.com/content/6bc839c0-ace6-11e7-aab9-abaa44b1e130.
69. Rush Doshi et al., "China as a "cyber great power: Beijing's two voices in telecommunications," Brookings Institution, April 2021, p. 8, at https://www.brookings.edu/wp-content/uploads/2021/04/FP_20210405_china_cyber_power.pdf.

70. "Made in China 2025 explanation 6: The manufacturing power 'three-step' strategy," Ministry of Industry and Information Technology, 19 May 2015, http://www.miit.gov.cn/n1146295/n1146562/n1146655/c3780688/content.html, in USTR p. 16.
71. Jannace and Tiffany, "A new world order: The rule of law, or the law of rulers?" *Fordham International Law Journal*, 42(5), 2019, p. 1394.
72. Jude Blanchette, "From 'China Inc.' to 'CCP Inc.': A new paradigm for Chinese state capitalism," Hinrich Foundation, February 2021, pp. 6–7.
73. Ibid., p. 7.
74. Office of the United States Trade Representative, "Findings of the investigation into China's acts, policies and practices related to technology transfer, intellectual property and innovation under section 301 of the Trade Act of 1974," (USTR Report) 22 March 2018, p. 65.
75. Karen Sutter, "Made in China 2025 industrial policies: Issues for Congress," Congressional Research Service, IF 10964, 11 August 2020, p. 2.
76. Ngor Luong, Zachary Arnold, and Ben Murphy, "Understanding Chinese government guidance funds: An analysis of Chinese-language sources," Georgetown Center for Security and Technology, March 2021, p. 1.
77. Yoko Kubota, "China sets up new $29 billion semiconductor Fund," *Wall Street Journal*, 25 October 2019.
78. Coco Liu, Lauly Li, Cheng Ting-Fang, and Kenji Kawase, "China bets on $2tn high-tech infrastructure plan to spark economy," *Nikkei Asia*, 1 June 2020, at https://asia.nikkei.com/Business/China-tech/China-bets-on-2tn-high-tech-infrastructure-plan-to-spark-economy.
79. Reuters, "China to set up national venture capital guidance fund, state planner says," 6 March 2025, at https://www.reuters.com/world/china/china-set-up-national-venture-capital-guidance-fund-state-planner-says-2025–03–06/.
80. Pierre Olivier Gourinchas et al., "Trade balances in China and the U.S. are largely driven by domestic macro forces," IMF Blog, 12 September 2024, at https://www.imf.org/en/Blogs/Articles/2024/09/12/trade-balances-in-china-and-the-us-are-largely-driven-by-domestic-macro-forces.
81. Gerard Dipippo et al., "Red ink: Estimating Chinese industrial policy spending in comparative perspective, CSIS, May 2022, at https://csis-website-prod.s3.amazonaws.com/s3fs-public/publication/220523_DiPippo_Red_Ink.pdf?VersionId=LH8ILLKWz4o.bjrwNS7csuX_C04FyEre.
82. International Energy Agency, "Global EV outlook 2024," at https://www.iea.org/reports/global-ev-outlook-2024.
83. Craig Allen, "Silicon showdown: The competition for technology leadership between the U.S. and China," Stanford SCCEI Briefs, 24 September 2024, at https://scceichinabriefs.substack.com/p/silicon-showdown-the-competition.
84. Mercator study cited in Office of the United States Trade Representative, "Findings of the investigation into China's acts, policies and practices related to technology transfer, intellectual property and innovation under section 301 of the Trade Act of 1974," (USTR Report) 22 March 2018, p. 91.

85. Ryan Fedasiuk, Emily Weinstein, and Anna Puglisi, "China's foreign technology wish list," Georgetown Center for Security and Emerging Technology, May 2021, p. 2.
86. Ibid., p. 2.
87. Ibid., p. 1.
88. Michael Brown and Pavneet Singh, "China's technology transfer strategy: How Chinese investments in emerging technology enable a strategic competitor to access the crown jewels of U.S. innovation," Defense Innovation Unit Experimental DIUx, Washington, DC, January 2018, p. 19.
89. Ibid., p. 16.
90. Ministry of Commerce, People's Republic of China, "MOFCOM, SAFE, NBS Jointly issue the annual communique of China's outward foreign direct investment," 18 September 2020 at http://english.mofcom.gov.cn/article/newsrelease/significantnews/202009/20200903004178.shtml.
91. China Brief, "China outbound direct investment tracker 2024–25," 31 October 2025, at https://www.china-briefing.com/news/china-outbound-direct-invest-odi-tracker-2024-25/.
92. Michael Brown and Pavneet Singh, "China's technology transfer strategy: How Chinese investments in emerging technology enable a strategic competitor to access the crown jewels of U.S. innovation," Defense Innovation Unit Experimental DIUx, Washington, DC, January 2018, p. 5.
93. Jeffrey Gordon and Curtis Milhaupt in "China as a 'national strategic buyer': Towards a multilateral regime for cross-border M&A," *Columbia Business Law Review*, 2019, p. 197, at SSRN: https://ssrn.com/abstract=3180250.
94. Elsa B. Kania, "China's threat to American government and private sector research and innovation leadership," Testimony before the House Permanent Select Committee on Intelligence, 19 July 2018, p. 5, at https://s3.us-east-1.amazonaws.com/files.cnas.org/documents/Elsa-B.-Kania-Testimony-before-the-House-Permanent-Select-Committee-on-Intelligence-July-192c-2018-final_v2.pdf?mtime=20180720095054&focal=none.
95. Office of the United States Trade Representative, "Findings of the investigation into China's acts, policies and practices related to technology transfer, intellectual property and innovation under section 301 of the Trade Act of 1974," (USTR Report) 22 March 2018, p. 100.
96. Michael Brown and Pavneet Singh, "China's technology transfer strategy: How Chinese investments in emerging technology enable a strategic competitor to access the crown jewels of U.S. innovation," Defense Innovation Unit Experimental DIUx, Washington, DC, January 2018, pp. 7, 9–10.
97. Michael Brown, Statement before the House Permanent Select Committee on Intelligence, 19 July 2018, p. 2, at https://docs.house.gov/meetings/IG/IG00/20180719/108561/HHRG-115-IG00-Wstate-BrownM-20180719.pdf.
98. Michael Brown and Pavneet Singh, "China's technology transfer strategy: How Chinese investments in emerging technology enable a strategic competitor to access

the crown jewels of U.S. innovation," Defense Innovation Unit Experimental DIUx, Washington, DC, January 2018, p. 5. Those numbers might be high. A report by the Rhodium Group reckons that from 2000 to May 2018, Chinese venture capital contributions in the US totaled $11 billion. In 2018, Chinese VC investment in the US reached a record $3.3 billion, up from $2.1 billion in 2017. It is cited in Sean O'Connor, "How Chinese companies facilitate technology transfer from the United States," U.S.-China Economic and Security Review Commission, 6 May 2019, p. 5.

99. Ibid., p. 6.
100. Ibid., p. 5.
101. Michael Brown, Statement before the House Permanent Select Committee on Intelligence, 19 July 2018, p. 2.
102. Ibid., p. 2–3. Emphasis in original.
103. Olga Petricevic and David Teece, "The structural reshaping of globalization: Implications for strategic sectors, profiting from innovation, and the multinational enterprise," *Journal of International Business Studies*, 50, 2019, p. 1497.
104. Emily de La Bruyère and Nathan Picarsic, "Made in Germany, Co-opted by China," Foundation for Defense of Democracies," 14 October 2020, p. 12, citing Cora Jongbluth, "Is China systematically buying up key technologies? Chinese M&A Transactions in Germany in the Context of 'Made in China 2025,'" *Bertelsmann Stiftung*, 2018, https://www.bertelsmann-stiftung.de/leadmin/les/BSt/Publikationen/GrauePublikationen/MT_Is_China_Systematically_Buying_Up_Key_Technologies.pdf.
105. Merics, "Chinese investment rebounds despite growing frictions: Chinese FDI in Europe—2024 update," 21 May 2025 at https://merics.org/en/report/chinese-investment-rebounds-despite-growing-frictions-chinese-fdi-europe-2024-update.
106. Cory Bennett and Bryan Bender, "How China acquires 'the crown jewels' of U.S. technology," *Politico*, 22 May 2018, at https://www.politico.com/story/2018/05/22/china-us-tech-companies-cfius-572413.
107. Camille Stewart, "Full court press: Preventing foreign adversaries from exfiltrating national security technologies through bankruptcy proceedings," *Journal of National Security Law and Policy*, 10, 2019, p. 278.
108. Cory Bennett and Bryan Bender, "How China acquires 'the crown jewels' of U.S. technology," *Politico*, 22 May 2018 at https://www.politico.com/story/2018/05/22/china-us-tech-companies-cfius-572413.
109. CFIUS Reform: Examining the essential elements: Hearing before the S. Comm. on Banking, Housing, & Urban Affairs, 115th Cong. (2018).
110. Cory Bennett and Bryan Bender, "How China acquires 'the crown jewels' of U.S. technology," *Politico*, 22 May 2018 at https://www.politico.com/story/2018/05/22/china-us-tech-companies-cfius-572413.
111. Jason Arterburn, "The party state in China's military industrial complex: Implications for U.S. National Security," Testimony before the U.S.-China Economic and Security Review Commission Hearing on "U.S. Investment in China's Capital Markets and Military-Industrial Complex," 19 March 2021, p. 8.

112. "China's quest for foreign technology," p. 39 (W. C. Hannas & D. K. Tatlow, eds.), cited in Arterburn, "The party state in China's military industrial complex," p. 8.
113. Sean O'Connor, "How Chinese companies facilitate technology transfer from the United States," U.S.-China Economic and Security Review Commission, 6 May 2019, p. 9
114. "Japan's thousand talents plan lured at least 44 Japanese researchers," *Yomiuri Shimbun*, 2 January 2021, at https://the-japan-news.com/news/article/0007036449.
115. Emily Weinstein, "Chinese talent program tracker," Georgetown Center for Security and Emerging Technology, November 2020, at https://chinatalent-tracker.cset.tech/static/cset_chinese_talent_program_tracker.pdf.
116. William Hannas and Huey-meei Chang, "China's access to foreign AI technology: An assessment," Georgetown Center for Security and Emerging Technology, September 2019, p. 5.
117. Ibid., p. 5.
118. Michael Brown and Pavneet Singh, "China's technology transfer strategy: How Chinese investments in emerging technology enable a strategic competitor to access the crown jewels of U.S. innovation," Defense Innovation Unit Experimental DIUx, Washington, DC, January 2018, p. 20.
119. Ibid., p. 20.
120. Ibid., p. 16.
121. Ibid., p. 18.
122. Sean O'Connor, "How Chinese companies facilitate technology transfer from the United States," U.S.-China Economic and Security Review Commission, 6 May 2019, p. 9.
123. Olga Petricevic and David Teece, "The structural reshaping of globalization: Implications for strategic sectors, profiting from innovation, and the multinational enterprise," *Journal of International Business Studies*, 50, 2019, p. 1494.
124. France is generally considered the most aggressive of Europe's mercantilists, but even the "French government is not pursuing a national industrial strategy of targeting foreign firms to obtain advanced technologies or regulating the volume of outbound deal flow in service of national economic policy," counters Jeffrey Gordon and Curtis Milhaupt in "China as a 'national strategic buyer': Towards a multilateral regime for cross-border M&A," *Columbia Business Law Review*, 2019, at https://journals.library.columbia.edu/index.php/CBLR/article/view/1692/718.
125. David Teece, "Response to Peter Buckley's comment on "fundamental issues in strategy," 5 May 2020, p. 5 at http://leeds-faculty.colorado.edu/jere1232/Teece%20Response%20to%20Buckley.pdf.
126. Mark Wu, "The 'China Inc.' challenge to global trade governance," *Harvard International Law Journal*, 57(2), 2016, p. 270.
127. Ibid., p. 264.

128. Olga Petricevic and David Teece, "The structural reshaping of globalization: Implications for strategic sectors, profiting from innovation, and the multinational enterprise," *Journal of International Business Studies*, 50, 2019, p. 1498.
129. Ibid., p. 1496.
130. Jeffrey Gordon and Curtis Milhaupt in "China as a 'national strategic buyer': Towards a multilateral regime for cross-border M&A," *Columbia Business Law Review*, 2019, p. 247.
131. Ibid., p. 198.
132. Ibid., p. 212, p. 218.
133. Ibid., p. 248.
134. Zongyuan Zoe Liu, *Sovereign Funds: How the Communist Party of China Finances Its Global Ambitions*, Belknap Press of Harvard University Press, 2023.
135. The best known are the Santiago Principles, officially the "Sovereign wealth funds: Generally accepted principles and practices," October 2008 at https://www.ifswf.org/sites/default/files/santiagoprinciples_0_0.pdf.
136. Curtis Milhaupt and Mariana Pargendler, "Governance challenges of listed state-owned enterprises around the world: National experiences and a framework for reform," *Cornell International Law Journal*, 50(73), 2017, pp. 519–24.
137. Olga Petricevic and David Teece, "The structural reshaping of globalization: Implications for strategic sectors, profiting from innovation, and the multinational enterprise," *Journal of International Business Studies*, 50, 2019, p. 1497 and Jude Blanchette, "Confronting the challenge of Chinese state capitalism," CSIS, 22 January 2021, at https://www.csis.org/analysis/confronting-challenge-chinese-state-capitalism.
138. David J. Teece, "Fundamental issues in strategy: Time to reassess?" *Strategic Management Review*, 1(1), 2020, p. 122.
139. Mark Wu, "The 'China Inc.' Challenge to global trade governance," *Harvard International Law Journal*, 57(2), 2016, p. 265. He acknowledges that the WTO is not utterly at sea. "There are situations in which the behavior of China, Inc. resembles that of other alternative economic structures. In such instances, a body of treaty text and jurisprudence exists to address the problematic behavior. These laws can be applied to the China, Inc. context, allowing China's trading partners to vindicate their rights." Wu, p. 295, 300. There are also widely accepted principles and best practices to guide the operation of SOEs, such as the Organization for Economic Cooperation and Development (OECD) Guidelines on Corporate Governance of State-Owned Enterprises, 2015, and the World Bank Group, Corporate Governance of State-Owned Enterprises: A Tool Kit, WB456286/91347, 2014, at http://documents.worldbank.org/. curated/en/228331468169750340/pdf/913470PUB097810B00PUBLIC00100602014.Pdf, https://perma.cc/2QRE-24P9.
140. Jeffrey Gordon and Curtis Milhaupt in "China as a 'national strategic buyer': Towards a multilateral regime for cross-border M&A," *Columbia Business Law Review*, 2019, p. 222.

141. BDI (Federation of German Industries), "Partner and systemic competitor: How do we deal with China's state-controlled economy," Policy Paper China, January 2019, p. 4.
142. Jeffrey Gordon and Curtis Milhaupt in "China as a 'national strategic buyer': Towards a multilateral regime for cross-border M&A," *Columbia Business Law Review*, 2019, p. 197.
143. Mark Wu, "The 'China Inc.' Challenge to global trade governance," *Harvard International Law Journal*, 57(2), 2016, p. 315–6.
144. Jeffrey Gordon and Curtis Milhaupt in "China as a 'national strategic buyer': Towards a multilateral regime for cross-border M&A," *Columbia Business Law Review*, 2019, p. 248.
145. Xi Jinping, "Certain major issues for our national medium- to long-term economic and social development strategy [国家中长期经济社会发展战略若干重大问题], *Qiushi*, 1 November 2020, https://web.archive.org/web/20201111020608/http://www.qstheory.cn/dukan/qs/2020-10/31/c_1126680390.htm, translated by Georgetown Center for Security and Emerging Technology, 10 November 2020, p. 3, at https://cset.georgetown.edu/wp-content/uploads/t0235_Qiushi_Xi_economy_EN-1.pdf.
146. Mo Yu and Yi-Hua Lee, "US conservatives find new merit in industrial policy," VOA, 22 September 2020, at https://www.voanews.com/usa/us-conservatives-find-new-merit-industrial-policy.
147. Regina Abrami, William C. Kirby and F. Warren McFarlan, "Why China can't innovate," *Harvard Business Review*, March 2014, Carl Benedikt Frey, "How culture gives the US an innovation edge over China," MIT Sloan Management Review, 8 February 2021 at https://sloanreview.mit.edu/article/how-culture-gives-the-us-an-innovation-edge-over-china/, and Andrew Gilli and Mauro Gilli, "Why China has not caught up yet: Military-technical superiority and the limits of imitation, reverse engineering, and cyber espionage," *International Security*, 43(3), Winter, 2018.
148. William Hannas and Huey-meei Chang, "China's access to foreign AI technology: An assessment," Georgetown Center for Security and Emerging Technology, September 2019, p. 7.
149. Rhodium Group, "Was Made in China 2025 successful?" 5 May 2025, pp. 4–5.
150. Hannas and Chang, "China's access to foreign AI technology," p. iii.
151. Ibid., p. 8.

4. GOVERNMENTS RESPOND: PLAYING DEFENSE

1. For one list of options, see Jim Balsillie, "Digital transformation: Pre- and post-Covid-19," Institute for New Economic Thinking webinar, presentation, 20 August 2020, at https://www.ineteconomics.org/uploads/papers/INET-Jim-Balsillie-Event-Slides.pdf.
2. OECD, "Cybersecurity policy making at a turning point: Analysing a new generation

of national cybersecurity strategies for the internet economy," OECD Digital Economy Papers, No. 211, Paris, 16 November 2012, p. 6.

3. International Telecommunications Union, "Countries strengthening cybersecurity efforts but increased action still required," 12 September 2024, at https://www.itu.int/en/mediacentre/Pages/PR-2024-09-10-Global-Cybersecurity-Index.aspx.
4. OECD, "Cybersecurity policy making at a turning point: Analysing a new generation of national cybersecurity strategies for the internet economy," OECD Digital Economy Papers, No. 211, Paris, 16 November 2012, pp. 6–7.
5. Sneha Dawda, "Exploring national cybersecurity strategies," Royal United Services Institute, London, Occasional Paper, February 2021, p. 12.
6. OECD, "Cybersecurity policy making at a turning point: Analysing a new generation of national cybersecurity strategies for the internet economy," OECD Digital Economy Papers, No. 211, Paris, 16 November 2012, p. 15.
7. Ibid., pp. 6–7.
7. Sneha Dawda, "Exploring national cybersecurity strategies," Royal United Services Institute, London, Occasional Paper, February 2021, p. 8.
8. A concise history is available at "Historical Background of Export Control Development in Selected Countries and Regions," Center for Information on Security Trade Control (CISTEC), Japan, April 2016 at https://www.cistec.or.jp/english/service/report/1605historical_background_export_control_development.pdf.
9. Jon Bateman provides an excellent summary of the various US controls in "US-China Technological 'Decoupling': A Strategic and Policy Framework," Carnegie Endowment for International Peace, 2022, pp. 17–20.
10. Elena Lazarou and Nicholas Lokker, "Briefing: United States: Export Control Reform Act (ECRA)," European Parliamentary Research Service, November 2019, p. 2.
11. White House, "Fact sheet: Executive order addressing the threat from securities investments that finance certain companies of the People's Republic of China," 3 June 2021.
12. Elena Lazarou and Nicholas Lokker, "Briefing: United States: Export Control Reform Act (ECRA)," European Parliamentary Research Service, November 2019, p. 2.
13. Kevin Wolf, "Prepared remarks of The Honorable Kevin J. Wolf: Perspectives on reform of the CFIUS review process," testimony before the United States House Committee on Energy and Commerce Subcommittee on Digital Commerce and Consumer Protection, 26 April 2018, p. 3, at https://docs.house.gov/meetings/IF/IF17/20180426/108216/HHRG-115-IF17-Wstate-WolfK-20180426.pdf.
14. Export Control Reform Act of 2018, 50 U.S.C. §4811(3).
15. Export Control Reform Act of 2018, Pub. L. No. 115–232, §§ 1758 (2018). The discussion that follows leans on Gibson Dunn, "New Controls for emerging technologies released, while commerce department comes under fire for delays," 27 October 2020, at https://www.gibsondunn.com/wp-content/uploads/2020/10/new-controls-on-emerging-technologies-released-while-us-commerce-department-comes-under-fire-for-delay.pdf.

16. Ian Fergusson, "Export controls: Key challenges," Congressional Research Service in Focus, 14 January 2021, p. 2.
17. US Department of Commerce, Bureau of Industry and Security, "Entity list," at https://www.bis.doc.gov/index.php/policy-guidance/lists-of-parties-of-concern/entity-list.
18. Emma Rafaelof, "Unfinished business: Export control and foreign investment reforms," U.S.-China Economic and Security Review Commission, Issue Brief, 1 June 2021, p. 5.
19. Ian Fergusson, "Export controls: Key challenges," Congressional Research Service in Focus, 14 January 2021, p. 2.
20. US Department of Commerce, Bureau of Industry and Security, "Rules affecting the export administration regulations," at https://www.bis.doc.gov/index.php/federal-register-notices#fr40921 (as of 6 June 2021).
21. Skadden, "New BIS rule expands export restrictions to entities 50% owned by parties on the entity and military end-user lists," 30 September 2025 at https://www.skadden.com/-/media/files/publications/2025/09/newbisruleexpandsexportrestrictionstoentities50ownedbypartiesontheentityandmilitaryenduserlists.pdf?rev=3c960178ac094ff3858c085793de03f7n.
22. Ian Fergusson, "Export controls: Key challenges," Congressional Research Service in Focus, 14 January 2021.
23. Gregory Allen, "Understanding the Biden administration's updated export controls," CSIS, 11 December 2024 at https://www.csis.org/analysis/understanding-biden-administrations-updated-export-controls.
24. Mayumi Negishi, "Japan plans to curb exports of chips and quantum-computing tech," *The Japan Times*, 1 February 2025.
25. William Reinsch et al., "Optimizing US export controls for critical and emerging technologies; Working with partners," CSIS, February 2024, p. 1.
26. The White House, "Remarks by national security advisor Jake Sullivan at the special competitive studies project global emerging technologies summit," 16 September 2022, at https://www.whitehouse.gov/briefing-room/speeches-remarks/2022/09/16/remarks-by-national-security-advisor-jake-sullivan-at-the-special-competitive-studies-project-global-emerging-technologies-summit/.
27. The Reagan Foundation, "Reagan national defense forum 2023 fireside chat," 13 December 2023 at https://www.reaganfoundation.org/media/361835/rndf-2023-fireside-chat-transcript-sec-raimondo.pdf.
28. Council Regulation 428/2009, 2009 O.J. (L 134) 1 (EC).
29. Noah Barkin, "Export controls and the US-China Tech War," MERICS China Monitor, 18 March 2020, p. 6. As Biden's June 2021 sanctions made clear, the US shares those concerns, but they are secondary in Washington. In Europe, they are the priority.
30. Reuters, "Netherlands to expand export controls on semiconductor equipment," 15 January 2025, at https://www.reuters.com/technology/netherlands-expand-export-controls-semiconductor-equipment-2025–01–15/.

31. Tobias Gehrke and Julian Ringhof, "Caught in the Crossfire: Why EU states should discuss strategic export controls," European Council on Foreign Relations, 11 January 2023, at https://ecfr.eu/article/caught-in-the-crossfire-why-eu-states-should-discuss-strategic-export-controls/.
32. European Commission, "Speech by President von der Leyen on EU-China relations," 29 March 2023 at https://ec.europa.eu/commission/presscorner/detail/en/speech_23_2063.
33. US Department of Commerce, Bureau of Industry and Security, China Export Control Information, at https://www.bis.doc.gov/index.php/enforcement/oee/220-eco-country-pages/1040-china-export-control-information, accessed 9 June 2021.
34. Lianzhong Pan and Jonathan Todd, "China's new export control law: A fact sheet with practical applications," Mondaq, 26 November 2020, at https://www.mondaq.com/china/export-controls-trade-investment-sanctions/1010194/china39s-new-export-control-law-a-fact-sheet-with-practical-applications.
35. Mercator Institute of China Studies, "New export controls give China new weapon in tech war," Merics Short Analysis, 22 October 2020, at https://merics.org/en/short-analysis/new-export-controls-give-china-new-weapon-tech-war.
36. Huan Zhu, "What to expect from China's new export control law," Cato Institute, 1 December 2020, at https://www.cato.org/blog/what-expect-chinas-new-export-control-law.
37. Karen Sutter, "China's issues new export control law and related policies," Congressional Research Service in Focus, 26 October 2020, p. 1.
38. Ibid., p. 2.
39. Ibid., p. 2.
40. Amber Wang and Wendy Wu, "China hurried work on anti-sanctions law after 'Joe Biden disappointed Beijing by continuing to take a tough stance,'" *South China Morning Post*, 10 June 2021 at https://www.scmp.com/news/china/diplomacy/article/3136676/china-speeded-work-anti-sanctions-law-after-joe-biden.
41. Julian Ku, Twitter, 11 June 2021 at https://twitter.com/julianku/status/1403012004961570826 (accessed 11 June 2021).
42. Amber Wang and Wendy Wu, "China hurried work on anti-sanctions law after 'Joe Biden disappointed Beijing by continuing to take a tough stance,'" *South China Morning Post*, 10 June 2021 at https://www.scmp.com/news/china/diplomacy/article/3136676/china-speeded-work-anti-sanctions-law-after-joe-biden.
43. Global Times, "Anti-sanctions law: A toolbox, deterrent: Global Times Editorial," 10 June 2021 at https://www.globaltimes.cn/page/202106/1225958.shtml.
44. Ministry of Economy, Trade and Industry, "Japan held consultations under the WTO Agreement with the Government of the Republic of Korea regarding Japan's update of licensing policies and procedures on exports to the ROK," 20 November 2019, at https://www.meti.go.jp/english/press/2019/1120_001.html.
45. See, for example, the OECD Code of Liberalization of Capital Movements (1961), the Code of Liberalization of Current Invisible Transactions (1961), the OECD

Declaration on International Investment and Multinational Treatment (1976) or the Guidelines for Recipient Country Investment Policies relating to National Security (2009).

46. "Transparency, predictability and accountability for investment screening mechanisms, research note by the OECD Secretariat," OECD, 27 May 2021, p. 5.
47. OECD, "Investment screening in times of COVID-19—and beyond," 7 July 2020, p. 5.
48. Sunny Mann, "Foreign investment scrutiny intensifies globally due to COVID-19," 21 September 2020, at https://foreigninvestment.bakermckenzie.com/2020/09/21/foreign-investment-scrutiny-intensifies-globally-due-to-covid-19/.
49. "Didi caught as China and US battle over data," *Financial Times*, 6 July 2021, at https://www.ft.com/content/00403ae5–7565–413e-907d-ad46549375ba, and UNCTAD, national security related screening mechanisms for foreign investment," p. 2.
50. OECD, "Investment screening in times of COVID-19—and beyond," 7 July 2020, p. 3.
51. Ibid., pp. 6–7.
52. Dechert, "The FDI and national security review," 2025 at https://www.dechert.com/knowledge/publication/the-evolving-global-foreign-direct-investment-and-national-secur.html.
53. Ana Novik et al., "Managing security implications of international investment," in *Economic Security in a Changing World*, OECD, 11 September 2025, at https://www.oecd.org/en/publications/economic-security-in-a-changing-world_4eac89c7-en/full-report/managing-security-implications-of-international-investment_65a8b23f.html.
54. UNCTAD, "National security related screening mechanisms for foreign investment," p. 9.
55. European Commission, "Report from the Commission to the European Parliament and the Council: Fifth annual report on the screening of foreign direct investments into the Union," 14 October 2025, https://ec.europa.eu/transparency/documents-register/detail?ref=COM(2025)632&lang=en.
56. CELIS, "Mapping FDI practices in advanced economies," 10 September 2024 at https://www.celis.institute/celis-blog/mapping-fdi-screening-practices-in-advanced-economies/.
57. UNCTAD, National security related screening mechanisms for foreign investment," p. 1. The authors caution that these figures only reflect cases and transactions for which information is publicly available.
58. This history is from Bennett and Bender, "How China acquires 'the crown jewels' of U.S. technology," *Politico*, 22 May 2018 at https://www.politico.com/story/2018/05/22/china-us-tech-companies-cfius-572413.
59. Author calculations from "Committee on foreign investment in the United States annual report to Congress," US Department of the Treasury, CFIUS Reports and Tables, at https://home.treasury.gov/policy-issues/international/the-committee-on-foreign-investment-in-the-united-states-cfius/cfius-reports-and-tables.

60. Bennett and Bender, "How China acquires 'the crown jewels' of U.S. technology," *Politico*, 22 May 2018 at https://www.politico.com/story/2018/05/22/china-us-tech-companies-cfius-572413.
61. Brown and Singh at https://admin.govexec.com/media/diux_chinatechnology-transferstudy_jan_2018_(1).pdf.
62. James Jackson, "CFIUS reform under FIRRMA," Congressional Research Service in Focus, 11 February 2020, p. 1.
63. Cathleen Cimino-Isaacs and Karen Sutter, "Committee on foreign investment in the United States," Congressional Research Service, 29 May 2025 at https://www.congress.gov/crs-product/IF10177.
64. US Department of Treasury, "CFIUS—annual report to Congress 2022," p. 39.
65. White House, "Executive order on ensuring robust consideration of evolving national security risks by the committee on foreign investment in the United States," 15 September 2022 at https://www.whitehouse.gov/briefing-room/presidential-actions/2022/09/15/executive-order-on-ensuring-robust-consideration-of-evolving-national-security-risks-by-the-committee-on-foreign-investment-in-the-united-states/.
66. "Regulation (EU) 2019/452 of the European Parliament and of the Council of 19 March 2019 establishing a framework for the screening of foreign direct investments into the Union," at https://eur-lex.europa.eu/eli/reg/2019/452/oj.
67. European Commission, "2020 strategic foresight report: Charting the course towards a more resilient Europe," 2020, p. 20, at https://ec.europa.eu/info/sites/info/files/strategic_foresight_report_2020_1.pdf.
68. Satake Minoru, "UK puts foreign investment under microscope to prevent tech leaks, *Nikkei Asia*, 2 May 2021, at https://asia.nikkei.com/Politics/UK-puts-foreign-investment-under-microscope-to-prevent-tech-leaks.
69. Government of Canada, "Guidelines on the National Security Review of Investments," 24 March 2021 at https://www.ic.gc.ca/eic/site/ica-lic.nsf/eng/lk81190.html.
70. "Significant changes to Australia's Foreign Investment Framework Commenced on 1 January 2021," Jones Day Commentaries, January 2021 at https://www.jonesday.com/en/insights/2021/01/significant-changes-to-australias-foreign-investment-framework-commenced-on-1-january-2021.
71. Japan Ministry of Finance, "Factors to be considered in authorities' screening of foreign direct investment," 8 May 2020, at https://www.mof.go.jp/english/policy/international_policy/fdi/gaitamehou_20200508.htm.
72. Kajimoto Tetsushi and Daniel Leussink, "Japan tightens rules on foreign stakes in 518 firms, citing national security," Reuters, 8 May 2020, at https://www.reuters.com/article/us-japan-investment-mof-idUSKBN22K0Z0 (accessed 14 June 2021).
73. Japan Ministry of Finance, "Factors to be considered in authorities' screening of foreign direct investment," 8 May 2020, at https://www.mof.go.jp/english/policy/international_policy/fdi/gaitamehou_20200508.htm.
74. Adam Chan, "CFIUS, team Telecom and China," *Lawfare*, 28 September 2021, https://www.lawfareblog.com/cfius-team-telecom-and-china.

75. Kyodo, "Japan and US plan to keep watch on Rakuten after Tencent investment," *The Japan Times*, 21 April 2021, https://www.japantimes.co.jp/news/2021/04/21/business/corporate-business/rakuten-tencent-investment/.
76. See http://www.people.cn/zixun/flfgk/item/dwjjf/falv/2/2-1-23.html (in Chinese).
77. Kong Tuan Yuen and Li Yao, "China's new national security screening rules on foreign investment," NUS-EAI Commentary, No. 24, 4 February 2021, pp. 2–4 at https://research.nus.edu.sg/eai/wp-content/uploads/sites/2/2021/02/EAIC-24-20210204-1.pdf.
78. Kong Tuan Yuen and Li Yao, "China's new national security screening rules on foreign investment," p. 5.
79. Lester Ross, Kenneth Zhou, and Tingting Zhou, "China's new foreign investment security review measures," WilmerHale, 22 December 2020, p. 3, at https://www.wilmerhale.com/en/insights/client-alerts/20201222-chinas-new-foreign-investment-security-review-measures.
80. Ibid., p. 3.
81. Alex Joske, "Picking flowers, making honey: The Chinese military's collaboration with foreign universities," The Australian Strategic Policy Institute, 2018, p. 9.
82. Centre for the Protection of National Infrastructure, "Trusted research guidance for academia," nd at https://www.cpni.gov.uk/trusted-research-guidance-academia#2 (accessed 11 June 2021).
83. Nidhi Subbaraman, "US universities call for clearer rules on science espionage amid China crackdown," *Nature*, 6 April 2021, at https://www.nature.com/articles/d41586-021-00901-7.
84. Eanna Kelly, "UK announces new unit to deal with risk of foreign espionage and theft of IP from universities," *Science Business*, 26 May 2021, at https://sciencebusiness.net/news/uk-announces-new-unit-deal-risk-foreign-espionage-and-theft-ip-universities.
85. Ibid.
86. Miki Reiko, "G7 plans to defend academic research with an eye to China," *Nikkei Asia*, 9 June 2021, at https://asia.nikkei.com/Business/Technology/G-7-plans-steps-to-defend-academic-research-with-eye-on-China.
87. Oikawa Akira, "Japan tightens rules on tech theft to safeguard research with US," *Nikkei Asia*, 28 April 2021, at https://asia.nikkei.com/Business/Technology/Japan-tightens-rules-on-tech-theft-to-safeguard-research-with-US.
88. Takenouchi Takehiro, "Guidelines drafted to screen researchers in sensitive fields," The Asahi Shimbun, 2 December 2025 at https://www.asahi.com/ajw/articles/16196327.
89. Government of Canada, "G7 Research Compact," 13 June 2021, at https://www.international.gc.ca/world-monde/international_relations-relations_internationales/g7/documents/2021006013-research_compact-pacte_recherche.aspx?lang=eng.
90. Eanna Kelly, "UK announces new unit to deal with risk of foreign espionage and

theft of IP from universities," *Science Business*, 26 May 2021, at https://science-business.net/news/uk-announces-new-unit-deal-risk-foreign-espionage-and-theft-ip-universities.

91. Evan Dyer, "Experts call on Canadian Universities to close off China's access to sensitive research," CBC, 15 September 2020 at https://www.cbc.ca/news/politics/china-canada-universities-research-waterloo-military-technology-1.5723846.
92. Saito Buntaro, "Tokyo police toughening espionage crackdowns amid intel threat from China, N. Korea," *Mainichi Shimbun*, 20 March 2021, at https://mainichi.jp/english/articles/20210320/p2a/00m/0na/018000c.
93. Bethan Staton, "'Dependency' of UK universities on China creates risks, report says," *Financial Times*, 4 March 2021, at https://www.ft.com/content/8cc66c7f-a80f-4d84–96dc-71c22f01b57f.
94. Christina Gallardo, "Commission seeks to block China from sensitive joint science projects," *Politico EU*, 30 March 2021, at https://www.politico.eu/article/commission-plans-to-limit-research-tie-ups-with-china/.
95. Yojana Sharma, "New data red tape could hamper international research," *University World News*, 20 July 2018 at https://www.universityworldnews.com/post.php?story=20180720072113906.
96. Ibid.
97. Elsa Kania, "Technological entanglement: Cooperation, competition and dual-use dilemma in artificial intelligence," Australia Strategic Policy Institute, Policy Brief, Report No. 7/2018, p. 6.
98. Ellen Nakashima and David Nakamura, "China Initiative aims to stop economic espionage. Is targeting academics over grant fraud 'overkill'?" *The Washington Post*, 15 September 2021, at https://www.washingtonpost.com/national-security/china-initiative-questions-dismissals/2021/09/15/530ef936-f482011eb-9738-8395ec2a44e7_story.html.
99. WilmerHale, "DOJ revives China Initiative tactics: investigating academic researchers by targeting universities under the False Claims Act," 13 November 2025 at https://www.wilmerhale.com/en/insights/client-alerts/20251113-doj-revives-china-initiative-tactics-investigating-academic-researchers-by-targeting-universities-under-the-false-claims-act.
100. Ibid., and Artie McConnell, "A refocused China Initiative: DOJs, new white collar enforcement strategy," 21 May 2025 at https://www.bakerlaw.com/insights/a-refocused-china-initiative-dojs-new-white-collar-enforcement-strategy/.
101. Xiaoli Jin, "The challenge and necessity of US-China technology cooperation," *Columbia/SIPA Journal of International Affairs*, 12 March 2020 at https://jia.sipa.columbia.edu/online-articles/challenge-and-necessity-us-china-technology-cooperation.
102. Working Group on Science and Technology in US–China Relations, "Task force report: Meeting the China challenge: A new American strategy for technology competition," UC San Diego, Twenty-First Century China Center and the Asia

Society Center for US-China Relations, 16 November 2020, p. 6, p. 17 at https://china.ucsd.edu/_files/meeting-the-china-challenge_2020_report.pdf.

103. Ellen Nakashima, "Justice Department shutters China Initiative, launches broader strategy to counter nation-state threats," *The Washington Post*, 23 February 2022, https://www.washingtonpost.com/national-security/2022/02/23/china-initiative-redo/. There was skepticism about the Biden administration's commitment to fixing all those problems, even if the name was changed. See Margaret Lewis, "After the China Initiative," *Lawfare*, 23 March 2022, at https://www.lawfareblog.com/after-china-initiative.

104. Justin Jacobs and Harry Dempsey, "Global supply chains face months of disruption from Texas storm," *Financial Times*, 24 March 2021, at https://www.ft.com/content/f15841b6-47c2-4fcc-98f6-15945a7fea84.

105. Harry Dempsey and Stefania Palma, "Global ports grapple with backlog from Suez canal blockage," *Financial Times*, 26 April 2021, at https://www.ft.com/content/da6c709d-29c3-4f8b-b37d-a3480c6938e7.

106. Cheng Ting-Fang and Lauly Li, "Taiwan chip supplier halts production due to COVID cluster," *Nikkei Asia*, 4 June 2021, at https://asia.nikkei.com/Business/Tech/Semiconductors/Taiwan-chip-supplier-halts-production-due-to-COVID-cluster.

107. Harry Dempsey, "Unvaccinated sailors risk deepening global supply chain crisis," *Financial Times*, 22 March 2021, at https://www.ft.com/content/72feaabd-5f94-40ad-9073-0178db969b23.

108. Associated Press, "Automotive chip maker says it's nearly recovered from blaze," *Washington Post*, 2 June 2021, at https://www.washingtonpost.com/business/automotive-chip-maker-says-its-nearly-recovered-from-blaze/2021/06/01/74b47cc8-c2f0–11eb-89a4-b7ae22aa193e_story.html.

109. Reuters, "EU to limit export control of virus protection gear to just masks," 15 April 2020, at https://www.reuters.com/article/us-health-coronavirus-eu-exports-idUSKCN21W2TX.

110. Richard Lough and Andreas Rinke, "'Act of modern piracy: U.S. slammed amid global scramble over face masks," Reuters, 3 April 2020, at https://globalnews.ca/news/6775423/coroanvirus-global-face-mask-competition/.

111. Japan Ministry of Economy, Trade and Industry, "ASEAN-Japan Economic Resilience Action Plan," 22 April 2020 at https://www.meti.go.jp/press/2020/07/20200729005/20200729005–1.pdf.

112. Hoyama Taisei, "US enlists Quad to strengthen critical supply chains," *Nikkei Asia*, 9 June 2021 at https://asia.nikkei.com/Politics/International-relations/US-China-tensions/US-enlists-Quad-to-strengthen-critical-supply-chains.

113. Japan Ministry of Economy, Trade and Industry, "Australia-India-Japan Economic Ministers' Joint Statement on Supply Chain Resilience," 1 September 2020, at https://www.meti.go.jp/press/2020/09/20200901008/20200901008–1.pdf.

114. Japan Ministry of Economy, Trade and Industry, "Joint Statement of the Supply Chain Resilience Initiative by Australian, Indian and Japanese Trade Ministers,"

27 April 2021 at https://www.meti.go.jp/press/2021/04/20210427004/20210427004-1.pdf.

115. White House, "Fact sheet: Quad Summit," 12 March 2021, at https://www.whitehouse.gov/briefing-room/statements-releases/2021/03/12/fact-sheet-quad-summit/.
116. White House, "Executive Order on America's supply chains, 24 February 2021, at https://www.whitehouse.gov/briefing-room/presidential-actions/2021/02/24/executive-order-on-americas-supply-chains/.
117. White House, "Fact sheet: Biden-Harris administration announces supply chain disruptions task force to address short-term supply chain discontinuities," 8 June 2021, at https://www.whitehouse.gov/briefing-room/statements-releases/2021/06/08/fact-sheet-biden-harris-administration-announces-supply-chain-disruptions-task-force-to-address-short-term-supply-chain-discontinuities/.
118. European Commission, "Updating the 2020 New Industrial Strategy building a stronger single market for Europe's recovery," 5 May 2021, p. 11.
119. Ibid., p. 13.
120. Ibid., p. 5.
121. Semiconductor Industry Association, "Written comments to the Bureau of Industry and Security, in the Matter of Executive Order 14017 of 24 February 2021 America's Supply Chains," 5 April 2021, p. 3, at https://www.semiconductors.org/wp-content/uploads/2021/04/4.5.21-SIA-supply-chain-submission.pdf.
122. Semiconductor Industry Association, "Strengthening the global semiconductor supply chain in an uncertain era," 1 April 2021, p. 5.
123. The White House, "Building resilient supply chains, revitalizing American manufacturing, and fostering broad-based growth: 100 days reviews under executive order 14017," June 2021, p. 17.
124. US Department of Commerce, "US Department of Commerce announces upcoming entry into force of IPEF supply chain agreement," 31 January 2024, at https://www.commerce.gov/news/press-releases/2024/01/us-department-commerce-announces-upcoming-entry-force-ipef-supply-chain.
125. Takeuchi Yasuo, "EU aims to be independent chip power with 20% global share," *Nikkei Asia*, 10 March 2021, at https://asia.nikkei.com/Business/Tech/Semiconductors/EU-aims-to-be-independent-chip-power-with-20-global-share.
126. Takeuchi Yasuo, "EU aims to be independent chip power with 20% global share," and Joyce Lee, "S. Korea seeks tax cuts for U.S. investment by firms such as Samsung," Reuters, 21 May 2021 at https://www.reuters.com/technology/samsung-elec-invest-17-bln-new-chip-foundry-us-blue-house-2021–05–21/.
127. Kyodo, "TSMC to build 2nd chip plant near 1st factory in Japan's Kumamoto," *Nikkei Asia*, 29 January 2024, at https://asia.nikkei.com/Business/Tech/Semiconductors/TSMC-to-build-2nd-chip-plant-near-1st-factory-in-Japan-s-Kumamoto#:~:text=TSMC%20to%20build%202nd%20chip%20plant%20near%201st%20factory%20in%20Japan%27s%20Kumamoto,-Formal%20announcement%20likely&text=KUMAMOTO%2C%20Japan%20(Kyodo)%20%2D%2D,prefecture%2C%20a%20lawmaker%20said%20Sunday.

128. Jeffrey Wilson, "Strategies for securing critical materials value chains," Perth USAsia Centre, April 2020, p. 14.
129. Ibid., pp. 15–16.
130. Jane Nakano, "The geopolitics of critical minerals supply chains," CSIS, March 2021, p. 10, and Takeuchi Yasuo, "EU aims to be independent chip power with 20% global share," and Hoyama Taisei, "US enlists Quad to strengthen critical supply chains," *Nikkei Asia*, 9 June 2021, at https://asia.nikkei.com/Politics/International-relations/US-China-tensions/US-enlists-Quad-to-strengthen-critical-supply-chains.
131. Ibid., p. 13.
132. European Commission, "Critical raw materials resilience: Charting a path towards greater security and sustainability," 3 September 2020, p. 8.
133. Jane Nakano, "The geopolitics of critical minerals supply chains," CSIS, March 2021, p. 10, and Takeuchi Yasuo, "EU aims to be independent chip power with 20% global share," and Hoyama Taisei, "US enlists Quad to strengthen critical supply chains," *Nikkei Asia*, 9 June 2021, at https://asia.nikkei.com/Politics/International-relations/US-China-tensions/US-enlists-Quad-to-strengthen-critical-supply-chains, p. 21.
134. White House, "Carbis Bay G7 Summit Communique," 13 June 2021, at https://www.whitehouse.gov/briefing-room/statements-releases/2021/06/13/carbis-bay-g7-summit-communique/.
135. "Science, supply chains and financial sector set as priorities for China's economy," Merics China Briefing, 14 January 2021, at https://merics.org/en/briefing/science-supply-chains-and-financial-sector-set-priorities-chinas-economy.
136. Jane Nakano, "The geopolitics of critical minerals supply chains," CSIS, March 2021, p. 10, and Takeuchi Yasuo, "EU aims to be independent chip power with 20% global share," and Hoyama Taisei, "US enlists Quad to strengthen critical supply chains," *Nikkei Asia*, 9 June 2021, at https://asia.nikkei.com/Politics/International-relations/US-China-tensions/US-enlists-Quad-to-strengthen-critical-supply-chains, pp. 5–6.
137. Liu Zhihua and Liu Yukun, "China to step up protection of rare earth resources," *China Daily*, 16 January 2021, at https://global.chinadaily.com.cn/a/202101/16/WS60023d3aa31024ad0baa3039.html.
138. Jing Zhang et al., "China imposes new export controls on two minerals critical to the manufacture of semiconductors," Mayer/Brown, 27 July 2023, at https://www.mayerbrown.com/en/insights/publications/2023/07/china-imposes-new-export-controls-on-two-minerals-critical-to-the-manufacture-of-semiconductors.
139. Yu Jing and Li Ruikang "Massive chip shortage exposes vulnerability in the global supply chain," *CGTV*, 2 April 2021, at https://news.cgtn.com/news/2021–04–02/Massive-chip-shortage-exposes-vulnerability-in-the-global-supply-chain-Z8oAJRhU0U/index.html.
140. Ibid.
141. Takeuchi Yasuo, "EU aims to be independent chip power with 20% global share."

142. Laura Rosenberger, "Making cyberspace safe for democracy," *Foreign Affairs*, May/June 2020, p. 150.
143. Thomas Lum, "Internet development and information control in the People's Republic of China," Congressional Research Service, CRS Report for Congress, 10 February 2006, p. 1.
144. Information Office of the State Council of the People's Republic of China, "The Internet in China White Paper," 8 June 2010 at http://www.china.org.cn/government/whitepaper/2010–06/08/content_20207978.htm.
145. Simon Denyer, "China's scary lesson to the world: Censoring the internet works," *The Washington Post*, 23 May 2016, at https://www.washingtonpost.com/world/asia_pacific/chinas-scary-lesson-to-the-world-censoring-the-internet-works/2016/05/23/413afe78-fff3-11e5-8bb1-f124a43f84dc_story.html.
146. OpenNet Initiative, "Internet filtering in China in 2004–2005: A Country Study," April 2005, p. 3, at http://www.opennetinitiative.net/studies/china/.
147. Laura Rosenberger, "Making cyberspace safe for democracy," *Foreign Affairs*, May/June 2020, p. 148.
148. European Commission, "Action plan against disinformation," June 2019, https://ec.europa.eu/info/sites/default/files/factsheet_disinfo_elex_140619_final.pdf.
149. Anna Polyakova and Daniel Fried, "Democratic defense against disinformation 2.0," The Atlantic Council, June 2019, p. 4.
150. Ibid., p. 4.
151. Nina Jankowicz, Statement on "United States Efforts to Counter Russian Disinformation and Malign Influence," House Committee on Appropriations on Stated Foreign Operations, and Related Programs, 10 July 2019, p. 332, at https://docs.house.gov/meetings/AP/AP04/20190710/109748/HHRG-116-AP04-Wstate-JankowiczN-20190710.pdf.
152. Alina Polyakova, Statement on "United States Efforts to Counter Russian Disinformation and Malign Influence," House Committee on Appropriations on Stated Foreign Operations, and Related Programs, 10 July 2019, p. 323, at https://www.govinfo.gov/content/pkg/CHRG-116hhrg39682/pdf/CHRG-116hhrg39682.pdf.
153. Laura Rosenberger, "Making cyberspace safe for democracy," *Foreign Affairs*, May/June 2020, p. 154.
154. The White House, US National Security Strategy Interim Guidance, March 2021, p. 7, 9, 19.
155. Theresa Hitchens, "New strategy aims to up DoD, IC Game to counter disinformation," *Breaking Defense*, 16 March 2021, at https://breakingdefense.com/2021/03/new-strategy-aims-to-up-dod-ic-game-to-counter-disinformation/.
156. Editorial Board, "The Disinformation Governance Board's collapse showcases the problem," *The Washington Post*, 23 May 2022, at https://www.washingtonpost.com/opinions/2022/05/23/disinformation-governance-boards-collapse-showcases-problem/.
157. Julian Barnes, "France steps up fight against disinformation as U.S. pulls back, offi-

cial says," *The New York Times*, 17 November 2025, at https://www.nytimes.com/2025/11/17/us/politics/france-disinformation.html.

158. Katie Rogers, "Trump dismisses warning of Russian 2020 meddling as a Democratic 'hoax,'" *The New York Times*, 21 February 2020, at https://www.nytimes.com/2020/02/21/us/politics/trump-and-russia.html.
159. "NATO's approach to countering disinformation: A focus on COVID-19," 17 July 2020, at https://www.nato.int/cps/en/natohq/177273.htm.
160. "Charlevoix commitment on defending democracy from foreign threats," 2 January 2019, https://international.gc.ca/world-monde/international_relations-relations_internationales/g7/documents/2018–06–09-defending_democracy-defense_democratie. aspx?lang=eng.
161. Noah Barkin, "Exclusive: Five Eyes intelligence alliance builds coalition to counter China," Reuters, 12 October 2018, https://www.reuters.com/article/us-china-fiveeyes/exclusive-five-eyes-intelligence-alliance-builds-coalition-to-counter-china-idUSKCN1MM0GH.
162. Julian Barnes, "France steps up fight against disinformation as U.S. pulls back, official says," *The New York Times*, 17 November 2025, at https://www.nytimes.com/2025/11/17/us/politics/france-disinformation.html.

5. GOVERNMENTS RESPOND: PLAYING OFFENSE

1. Michael Sulmeyer, "How the U.S. can play cyber-offense: Deterrence isn't enough," *Foreign Affairs*, 22 March 2018, at https://www.foreignaffairs.com/articles/world/2018-03-22/how-us-can-play-cyber-offense.
2. Danny Vinik, "America's Secret Arsenal," *Politico*, 9 December 2015, at https://www.politico.com/agenda/story/2015/12/defense-department-cyber-offense-strategy-000331/.
3. Nicole Perlroth explains how "going on offense" in cyber is dangerous and capable of causing all sorts of unanticipated (?) consequences in *This Is How They Tell Me the World Ends: The Cyberweapons Arms Race*, Bloomsbury Publishing, 2021. Prepare for sleepless nights after reading it.
4. Herb Lin and Amy Zegart, "Introduction to the special issue on strategic dimensions of offensive cyber operations," *Journal of Cybersecurity*, 2 March 2017, p. 1.
5. Zachary Goldman and Damon McCoy, "Deterring financially motivated cybercrime," *Journal of National Security Law and Policy*, 8, 2016, p. 597.
6. In some cases this is easy to distinguish: cybercrime, even if state supported, is not an act of war. But things get murky when infrastructure is attacked: is extortion an act of war?
7. Jason Healey, "The Implications of persistent (and permanent) engagement in cyberspace," *Journal of Cybersecurity*, 5(1), 2019, p. 2.
8. Cited in Jacquelyn Schneider, "Persistent engagement: Foundation, evolution and evaluation of a strategy," *Lawfare*, 10 May 2019, at https://www.lawfareblog.com/persistent-engagement-foundation-evolution-and-evaluation-strategy.

9. Josh Gold, "The Five Eyes and offensive cyber capabilities: Building a cyber deterrence initiative," NATO Cooperative Cyber Defence Center of Excellence, Tallinn, 2020, p. 10.
10. The White House of President Barack Obama, "Fact sheet: President Xi Jinping's State Visit to the United States," 25 September 2015, at https://obamawhitehouse.archives.gov/the-press-office/2015/09/25/fact-sheet-president-xi-jinpings-state-visit-united-states.
11. Adam Segal, "The U.S.-China cyber espionage deal one year later," Council on Foreign Relations blog, 28 September 2016, at https://www.cfr.org/blog/us-china-cyber-espionage-deal-one-year-later and Perlroth, *This Is How They Tell Me the World Ends*, p. 352.
12. David Sanger, "White house used hotline to warn Russia over hacking before election," *The New York Times*, 17 November 2016.
13. Reuters staff, "US accuses China of violating bilateral anti-hacking deal," Reuters, 9 November 2018, at https://www.reuters.com/article/us-usa-china-cyber-idUSKCN1NE02E. Chinese officials rejected the allegations, saying they lacked "factual basis."
14. The White House, "Fact sheet: Imposing costs for harmful foreign activities by the Russian Government," 15 April 2021, at https://www.whitehouse.gov/briefing-room/statements-releases/2021/04/15/fact-sheet-imposing-costs-for-harmful-foreign-activities-by-the-russian-government/.
15. Raphael Satter and Mike Corder, Report: Dutch spies caught Russian hackers on tape," *Associated Press*, 27 January 2018, at https://apnews.com/article/hacking-elections-international-news-security-services-technology-ef3b036949174a9b98d785129a93428b.
16. Ellen Nakashima, "Feds recover more than $2 million in ransomware payments from Colonial Pipeline hackers," *The Washington Post*, 8 June 2021, at https://www.washingtonpost.com/business/2021/06/07/colonial-pipeline-ransomware-payment-recovered/.
17. While this may seem far-fetched, the scale of the Kaseya ransomware attack may change thinking about this policy. See Matt Tait, "The Kaseya ransomware attack is a really big deal," *Lawfare*, 5 July 2021 at https://www.lawfareblog.com/kaseya-ransomware-attack-really-big-deal for details of the attack, and Peter Pascucci and Kurt Sanger, "Revisiting a framework for military takedowns of cybercriminals," *Lawfare*, 2 July 2021, at https://www.lawfareblog.com/revisiting-framework-military-takedowns-against-cybercriminals.
18. National Cyber Strategy of the United States of America, pp. 20–21.
19. James R. Clapper, Marcel Lettre, and Michael S. Rogers, "Foreign cyber threats to the United States," 5 January 2017, https://www.armed-services.senate.gov/imo/media/doc/Clapper-Lettre-Rogers_01-05-16.pdf?fbclid=IwAR3zLlXbhy-2ggxkXk4FEmJD4JkEfvWWK_-RDf3ONtesa80k9ypTEeTV90g.
20. Christopher Ford, "Responding to modern cyber threats with deterrence and diplomacy," US Department of State, 19 October 2020, at https://2017–2021.state.

gov/responding-to-modern-cyber-threats-with-diplomacy-and-deterrence/index.html.

21. UK Parliament Intelligence and Security Committee (ISC), *Annual Report 2016–2017*, December 2017, p. 35, http://isc.independent.gov.uk/les/2016–2017_ISC_AR.pdf.
22. Australian Government, "Australia's cyber security strategy," 2016, https://cybersecuritystrategy.homeaffairs.gov.au/AssetLibrary/dist/assets/images/PMC-Cyber-Strategy.pdf, p. 28.
23. National Defence (Canada), "Strong, secure, engaged: Canada's defence policy," 2017, https://www.canada.ca/content/dam/dnd-mdn/documents/reports/2018/strong-secure-engaged/canada-defence-policy-report.pdf, p. 41.
24. Josh Gold, "The Five Eyes and offensive cyber capabilities: Building a cyber deterrence initiative," NATO Cooperative Cyber Defence Center of Excellence, Tallinn, 2020, p. 20.
25. NATO, Statement by the North Atlantic Council concerning Malicious Cyber Activities, 3 June 2020, https://www.nato.int/cps/en/natohq/o cial_texts_176136.htm.
26. Alessandro Marrone and Ester Sabatino, "Cyber defense in NATO Countries: Comparing models," Instituto d'Affari Internazionale, IAI Papers 21/05 February 2021, pp. 26–7.
27. International Institute of Strategic Studies, "Cyber capabilities and national power: a net assessment," London, June 2021, p. 7.
28. Paris Call for Trust and Security in Cyberspace, at https://parispeaceforum.org/initiatives/paris-call-for-trust-and-security-in-cyberspace/. As of 1 April 2021, there were 1,209 signatories.
29. US Cyber Command, "Command Vision for US Cyber Command,"14 June 2018, https://www.cybercom.mil/Portals/56/Documents/USCYBERCOM%20Vision%20April%202018.pdf?ver=2018-06-14-152556-010, 10.
30. Thomas Franck, "Trump doubles down: 'Trade wars are good, and easy,'" *CNBC*, 2 March 2018 at https://www.cnbc.com/2018/03/02/trump-trade-wars-are-good-and-easy-to-win.html.
31. (19 U.S.C. §1862) "Section 232 Investigations: Overview and Issues for Congress," Congressional Research Service, 18 May 2021, p. 1, at https://fas.org/sgp/crs/misc/R45249.pdf.
32. This history is from "Section 232 Investigations," pp. 1–4.
33. Jing Zhang et al., "Biden administration concludes section 232 investigation into imports of NDFEB magnets but does not impose tariffs," Mayer-Brown, 26 October 2022 at https://www.mayerbrown.com/fr/insights/publications/2022/10/biden-administration-concludes-section-232-investigation-into-imports-of-ndfeb-magnets-but-does-not-impose-tariffs.
34. William Burkhart and Keigh Hammond, "Presidential 2025 tariff actions: Timeline and Status," Library of Congress, 16 September 2025 at https://www.congress.gov/crs-product/R48549.

35. Andres Schwarzenberg, "Section 301 of the Trade Act of 1974: Origin, evolution and use," Congressional Research Service, 14 December 2020, p. i-5, at https://fas.org/sgp/crs/misc/R46604.pdf.
36. William Burkhart and Keigh Hammond, "Presidential 2025 tariff actions: Timeline and status," Library of Congress, 16 September 2025 at https://www.congress.gov/crs-product/R48549.
37. Office of the USTR, "Findings of the investigation into China's acts, policies, and practices related to technology transfer, intellectual property, and innovation under Section 301 of the Trade Act of 1974," 22 March 2018.
38. Andres Schwarzenberg, "Section 301 of the Trade Act of 1974," p. 29.
39. White House, "Fact sheet: Executive order addressing the threat from securities investments that finance certain companies of the People's Republic of China," 3 June 2021 at https://www.whitehouse.gov/briefing-room/statements-releases/2021/06/03/fact-sheet-executive-order-addressing-the-threat-from-securities-investments-that-finance-certain-companies-of-the-peoples-republic-of-china/.
40. Jeanne Whalen and Ellen Nakashima, "Biden expands Trump order by banning U.S. investment in Chinese companies linked to the military or surveillance technology, *The Washington Post*, 4 June 2021 at https://www.washingtonpost.com/technology/2021/06/03/investment-ban-chinese-surveillance-tech/.
41. Xavier Espinoza and James Politi, "US warns EU against anti-American tech policy," *Financial Times*, 15 June 2021 at https://www.ft.com/content/2036d7e9-daa2-445d-8f88-6fcee745a259.
42. Department of the Treasury, "Treasury issues regulations to implement executive order addressing U.S. investments in certain national security technologies and products in countries of concern," 28 October 2024 at https://home.treasury.gov/news/press-releases/jy2687.
43. UNCTAD, "Outward FDI policies: Promotion and facilitation—regulation and screening," Investment Policy Monitor," February 2024 at https://unctad.org/system/files/official-document/diaepcbinf2024d1_en.pdf.
44. James Killick et al., "Two years of the EU's foreign subsidies regulation—what dealmakers need to know," White & Case, 1 July 2025, https://www.whitecase.com/insight-alert/two-years-eu-foreign-subsidies-regulation-what-dealmakers-need-know.
45. Robert Atkinson and Hilal Aka, "Defending American tech in global markets," Information Technology & Innovation Foundation," 1 December 2025, p. 3.
46. While the meaning of infrastructure is the subject of intense debate in the United States, in most discussions of economic development the term refers to "hard" items like roads, railways, ports, airports, and energy and communications networks. Those last items open the door to a wide variety of services, many of them digital.
47. OECD, "The Belt and Road Initiative in the global trade, investment and finance landscape," in *OECD Business and Finance Outlook 2018*, OECD Publishing, p. 5.
48. Ibid., p. 6.
49. Jacob Lew and Gary Roughead, China's Belt and Road: Implications for the United

States, Council on Foreign Relations Independent Task Force Report No. 79 (CFR Task Force), March 2021, p. 12.

50. Michele Ruta et al., "Belt and Road economics: Opportunities and risks of transport corridors," World Bank Group, Washington, DC, 2019, p. xiii.
51. G20 Ministerial Statement on Trade and Digital Economy, 9 June 2019, p. 5, at https://www.mofa.go.jp/files/000486596.pdf.
52. Heather Johnson, "About 2.5 billion lack internet connectivity: How connectivity can unlock their potential," World Economic Forum, 25 September 2024, at https://www.weforum.org/stories/2024/09/2–5-billion-people-lack-internet-access-how-connectivity-can-unlock-their-potential/.
53. International Telecommunications Union, "Connecting humanity: Assessing investment needs of connecting humanity to the internet by 2030," Geneva, 2020, p. iii.
54. Ibid., p. 4.
55. There is speculation the name was changed to deflect suspicion that BRI is a China-centered model. In Chinese, the name remains the same.
56. Michele Ruta et al., "Belt and Road economics: Opportunities and risks of transport corridors," World Bank Group, Washington, DC, 2019, p. 13.
57. Ouyang Shijia, "China's top economic planner; BRI delivers tangible gains, boosts global growth," China Daily, 24 June 2025 at https://www.chinadailyhk.com/hk/article/614636.
58. Deutsche Bank, "China's Belt and Road initiative: A guide to market participation," November 2019, p. 7.
59. Oxford Business Group, "Has COVID-19 prompted the Belt and Road to go green?" 28 January 2021 at https://oxfordbusinessgroup.com/news/has-covid-19-prompted-belt-and-road-initiative-go-green.
60. Christoph Nedopil Wang, "China Belt and Road Initiative investment report 2025 H1, 17 July 2025 at https://greenfdc.org/china-belt-and-road-initiative-bri-investment-report-2025-h1/.
61. Deutsche Bank, "China's Belt and Road Initiative," p. 6.
62. Brenda Goh and Ryan Woo, "China President Xi says goal of Belt and Road is advance 'win-win cooperation,'" Reuters, 26 April 2019, https://www.reuters.com/article/us-china-silkroad-xi/china-president-xi-says-goal-of-belt-and-road-is-advance-win-win-cooperation-idUSKCN1S205Z.
63. Jacob Lew and Gary Roughead, China's Belt and Road: Implications for the United States, Council on Foreign Relations Independent Task Force Report No. 79, (CFR Task Force), March 2021, pp. 62–5.
64. "Speech by Executive Vice Foreign Minister Zhang Yesui at the luncheon of the Sixth World Peace Forum," 24 June 2017 at http://www.china-un.ch/eng/zywjyjh/t1479002.htm.
65. Daniel Russel and Blake Berger, "Weaponizing the Belt and Road Initiative," Asia Society Policy Institute, September 2020, pp. 10–12.
66. Kerry Brown, "The Belt and Road: Security dimensions," *The Asia Europe Journal*, 16(3), April 2018, p. 5.

67. "习近平在周边外交工作座谈会上发表重要讲话 [Xi Jinping's important speech at the Peripheral Diplomacy Work Conference]," 新华网 [*Xinhua*], 25 October 2013, http://news.xinhuanet.com/politics/2013-10/25/c_117878897.htm, cited in Daniel Russel and Blake Berger, "Weaponizing the Belt and Road Initiative," Asia Society Policy Institute, September 2020, p. 14.
68. Kerry Brown, "The Belt and Road: Security dimensions," *The Asia Europe Journal*, 16(3), April 2018, p. 3.
69. While both sides benefit from BRI projects, China gains a great deal. In addition to the benefits below, BRI absorbs China's excess manufacturing capacity, puts its excess savings to use, secures cheap inputs, and helps reorient global commerce away from the US and Western Europe toward China. See the CFR Task Force report, pp. 9–10.
70. CFR Task Force, p. 21.
71. Michele Ruta et al., "Belt and Road economics: Opportunities and risks of transport corridors," World Bank Group, Washington, DC, 2019, pp. 4–5.
72. Jacob Lew and Gary Roughead, China's Belt and Road: Implications for the United States, Council on Foreign Relations Independent Task Force Report No. 79 (CFR Task Force), March 2021, p. 34.
73. Ibid., p. 50.
74. Ibid., p. 22.
75. Ibid., p. 23.
76. Sarina Patterson, "The BRI at 10: a report card from the global South," 26 March 2024 at https://www.aiddata.org/blog/the-bri-at-10-a-report-card-from-the-global-south.
77. Robert Greene and Paul Triolo, "Will China control the global internet through its Digital Silk Road?" Carnegie Endowment for International Peace, 8 May 2020 at https://carnegieendowment.org/2020/05/08/will-china-control-global-internet-via-its-digital-silk-road-pub-81857.
78. Ibid.
79. Daniel Russel and Blake Berger, "Weaponizing the Belt and Road Initiative," Asia Society Policy Institute, September 2020, p 21.
80. OECD, "The Belt and Road Initiative in the global trade, investment and finance landscape," in *OECD Business and Finance Outlook 2018*, OECD Publishing, Paris, p. 27.
81. Jacob Lew and Gary Roughead, China's Belt and Road: Implications for the United States, Council on Foreign Relations Independent Task Force Report No. 79 (CFR Task Force), March 2021, p. 17.
82. Tim Ruhlig, "China, Europe and the new power competition over technical standards," Swedish Institute of International Affairs, UI Brief, 1/2021, p. 8.
83. Yingqiu Kuang, "China in global technology governance: Experimentation, achievements and uncertainties," in Alessia Almighini, ed., *China: Champion of (which) Globalization?* Italian Institute for International Political Studies, Milano, 2018, p. 88.

84. Jacob Lew and Gary Roughead, China's Belt and Road: Implications for the United States, Council on Foreign Relations Independent Task Force Report No. 79 (CFR Task Force), March 2021, p. 17.
85. Akane Okutsu, Cliff Venzon, and CK Tan, "China's Belt and Road power grids keep security critics awake," *Financial Times*, 10 March 2020, http://ft.com/content/f2b6e395-f1ee-4a6d-8839-009a35555b52.
86. Phillip Cornell, "Energy governance and China's bid for global grid integration," *Energy Source* (blog), Atlantic Council, 30 May 2019, http://atlanticcouncil.org/blogs/energysource/energy-governance-and-china-s-bid-for-global-grid-integration, and Jacob Lew and Gary Roughead, China's Belt and Road: Implications for the United States, Council on Foreign Relations Independent Task Force Report No. 79 (CFR Task Force), March 2021, p. 67.
87. Ibid., and Jacob Lew and Gary Roughead, China's Belt and Road: Implications for the United States, Council on Foreign Relations Independent Task Force Report No. 79 (CFR Task Force), March 2021, p. 69.
88. Jacob Lew and Gary Roughead, China's Belt and Road: Implications for the United States, Council on Foreign Relations Independent Task Force Report No. 79 (CFR Task Force), March 2021, pp. 42–4.
89. Karen Sutter, Andres Scwarzenberg and Michael Sutherland, "China's 'One Belt, One Road' initiative: Economic issues," Congressional Research Service, In Focus, 22 January 2021, p. 2.
90. Anna Gelpern et al., "How China lends: A rare look into 100 debt contracts with foreign governments," Center for Global Development, March 2021.
91. Lee Jones and Shahar Hameiri, "Debunking the myth of debt-trap diplomacy," Chatham House, August 2020.
92. Daniel Russel and Blake Berger, "Weaponizing the Belt and Road Initiative," Asia Society Policy Institute, September 2020, p. 22.
93. Ibid., p. 36.
94. Jacob Lew and Gary Roughead, China's Belt and Road: Implications for the United States, Council on Foreign Relations Independent Task Force Report No. 79 (CFR Task Force), March 2021, p. 4.
95. Ibid., p. 85.
96. Steve Herman, "US seen falling short countering China's rising political clout," *VOA*, 23 March 2021 at https://www.voanews.com/usa/us-politics/us-seen-falling-short-countering-chinas-rising-geopolitical-clout.
97. Tobias Harris, "'Quality' Infrastructure: Japan's robust challenge to China's Belt and Road," War on the Rocks, 9 April 2019 at https://warontherocks.com/2019/04/quality-infrastructure-japans-robust-challenge-to-chinas-belt-and-road/.
98. Asia Africa Growth Corridor, at https://aagc.ris.org.in/about-aagc.
99. Alex Fang, "Biden pushes Belt and Road rival 'Blue Dot' with Japan and Australia," *Nikkei Asia*, 9 June 2021, at https://asia.nikkei.com/Spotlight/Belt-and-Road/Biden-pushes-Belt-and-Road-rival-Blue-Dot-with-Japan-and-Australia.

100. Michael Peel and Sam Fleming, "West and allies relaunch push for own version of China's Belt and Road," *Financial Times*, 2 May 2021 at https://www.ft.com/content/2c1bce54-aa76-455b-9b1e-c48ad519bf27.
101. OECD, "OECD and global executives engage with the Blue Dot Network to develop certification for quality infrastructure investment," 7 June 2021, at https://www.oecd.org/newsroom/oecd-and-global-executives-engage-with-the-blue-dot-network-to-develop-certification-for-quality-infrastructure-investment.htm.
102. Matthew Goodman, Daniel Runde, and Jonathan Hillman, "Connecting the Blue Dots," CSIS, 26 February 2020, at https://www.csis.org/analysis/connecting-blue-dots.
103. Sam Fleming, "EU to develop investment initiative to counter Chinese influence," *Financial Times*, 5 July 2021 at https://www.ft.com/content/d8e1747d-5340–47ce-b5e2-dd6002a74bb7.
104. Ibid.
105. Stuart Lau, "China direct from Politico," 2 December 2021, at https://www.politico.eu/newsletter/china-direct/beijings-envoy-tours-europe-belt-vs-gateway-mi6-warning/.
106. White House, "Memorandum on the Partnership for Global Infrastructure and Investment," 26 June 2022 at https://www.whitehouse.gov/briefing-room/presidential-actions/2022/06/26/memorandum-on-the-partnership-for-global-infrastructure-and-investment/?utm_source=substack&utm_medium=email.
107. Jonathan Holslag, "The EU's 'global gateway'—an answer to China or a 'dead end'?" *The EU Observer*, 1 December 2021, https://euobserver.com/opinion/153678.
108. Michael Peel and Sam Fleming, "West and allies relaunch push for own version of China's Belt and Road," *Financial Times*, 2 May 2021, at https://www.ft.com/content/2c1bce54-aa76-455b-9b1e-c48ad519bf27.
109. Alex Fang, "Biden pushes Belt and Road Rival 'Blue Dot' with Japan and Australia," *Nikkei Asia*, 9 June 2021 at https://asia.nikkei.com/Spotlight/Belt-and-Road/Biden-pushes-Belt-and-Road-rival-Blue-Dot-with-Japan-and-Australia.
110. Michael Peel and Sam Fleming, "West and allies relaunch push for own version of China's Belt and Road," *Financial Times*, 2 May 2021 at https://www.ft.com/content/2c1bce54-aa76-455b-9b1e-c48ad519bf27.
111. Marianne Schneider-Petsinger, Jue Wang, Yu Jie, and James Crabtree, The US–China strategic competition: The quest for global technological leadership," Chatham House, November 2019, p. 29 at https://www.chathamhouse.org/sites/default/files/publications/research/CHHJ7480-US-China-Competition-RP-WEB.pdf.
112. Ray Bowen, "A 'China model?' Beijing's Promotion of Alternative Global Norms and Standards Panel III: Technological Competition and Driving New Standards Beijing's Promotion of PRC Technical Standards," Testimony for The United States-China Economic and Security Review Commission, 13 March 2020, p. 5,

at https://www.uscc.gov/sites/default/files/testimonies/March%2013%20Hearing_Panel%203_Ray%20Bowen%20Pointe%20Bello.pdf.

113. US–China Business Council, "China in international standards setting," February 2020, p. 2.
114. Cited in Tim Ruhlig, "China, Europe and the new power competition over technical standards," Swedish Institute of International Affairs, UI Brief, 1/2021, p. 5.
115. Deputy Director of the Policy Research Office of the Zhejiang Provincial Party Committee Guo Zhanheng, "习近平标准化思想与浙江实践 [Xi Jinping's standardization thought and Zhejiang practice]," *Zhejiang Daily* (China), 25 September 2015, in Nathan Picarsic and Emily de la Bruyere, "International Telecommunication Union," Foundation for Defense of Democracies, 30 June 2021 at https://www.fdd.org/analysis/2021/06/30/international-telecommunication-union/#easy-footnote-3-119135.
116. Anu Bradford, "When it comes to markets, Europe is no fading power," *Foreign Affairs*, 3 February 2020, at https://www.foreignaffairs.com/articles/europe/2020-02-03/when-it-comes-markets-europe-no-fading-power.
117. Anu Bradford, "The Brussels effect," 107 *Northwestern University Law Review*, 1, 2012.
118. Ibid., p. 3.
119. Tim Ruhlig, "China, Europe and the new power competition over technical standards," Swedish Institute of International Affairs, UI Brief, 1/2021, p. 4.
120. Julia Voo, "Shaping global technology governance: Why the U.S. must adopt a proactive approach to technical standards for long term security," Working Paper for the Penn Project on the Future of U.S.-China Relations, 2020, p. 9, at https://cpb-us-w2.wpmucdn.com/web.sas.upenn.edu/dist/b/732/files/2020/10/Julia-Voo_Shaping-Global-Technology-Governance_Final.pdf.
121. Marianne Schneider-Petsinger, Jue Wang, Yu Jie, and James Crabtree, "The US–China Strategic Competition: The quest for global technological leadership," Chatham House, November 2019, p. 30 at https://www.chathamhouse.org/sites/default/files/publications/research/CHHJ7480-US-China-Competition-RP-WEB.pdf.
122. Tim Ruhlig, "China, Europe and the new power competition over technical standards," Swedish Institute of International Affairs, UI Brief, 1/2021, p. 6.
123. T. Loya and J. Boli, "Standardization in the world polity. Technical rationality over power," in Boli, J., and Loya, T., (eds.) *Constructing World Culture*, Stanford University Press, 1999, p. 188.
124. Filippo Blancato, "Regulate to dominate: The geopolitics of standard-setting in digital technologies and its strategic implications for the EU," UNU-CRIS Policy Briefs. United Nations University Institute on Comparative Regional Integration Studies, 2019, p. 3.
125. Xi Jinping "Accelerating from factor driven, investment scale driven development to innovation driven development [加快从 要素驱动、投资规模驱动发展为主向以创新驱动发展为主的转变]," 9 June 2014, selected works of important documents since

the Eighteenth National Congress, Central Literature Publishing House 2016 Year edition, pp. 22–3.

126. Doshi et al., "China as a 'cyber great power,'" p. 19, citing 杨震 [Yang Zhen], "物联网:引领新一轮信息技术革命" [Internet of Things: Leading a new round of information technology revolution], 江苏通信 [*Jiangsu Communications*] 3, 2010, 12113, CNKI: F49;F426.6, at https://www.brookings.edu/wp-content/uploads/2021/04/FP_20210405_china_cyber_power.pdf.
127. National medium- and long-term program for scientific and technological development (2006–2020), State Council of the People's Republic of China, 2006, p. 48
128. Doshi et al., "China as a "cyber great power," p. 16, citing 习近平 [Xi Jinping], "中共中央政治局就实施网络强国战略进行第三十六次集体 学习" [The Political Bureau of the CPC Central Committee Conducts the 36th Collective Study on the Implementation of the Strategy of Network Power], 新华 [Xinhua], 9 October 2016, http://www.gov.cn/xinwen/2016–10/09/content_5116444.htm, at https://www.brookings.edu/wp-content/uploads/2021/04/FP_20210405_china_cyber_power.pdf.
129. Ibid., p. 16.
130. Science and Technology Division of the Ministry of Industry and Information Technology (MIIT [工业和信息化部]) "MIIT Document (2018) 231 Ministry of Industry and Information Technology Opinion on the Implementation of Standardization of the Industrial Communication Industry for Construction of the 'Belt and Road' [工信部科〔2018〕231 号 工业和信息化部关于工业通信业标准化工作服务于"一带一路"建设的实施意见]," published 12 November 2018, at http://www.miit.gov.cn/n1146295/n1652858/n1652930/n3757016/c6480388/content.html, in Ray Bowen, "A 'China Model?' Beijing's Promotion of Alternative Global Norms and Standards Panel III: Technological Competition and Driving New Standards Beijing's Promotion of PRC Technical Standards," Testimony for The United States-China Economic and Security Review Commission (USCC), 13 March 2020, p. 6, at https://www.uscc.gov/sites/default/files/testimonies/March%2013%20Hearing_Panel%203_Ray%20Bowen%20Pointe%20Bello.pdf.
131. Ray Bowen, "A 'China model?' Beijing's promotion of alternative global norms and standards panel III: Technological Competition and Driving New Standards Beijing's Promotion of PRC Technical Standards," Testimony for The United States-China Economic and Security Review Commission (USCC), 13 March 2020, p. 13, at https://www.uscc.gov/sites/default/files/testimonies/March%2013%20Hearing_Panel%203_Ray%20Bowen%20Pointe%20Bello.pdf.
132. Julia Voo and Roger Creemers, "China's role in digital standards for emerging technologies—Impacts on The Netherlands and Europe," Leiden Asia Center, May 2021, p. 9.
133. US–China Business Council, "China in international standards setting," February 2020, pp. 2–3.

134. Tim Ruhlig, "China, Europe and the new power competition over technical standards," Swedish Institute of International Affairs, UI Brief, 1/2021, p. 8.
135. Ray Bowen, "A 'China model?' Beijing's promotion of alternative global norms and standards panel III: Technological Competition and Driving New Standards Beijing's Promotion of PRC Technical Standards," Testimony for The United States-China Economic and Security Review Commission (USCC), 13 March 2020, p. 3.
136. Tim Ruhlig, "China, Europe and the new power competition over technical standards," Swedish Institute of International Affairs, UI Brief, 1/2021, p. 11.
137. See, for example, COM (2020) 102 final, A New Industrial Strategy for Europe; SWD (2020)98 final, Identifying Europe's recovery needs, p. 15, https://ec.europa.eu/info/sites/info/files/strategic_foresight_report_2020_1.pdf and the other reports referenced in this discussion.
138. Alan Beattie, "Technology: How the US, EU and China compete to set industry standards," *Financial Times*, 24 July 2019, at https://www.ft.com/content/0c91b884-92bb-11e9-aea1–2b1d33ac3271.
139. European Commission, "Communication from the Commission to the European Parliament, the European Council, the Council, the European Economic and Social Committee and the Committee of the Regions, an EU strategy on standardisation," 2 February 2022.
140. European Commission, "Communication from the Commission to the European Parliament, the European Council, the Council, the European Economic and Social Committee and the Committee of the Regions, Updating the 2020 New Industrial Strategy: Building a stronger single market for Europe's recovery," COM (2021) 350 final, p. 15.
141. John Chen et al., "China's Internet of Things," Research Report Prepared on Behalf of the US–China Economic and Security Review Commission, October 2018, pp. 1–3.
142. Ibid., p. 3.
143. The White House, "United States Government national standards strategy for critical and emerging technology," May 2023, p. 3, https://bidenwhitehouse.archives.gov/wp-content/uploads/2023/05/US-Gov-National-Standards-Strategy-2023.pdf.
144. Rush Doshi et al., 'China as a "cyber great power': Beijing's two voices in telecommunications," Brookings Institution, April 2021, p. 16, citing Xi Jinping, The Full Text of Xi Jinping's Speech at the Forum on Cybersecurity and Informatization Work, Beijing, 25 April 2016, http://www.xinhuanet.com//politics/2016–04/25/c_1118731175.htm, at https://www.brookings.edu/wp-content/uploads/2021/04/FP20210405_china_cyber_power.pdf.
145. Ibid., p. 18.
146. Eurasia Group White Paper, "The geopolitics of 5G," 15 November 2018, p. 4, and Defense Innovation Board, "The 5G ecosystem: Risks and opportunities for DoD," April 2019, p. 7.

147. 郭超 [Guo Chao], 于川信 [Yu Chuanxin], and 王景芳 [Wang Jingfang], "对第五代移动通信技术军事应用的几点认识" [Some Understandings on the military application of the fifth-generation mobile communication technology] in Doshi et al., "China as a "cyber great power: Beijing's two voices in telecommunications," Brookings Institution, April 2021, p. 17.
148. Defense Innovation Board, "The 5G ecosystem: Risks and opportunities for DoD," April 2019, p. 2 and Arjun Kharpal, "Huawei 2024 revenue surges to near-record high as China smartphone comeback takes hold," 31 March 2025, https://www.cnbc.com/2025/03/31/huawei-2024-revenue-surges-to-near-record-high-on-smartphone-comeback.html.
149. Kadri Kaska, Henrik Beckvard, and Tomas Minarik, "Huawei, 5G and China as a security threat," NATO Cooperative Cyber Defence Center of Excellence, 2019, Talinn, p. 7.
150. Defense Innovation Board, "The 5G ecosystem: Risks and opportunities for DoD," April 2019, p. 13.
151. "Huawei named a leader in the Gartner Magic Quadrant for 5G core network infrastructure solutions," Huawei.com, 13 October 2025 at https://www.huawei.com/en/news/2025/10/5gc-mqleader.
152. US–China Business Council, "China in International Standards Setting," February 2020, p. 3.
153. Voo and Creemers, "China's role in digital standards for emerging technologies—Impacts on The Netherlands and Europe," Leiden Asia Center, May 2021, p. 8.
154. Tim Ruhlig, "China, Europe and the new power competition over technical standards," Swedish Institute of International Affairs, UI Brief, 1/2021, p. 9 and LexisNexis, "Unpacking the 5G revolution," IP Blog, 23 January 2025 at https://www.lexisnexisip.com/resources/5g-revolution/.
155. Alan Beattie, "Technology: How the US, EU and China compete to set industry standards," *Financial Times*, 24 July 2019, at https://www.ft.com/content/0c91b884–92bb-11e9-aea1–2b1d33ac3271.
156. Chinese government yields discounts of 25 per cent on 5G technology and services and loss-leading financing terms. John Watts, "A framework for an open, trusted, and resilient 5G global telecommunications network," The Atlantic Council, March 2020, p. 1.
157. James Lewis, "How 5G will shape innovation and security: A primer," CSIS, December 2018, p. 2.
158. Readiness to back the US position has varied, and Washington has resorted to pressure and threats to get partners to toe the line. Jeanne Whalen et al., "Huawei supply ban roils stocks as US tech companies begin to cut off China tech giant," *The Washington Post*, 21 May 2019 at https://www.washingtonpost.com/technology/2019/05/20/google-cuts-off-huawei-after-trump-administration-crack-down/.
159. Jeanne Whalen, "A remote corner of Idaho has become the best hope for the US challenge to Huawei," *The Washington Post*, 30 June 2020 at https://www.washingtonpost.com/business/2020/06/29/huawei-alternative-oran-idaho/.

160. The White House, "Fact sheet: US-Japan competitiveness and resilience (CoRe) partnership," 16 April 2021 at https://www.whitehouse.gov/briefing-room/statements-releases/2021/04/16/fact-sheet-u-s-japan-competitiveness-and-resilience-core-partnership/.
161. Clean Network, "Clean Time Network," nd, at https://techdiplomacy.org/tech-statecraft/timeline/.
162. Gideon Rachman, "US-China rivalry drives the retreat of market economics," *Financial Times*, 10 May 2021, at https://www.ft.com/content/1e749857-3cd6-453d-8cee-2c501cbfd53b.
163. World Trade Organization "World trade report 2020: Government policies to promote innovation in the digital age," Geneva, 2020, endnote 1, p. 21.
164. Stephen Ezell, Frank Spring and Katarzyna Bitka, "The global flourishing of national innovation foundations," Information Technology and Innovation Foundation, April 2015, at https://www2.itif.org/2015-flourishing-national-innovation.pdf.
165. Simon Evenett et al., "The return of industrial policy in data," International Monetary Fund, WP/24/1, January 2024, p. 7.
166. Simon Evanett et al., "Industrial policy since the great financial crisis," International Monetary Fund, WP/25/222, October 2025, p. 17.
167. David Bailey et al., "Industrial policy: New technologies and transformative innovation policies," *Cambridge Journal of Regions, Economy and Society*, 12, 2019, 169–70, and Martin Sandbu, "What is industrial policy for?" *Financial Times*, 8 July 2021, at https://www.ft.com/content/8a355fbb-2f09-48eb-b9b7-e6780228083a.
168. World Trade Organization "World trade report 2020: Government policies to promote innovation in the digital age," Geneva, 2020, p. 16.
169. David Bailey et al., "Industrial policy: New technologies and transformative innovation policies," *Cambridge Journal of Regions, Economy and Society*, 12, 2019, p. 171.
170. World Trade Organization "World trade report 2020: Government policies to promote innovation in the digital age," Geneva, 2020, p. 17.
171. Ibid.
172. Mariana Mazzucato and George Dibb, "Innovation policy and industrial strategy for post-covid economic recovery," Institute for Innovation and Public Purpose, Issue Brief 10, September 2020, p. 2.
173. Congressional Budget Office, "Federal policies and innovation," November 2014, p. 13.
174. Melissa Flagg, Autumn Toney, and Paul Harris, "Research security, collaboration, and the changing map of global R&D," Center for Security and Emerging Technology, June 2021, p. 2.
175. Jimmy Goodrich, "China's surge in basic research threatens U.S. leadership," LinkedIn, July 2025, at https://www.linkedin.com/posts/jimmygoodrich_scienceleadership-nationalsecurity-uschinacompetition-activity-7349153282629455874-_9Mc/?rcm=ACoAAACvAlUBENsDLSosV4eSXpA4mbZViWPmj54&utm_source=substack&utm_medium=email.

176. Council on Foreign Relations, "Innovation and national security: Keeping our edge," Task Force Report No. 77, 2019, p. vi.
177. Dylan Gerstel and Matthew Goodman, "From industrial policy to innovation strategy: Lessons from Japan, Europe and the United States," CSIS, Washington, DC, September 2020, p. 18.
178. James Politi and Aime Williams, "Top Biden aide calls for US to embrace 'industrial strategy,'" *Financial Times*, 23 June 2021, at https://www.ft.com/content/b59ef9df-c78b-4afa-922e-676547d68c26.
179. Gideon Rachman, "US-China rivalry drives the retreat of market economics," *Financial Times*, 10 May 2021, at https://www.ft.com/content/1e749857–3cd6–453d-8cee-2c501cbfd53b.
180. Dylan Gerstel and Matthew Goodman, "From industrial policy to innovation strategy: Lessons from Japan, Europe and the United States," CSIS, Washington, DC, September 2020, p. 18.
181. Council on Foreign Relations, "Innovation and national security: Keeping our edge," Task Force Report No. 77, 2019, pp. 16–17.
182. Gideon Rachman, "US–China rivalry drives the retreat of market economics," *Financial Times*, 10 May 2021, at https://www.ft.com/content/1e749857–3cd6–453d-8cee-2c501cbfd53b.
183. James Politi and Aime Williams, "Top Biden aide calls for US to embrace 'industrial strategy,'" *Financial Times*, 23 June 2021, at https://www.ft.com/content/b59ef9df-c78b-4afa-922e-676547d68c26.
184. US Congress, "Strategic Competition Act of 2021," pp. 6–7, at https://www.foreign.senate.gov/imo/media/doc/DAV21598%20-%20Strategic%20Competition%20Act%20of%202021.pdf, and Gideon Rachman, "US–China rivalry drives the retreat of market economics," *Financial Times*, 10 May 2021, at https://www.ft.com/content/1e749857-3cd6-453d-8cee-2c501cbfd53b.
185. David Sanger et al., "Senate poised to pass huge industrial policy bill to counter China," *The New York Times*, 7 June 2021 at https://www.nytimes.com/2021/06/07/us/politics/senate-china-semiconductors.html.
186. Demetri Sevastopulo, "US Senate passes bill aimed at countering China's tech sector," *Financial Times*, https://www.ft.com/content/2429f43f-f1bf-4131-aefa-0d20a64b9d17.
187. Kathryn Rattigan, "The endless frontier act: Shifting the focus from defense to offense," *National Law Review*, 15 April 2021 at https://www.natlawreview.com/article/endless-frontier-act-shifting-focus-defense-to-offense.
188. Tripp Mickle and Ana Swanson, "Trump's call to scrap 'horrible' chip program spreads panic," *The New York Times*, 11 March 2025, at https://www.nytimes.com/2025/03/10/technology/trump-chips-act.html.
189. A very quick review is at Andrea Renda, "The EU industrial strategy: Towards a post-growth agenda?" *Intereconomics*, 56, 2021, p. 133 at https://doi.org/10.1007/s10272–021–0968–7.
190. Kevin Komer, "Industrial policy in times of COVID-19 and its aftermath," EU Monitor, Deutsche Bank Research, 31 August 2020, p. 16.

191. Ibid., p. 3.
192. Janka Oertel, "The new China consensus: How Europe is growing wary of Beijing," European Council on Foreign Relations, Policy Brief, September 2020, pp. 2–4.
193. Kevin Komer, "Industrial policy in times of COVID-19 and its aftermath," EU Monitor, Deutsche Bank Research, 31 August 2020, p. 3.
194. Janka Oertel, "The new China consensus: How Europe is growing wary of Beijing," European Council on Foreign Relations, Policy Brief, September 2020, pp. 2–4.
195. Michael Peel, "Europe gets squeezed between great powers," *Financial Times*, 17 February 2020, at https://www.ft.com/content/dafa0c7c-5132–11ea-90ad-25e377c0ee1f.
196. Brigitte Dekker and Anna-Lena Rhiem, "The rise of geoeconomics and the need for a resilient European semiconductor industry," Mercator Institute, 23 February 2021, at https://merics.org/en/short-analysis/rise-geoeconomics-and-need-resilient-european-semiconductor-industry.
197. European Commission, "A new industrial strategy for Europe," COM(2020)102 final, 10 March 2020 at https://eur-lex.europa.eu/legal-content/EN/TXT/?uri=CELEX:52020DC0102.
198. Ibid.
199. Kevin Komer, "Industrial policy in times of COVID-19 and its aftermath," EU Monitor, Deutsche Bank Research, 31 August 2020, p. 11.
200. Andrea Renda, "The EU industrial strategy: Towards a post-growth agenda?" *Intereconomics*, 56, 2021, p. 134, at https://doi.org/10.1007/s10272-021-0968-7.
201. European Commission, "2020 strategic foresight report: Charting the course towards a more resilient Europe," Brussels, p. 22, https://ec.europa.eu/info/sites/info/files/strategic_foresight_report_2020_1.pdf.
202. European Commission, "State aid: Commission adopts new rules to support important projects of common European interest," 13 June 2014, at https://ec.europa.eu/commission/presscorner/detail/en/IP_14_673.
203. European Commission, "Important projects of common European interest (IPCEI), nd, https://ec.europa.eu/competition-policy/state-aid/legislation/modernisation/ipcei_en.
204. Brigitte Dekker and Anna-Lena Rhiem, "The rise of geoeconomics and the need for a resilient European semiconductor industry," Mercator Institute, 23 February 2021, at https://merics.org/en/short-analysis/rise-geoeconomics-and-need-resilient-european-semiconductor-industry.
205. The future of European competitiveness: Part A | A competitiveness strategy for Europe," September 2024, p. 5.
206. European Commission, "The Draghi Report: One year on," 16 September 2025 at https://commission.europa.eu/topics/competitiveness/draghi-report/one-year-after_en.
207. Mario Draghi, "High-level conference—One year after the Draghi report: what

has been achieved, what has changed," European Commission, 16 September 2025, p. 1.

208. Dylan Gerstel and Matthew Goodman, "From industrial policy to innovation strategy: Lessons from Japan, Europe and the United States," CSIS, Washington, DC, September 2020, p. 5.
209. Ibid., p. 5.
210. Ibid., p. 8.
211. Brad Glosserman, *Peak Japan: The End of Great Ambitions*, Georgetown University Press, 2019.
212. Japan Cabinet Office, "Society 5.0," nd, at https://www8.cao.go.jp/cstp/english/society5_0/index.html.
213. Subcommittee on Security Export Control Policy, Trade Committee, Industrial Structure Council, Interim Report (Overview), 8 October 2019, https://www.meti.go.jp/english/policy/external_economy/trade_control/pdf/191008a.pdf.
214. Cabinet Office of Japan, "Integrated innovation strategy," 2019, p. 3 at https://www8.cao.go.jp/cstp/english/doc/integrated_main.pdf.
215. Cabinet Office of Japan, "Integrated innovation strategy," 2019, at https://www8.cao.go.jp/cstp/english/doc/integrated_main.pdf and "Integrated innovation strategy 2020," 17 July 2020, footnotes 12–13, at https://www8.cao.go.jp/cstp/english/strategy_2020.pdf.
216. Subcommittee on Security Export Control Policy, Trade Committee, Industrial Structure Council, Interim Report (overview), 8 October 2019, fn. 5 at https://www.meti.go.jp/english/policy/external_economy/trade_control/pdf/191008a.pdf.
217. Kyodo, "Japan to strengthen domestic chip production under growth strategy," *The Mainichi*, 2 June 2021 at https://mainichi.jp/english/articles/20210602/p2g/00m/0bu/082000c and Tsuji Takashi and Takeuchi Kosuke, "Chip supply chain a top priority in Suga's first economic plan," *Nikkei Asia*, 19 June 2021 at https://asia.nikkei.com/Business/Tech/Semiconductors/Chip-supply-chain-a-top-priority-in-Suga-s-first-economic-plan.
218. Kyodo, "Japan to strengthen domestic chip production under growth strategy," *The Mainichi*, 2 June 2021 at https://mainichi.jp/english/articles/20210602/p2g/00m/0bu/082000c.
219. Marianna Mazzucato, Rainer Kattel and Josh Ryan-Collins, "Challenge-driven innovation policy: Towards a new policy toolkit," *Journal of Industry, Competition and Trade*, 20, 2020, pp. 421–37.
220. Japan Science and Technology Agency, "Moonshot research and development program," nd, at https://www.jst.go.jp/moonshot/en/about.html (accessed 16 July 2021).
221. Cabinet Office of Japan, "Integrated innovation strategy 2020," 17 July 2020, p. 16, at https://www8.cao.go.jp/cstp/english/strategy_2020.pdf.
222. Makiko Yamazaki and Kentaro Sugiyama, "Japan PM Takaichi launches economic HG, gears up public investments," Reuters, 3 November 2025 at https://www.

reuters.com/world/asia-pacific/japan-pm-takaichi-creates-new-economic-growth-panel-gears-up-proactive-2025–11–04/.

223. Jimmy Goodrich, "Beijing calls for 'extraordinary measures' to boost tech self-sufficiency," IGCC, 20 November 2025, at https://ucigcc.org/blog/beijing-calls-for-extraordinary-measures-to-boost-tech-self-sufficiency/.
224. Adam Segal, "Seizing core technologies: China responds to US technology competition," China Leadership Monitor, 1 June 2019, p. 3 https://3c8314d6–0996-4a21-9f8a-a63a59b09269.filesusr.com/ugd/10535f_4b07cc3458014c47a03b3f23cae5fde1.pdf.
225. Henrique Choer Moraes and Mikaek Wigell, "The Emergence of Strategic Capitalism: Geoeconomics, Corporate Statecraft and the Repurposing of the Global Economy," Finnish Institute of International Affairs, FIIA Working Paper, 117, September 2020, p. 4.

6. THE NEW NORMAL FOR BUSINESS

1. "World Economic Forum Annual Meeting 2022 rescheduled to 22–26 May," World Economic Forum, accessed 28 May 2022, at https://www.weforum.org/agenda/2022/01/davos-2022-world-economic-forum-annual-meeting-may/.
2. Gideon Rachman, "The Ukraine War casts its shadow over Davos," *Financial Times*, 24 May 2022 at https://www.ft.com/content/4db632df-ff13–4c22–8808-f7ddb047aec9?emailId=628f480d1f6d600023306af6&segmentId=c393f5a6-b640-bff3-cc14-234d058790ed.
3. Gideon Rachman, *Zero-Sum World: Politics, Power and Prosperity After the Crash*, Atlantic Books, 2010.
4. Kaye Wiggins, Antoine Gara, and Jamie Smyth, "Business leaders warn that three-decade era of globalisation is ending," *Financial Times*, 22 May 2022 at https://www.ft.com/content/0599878e-a820–4657–8e52-f069bb10d512.
5. Chad Bray, "HSBC boss: Complex geopolitical landscape 'fact of life' for global banks," *South China Morning Post*, 11 October 2021 at https://www.scmp.com/business/banking-finance/article/3151963/hsbc-boss-noel-quinn-complex-geopolitical-landscape-fact.
6. "Huawei CFO Meng returns to China after striking U.S. Deal," *Nikkei Asia*, 25 September 2021 at https://asia.nikkei.com/Spotlight/Huawei-crackdown/Huawei-CFO-Meng-returns-to-China-after-striking-U.S.-deal.
7. Sophie Heading, "New world: New rules or no rules," *KPMG*, accessed 27 May 2022 at https://home.kpmg/xx/en/home/insights/2018/03/new-world-new-rules-or-no-rules.html.
8. Ibid.
9. Francis Fukuyama, *The End of History and the Last Man*, Free Press, 1992.
10. Francis Fukuyama, "The End of History?" *The National Interest*, 16, 1989, pp. 3–18, https://doi.org/http://www.jstor.org/stable/24027184.
11. Thomas L. Friedman, *The Lexus and the Olive Tree: Understanding Globalization*, Farrar, Straus and Giroux, 1999.

12. Thomas L. Friedman, *The World Is Flat: A Brief History of the Twenty-First Century*, Farrar, Straus and Giroux, 2005.
13. Joseph S. Nye and John D. Donahue, "Governance in a Globalizing World—a World Interconnected by 'Thick Globalism,'" *HBS Working Knowledge*, accessed 27 May 2022 at https://hbswk.hbs.edu/archive/governance-in-a-globalizing-world-a-world-interconnected-by-thick-globalism.
14. Ibid.
15. Bill Clinton, full text of Clinton's speech on China Trade Bill, 8 March 2000, Paul H. Nitze School of Advanced International Studies of the Johns Hopkins University, accessed 30 May 2022 at https://archive.nytimes.com/www.nytimes.com/library/world/asia/030900clinton-china-text.html.
16. Shaun Rein's *The End of Cheap China, Revised and Updated: Economic and Cultural Trends that Will Disrupt the World*, Wiley, 2014 and *The End of Copycat China: The Rise of Creativity, Innovation, and Individualism in Asia*, Wiley, 2014 provide insight on this transition.
17. Robert S. Kaplan and Anette Mikes, "Managing risks: A new framework," *Harvard Business Review*, June 2012 at https://hbr.org/2012/06/managing-risks-a-new-framework.
18. Robert S. Kaplan, Herman B. Leonard, and Anette Mikes, "The risks you can't foresee," *Harvard Business Review*, November–December 2020 at https://hbr.org/2020/11/the-risks-you-cant-foresee.
19. Ibid.
20. Eurasia Group at https://www.eurasiagroup.net/our-story.
21. Rohitesh Dhawan and Sean West, "The CEO as Chief Geopolitical Officer," *KPMG*, accessed 27 May 2022 at https://assets.kpmg/content/dam/kpmg/xx/pdf/2018/03/the-ceo-as-chief-geopolitical-officer.pdf.
22. Ibid.
23. John Chipman, "Your company needs a foreign policy," *Harvard Business Review*, 8 August 2016 at https://hbr.org/2016/09/why-your-company-needs-a-foreign-policy.
24. World Economic Forum White Paper, "Our Shared Digital Future: Responsible Digital Transformation-Board Briefing," 6 February 2019, at https://www.weforum.org/whitepapers/our-shared-digital-future-responsible-digital-transformation-board-briefing-9ddf729993/.
25. Aon and Ponemon Institute, "2019 intangible assets financial statement impact comparison report," April 2019 at https://www.aon.com/unitedkingdom/insights/2019-emea-ponemon-intangible-assets-report.jsp.
26. Somini Sengupta and Evelyn M. Rusli, "Personal data's value? Facebook is set to find out," *The New York Times*, 1 February 2012 at https://www.nytimes.com/2012/02/01/technology/riding-personal-data-facebook-is-going-public.html.
27. Nicholas Confessore, "Cambridge Analytica and Facebook: The scandal and the fallout so far," *The New York Times*, 4 April 2018, at https://www.nytimes.com/2018/04/04/us/politics/cambridge-analytica-scandal-fallout.html.

28. Carole Cadwalladr, "The great British Brexit robbery; how our democracy was hijacked," *The Guardian*, 7 May 2017, at https://www.theguardian.com/technology/2017/may/07/the-great-british-brexit-robbery-hijacked-democracy.
29. Sha Hua, "China orders ride-hailing firm Didi's app removed from app stores," *The Wall Street Journal*, 5 July 2021 at https://www.wsj.com/articles/chinas-internet-regulator-finds-serious-problems-with-didi-china-units-app-11625402263.
30. Lizhi Liu, "The rise of data politics: Digital China and the world," *Studies in Comparative International Development*, 56, 2021, 45–67.
31. Hugh Son, "Jamie Dimon says risk of cyberattacks 'may be biggest threat to the US financial system,'" *CNBC*, 4 April 2019 at https://www.cnbc.com/2019/04/04/jp-morgan-ceo-jamie-dimon-warns-cyber-attacks-biggest-threat-to-us.html.
32. Matt Egan, "Exclusive: 'Cyber is the most dangerous weapon in the world,' JPMorgan Council Warns," *CNN*, 16 December 2021 at https://edition.cnn.com/2021/12/16/business/cyber-security-hacking-jpmorgan/index.html.
33. "Over 60% of the Fortune 1000 had at least one public breach over the last decade," *Help Net Security*, 18 March 2020 at https://www.helpnetsecurity.com/2020/03/18/fortune-1000-breach.
34. Michel Cukier, "Study: Hackers attack every 39 seconds," *A. James Clark School of Engineering, University of Maryland*, 2007 at https://eng.umd.edu/news/story/study-hackers-attack-every-39-seconds.
35. "Gartner survey finds 88% of boards of directors view cybersecurity as a business risk," *Gartner*, 18 November 2021 at https://www.gartner.com/en/newsroom/press-releases/2021-11-18-gartner-survey-finds-88-percent-of-boards-of-directors-view-cybersecurity-as-a-business-risk.
36. Tara Siegel Bernard et al., "Equifax says cyberattack may have affected 143 million in the U.S.," *The New York Times*, 7 September 2017 at https://www.nytimes.com/2017/09/07/business/equifax-cyberattack.html.
37. John McCrank and Aparajita Saxena, "Equifax clears executives who sold shares after hack," Reuters, 3 November 2017 at https://www.reuters.com/article/us-equifax-cyber-idUSKBN1D31EK.
38. "Equifax Data Breach Settlement," *Federal Trade Commission*, 18 March 2022 at https://www.ftc.gov/enforcement/refunds/equifax-data-breach-settlement.
39. "Chinese military personnel charged with computer fraud, economic espionage and wire fraud for hacking into credit reporting agency Equifax," *The United States Department of Justice*, 13 February 2020 at https://www.justice.gov/opa/pr/chinese-military-personnel-charged-computer-fraud-economic-espionage-and-wire-fraud-hacking.
40. TRM Team, "North Korea and the industrialization of cryptocurrency theft," TRM Labs, 18 December 2025 at https://www.trmlabs.com/resources/blog/north-korea-and-the-industrialization-of-cryptocurrency-theft.
41. Hyonhee Shin, "Mystery hacker steals data on 1,000 North Korean defectors in South," Reuters, 28 December 2018 at https://www.reuters.com/article/us-northkorea-southkorea-idUSKCN1OR0CS.

42. Seulkee Jang, "North Korea recently hacked Pfizer to steal vaccine development-related secrets," *Daily NK*, 24 February 2021 at https://www.dailynk.com/english/north-korea-recently-hacked-pfizer-steal-vaccine-development-secrets/.
43. "Cash-strapped North Korea believed to be stepping up cyber thefts against the South," *The Straits Times*, 2 February 2021 at https://www.straitstimes.com/asia/east-asia/cash-strapped-north-korea-believed-to-be-stepping-up-cyberthefts-against-the-south.
44. Andrei Hagiu, *MIT Sloan Management Review*, "Strategic decisions for multisided platforms," 19 December 2013 at https://sloanreview.mit.edu/article/strategic-decisions-for-multisided-platforms/.
45. Andrew Browne, "How the coronavirus is accelerating deglobalization," *Bloomberg*, 29 February 2020 at https://www.bloomberg.com/news/newsletters/2020-02-29/why-deglobalization-is-accelerating-bloomberg-new-economy.
46. Larry Fink, "Larry Fink's annual 2022 letter to CEOS," *BlackRock*, 18 January 2022 at https://www.blackrock.com/corporate/investor-relations/larry-fink-ceo-letter.
47. Joe Nocera, "Why American mask makers are going out of business," *The New York Times*, 5 March 2022 at https://www.nytimes.com/2022/03/05/business/dealbook/american-mask-makers.html.
48. Ibid.
49. Ibid.
50. Susan Lund et al., "Globalization in transition: The future of trade and value chains," McKinsey & Company, 2 April 2021 at https://www.mckinsey.com/featured-insights/innovation-and-growth/globalization-in-transition-the-future-of-trade-and-value-chains.
51. Christos Cabolis, "Strategic trade-offs: Beyond the old playbook," *Hinrich Foundation*, 14 October 2021 at https://www.hinrichfoundation.com/research/article/sustainable/beyond-the-old-playbook?utm_campaign=article-cabolis-sti-2025-strategic-trade-offs&utm_medium=email&_hsenc=p2ANqtz-8avLHUVCxpiluWvHZ8i_49cUXKf5tKTCva2RFv9CyBD8LDiQpUwina8W31Kvx5--q5o8PMYV0ORmwUYX7MifbTBMFxxA&_hsmi=385018663&utm_content=20251014-weekly-research-&utm_source=hinrich-thought-leadership.
52. Kevin Roose, "Can Ford turn itself into a tech company?" *The New York Times*, 9 November 2017 at https://www.nytimes.com/interactive/2017/11/09/magazine/tech-design-autonomous-future-cars-detroit-ford.html.
53. Dirk A. Zetzsche, Ross P. Buckley, Douglas W. Arner & Janos N. Barberis, "From fintech to techfin: The regulatory challenges of data-driven finance," *NYU Journal of Law and Business*, 14(2), Spring 2018, pp. 393–446
54. Hannah Murphy and Kiran Stacey, "Facebook Libra: The inside story of how the company's cryptocurrency dream died," *Financial Times*, 10 March 2022, at https://www.ft.com/content/a88fb591–72d5–4b6b-bb5d-223adfb893f3.
55. Stephanie Bodoni, "Amazon gets record $888 million EU fine over data violations," *Bloomberg*, 30 July 2021 at https://www.bloomberg.com/news/articles/2021-07-30/amazon-given-record-888-million-eu-fine-for-data-privacy-breach.

56. European Commission, "Commission fines X €120 million under the Digital Services Act," 5 December 2025, at https://digital-strategy.ec.europa.eu/en/news/commission-fines-x-eu120-million-under-digital-services-act.
57. Chang Che and Jeremy Goldkorn, "China's 'big tech crackdown': A guide," *SupChina*, 3 December 2021 at https://supchina.com/2021/08/02/chinas-big-tech-crackdown-a-guide/.
58. "China introduces anti-foreign sanctions law," *Clifford Chance*, July 2021 at https://www.cliffordchance.com/briefings/2021/07/china-introduces-anti-foreign-sanctions-law.html.
59. Ji Li, "Superpower rivalry and the global compliance dilemma," 45 *University of Pennsylvania Journal of International Law*, 891, 2024.
60. Guogang Wu, Asia Society Policy Institute Report, *China's Common Prosperity Program: Causes, Challenges, and Implications*, 30 May 2022 at https://asiasociety.org/policy-institute/chinas-common-prosperity-program-causes-challenges-and-implications.
61. Kaye Wiggins, Antoine Gara, and Jamie Smyth, "Business leaders warn that three-decade era of globalisation is ending," *Financial Times*, 22 May 2022 at https://www.ft.com/content/0599878e-a820–4657–8e52-f069bb10d512.
62. Angela Zhang, *Chinese Antitrust Exceptionalism: How the rise of China challenges global regulation*. Oxford University Press, 2021.
63. "Case M.7850—EDF / CGN / NNB GROUP OF COMPANIES," *European Commission*, 3 October 2016 at https://ec.europa.eu/competition/mergers/cases/decisions/m7850_429_3.pdf.

7. FRAMING THE BUSINESS RESPONSE

1. "Lotte Mart to exit the Chinese market," *China Daily*, 18 October 2018 at https://www.chinadaily.com.cn/a/201810/18/WS5bc7f9bfa310eff303283238.html.
2. Dal Yong Jin, "The Korean wave," *Georgetown Journal of International Affairs*, 26 May 2020 at https://gjia.georgetown.edu/2020/05/26/the-korean-wave/.
3. Hae-rin Lee, "China allows group tours to Korea after more than 6 years of restrictions," *The Korea Times*, 10 August 2023 at https://www.koreatimes.co.kr/lifestyle/travel-food/20230810/china-allows-group-tours-to-korea-after-more-than-6-years-of-restrictions.
4. Bo-eun Kim, "South Korea looks to break China import dependence and establish 'supply chain alliances,'" *South China Morning Post*, 20 May 2022 at https://www.scmp.com/economy/global-economy/article/3178370/south-korea-looks-break-china-import-dependence.
5. Sang-hun Choe, "South Korea and the U.S. agree to deploy missile defense system," *The New York Times*, 7 July 2016 at https://www.nytimes.com/2016/07/08/world/asia/south-korea-and-us-agree-to-deploy-missile-defense-system.html.
6. Ethan Meick and Nargiza Salidjanova, U.S.-China Economic and Security Review Commission, "China's response to U.S.-South Korean missile defense system

deployment and its implication," 26 July 2017 at https://www.uscc.gov/sites/default/files/Research/Report_China%27s%20Response%20to%20THAAD%20Deployment%20and%20its%20Implications.pdf.

7. Erika Na, "As South Korean dramas and films ride wave of international success, how long will China remain off-limits?" *South China Morning Post*, 27 January 2022 at https://www.scmp.com/economy/china-economy/article/3164822/south-korean-dramas-and-films-ride-wave-international-success.
8. Bryan Harris and Jung-a Song, "China bans tour groups to South Korea as defence spat worsens," *Financial Times*, 3 March 2017 at https://www.ft.com/content/9fc4b1b4-ffb1-11e6-96f8-3700c5664d30.
9. "South Korea's Lotte seeks to exit China after investing $9.6 billion, as Thaad fall-out ensues," *The Straits Times*, 13 March 2019 at https://www.straitstimes.com/asia/east-asia/south-koreas-lotte-seeks-to-exit-china-after-investing-96-billion.
10. Ibid.
11. Cynthia Kim and Hyunjoo Jin, "With China dream shattered over missile land deal, Lotte faces costly overhaul," Reuters, 25 October 2017, at https://www.reuters.com/article/us-lotte-china-analysis-idUSKBN1CT35Y.
12. Warren G. Bennis and Burt Nanus, *Leaders: The Strategies for Taking Charge*, Harper and Row, 1985.
13. Nate Bennett and G. James Lemoine, "What VUCA really means for you," *Harvard Business Review*, January–February 2014, at https://hbr.org/2014/01/what-vuca-really-means-for-you.
14. Ibid.
15. "DoD news briefing—Secretary Rumsfeld and Gen. Myers," *U.S. Department of Defense*, 12 February 2002 at https://archive.ph/20180320091111/http://archive.defense.gov/Transcripts/Transcript.aspx.
16. Geoff Colvin, "Inside McDonald's-months long decision to sell all 853 stores in Russia—and lose a significant chunk of global revenue with them," *Fortune*, 28 July 2022, at https://fortune.com/2022/07/27/mcdonalds-closing-all-stores-russia-revenue-employees/.
17. When the first McDonald's opened in Moscow's Pushkin Square in 1990, it was reported that thousands queued for hours.
18. Anthony Elson, "The sovereign wealth funds of Singapore," *World Economics*, 9(3), 2008, pp. 73–96.
19. Yuna Park, "S. Korean state pension funds invest $1.2 bn in equities in August," Reuters, 7 August 2019 at https://www.reuters.com/article/southkorea-markets-pension-idUSL4N2525A6.
20. Press Release, "Intel and Trump administration reach historic agreement to accelerate American technology and manufacturing leadership," 22 August 2025 at https://newsroom.intel.com/corporate/intel-and-trump-administration-reach-historic-agreement.
21. Gregory Ip, "CEOs are learning to live with Trump's turn to state capitalism," *The Wall Street Journal*, 15 December 2025 at https://www.wsj.com/politics/policy/ceos-trump-state-capitalism-f7111bf3?mod=djem10point.

22. *Citizens United v. Federal Election Commission*, 558 U.S. 310 (2010).
23. Dorothy S. Lund and Leo E. Strine, "Corporate political spending is bad business," *Harvard Business Review*, January–February 2022 at https://hbr.org/2022/01/corporate-political-spending-is-bad-business.
24. John Chipman, "Why your company needs a foreign policy," *Harvard Business Review*, September 2016 at https://hbr.org/2016/09/why-your-company-needs-a-foreign-policy.
25. "South Korea pardons SK Group leader amid growing discontent over chaebol," Reuters, 13 August 2015 at https://www.reuters.com/article/us-southkorea-pardons-idUSKCN0QI0D020150813.
26. Ibid.
27. David S. Lee, "Pardon for Samsung's Lee is business as usual for South Korea," *Nikkei Asia*, 12 August 2022, at https://asia.nikkei.com/Opinion/Pardon-for-Samsung-s-Lee-is-business-as-usual-for-South-Korea.
28. Meta, "Responsible innovation," at https://about.facebook.com/metaverse/responsible-innovation/.
29. Responsible Innovation Labs, "Charter," at https://www.rilabs.org/charter#members.
30. Business Roundtable, "Business roundtable launches responsible AI initiative," 26 January 2022 at https://www.businessroundtable.org/business-roundtable-launches-responsible-ai-initiative.
31. Press Release, "Magnachip enters into definitive agreement with wise road capital in a take private transaction valued at $1.4 billion," at https://www.prnewswire.com/news-releases/magnachip-enters-into-definitive-agreement-with-wise-road-capital-in-a-take-private-transaction-valued-at-1-4-billion-301256607.html.
32. Brandon L. Van Grack and James Brower, "CFIUS's expanding jurisdiction in the magnachip acquisition," *Lawfare*, at https://www.lawfareblog.com/cfiuss-expanding-jurisdiction-magnachip-acquisition.
33. Yahoo Finance, "Magnachip to be acquired by wise road capital for $1.45B, shares pop 27%," at https://finance.yahoo.com/news/magnachip-acquired-wise-road-capital-074655554.html.
34. Ibid., at 31.
35. Press Release, "Magnachip and wise road capital announce withdrawal of CFIUS filing and mutual termination of merger agreement," at https://www.prnewswire.com/news-releases/magnachip-and-wise-road-capital-announce-withdrawal-of-cfius-filing-and-mutual-termination-of-merger-agreement-301443561.html.
36. Vivek Astvansh, Wesley Ding and Adnan Habib, "Research: When geopolitical risk rises, innovation stalls," *Harvard Business Review*, 3 March 2022 at https://hbr.org/2022/03/research-when-geopolitical-risk-rises-innovation-stalls.

8. BUSINESS PURPOSE IN THE NEW NATIONAL SECURITY ECONOMY

1. Steven Pifer, "Watch out for little green men," Brookings Institute, 7 July 2014 at https://www.brookings.edu/opinions/watch-out-for-little-green-men/.

2. Helen Thomas, "Investors can no longer afford to ignore geopolitical risk," *Financial Times*, 2 March 2022 at https://www.ft.com/content/1f2c3627-120b-4a09-93e5-4bf265eb76e1.
3. Jeffrey Sonnenfeld et al., "Yale CELI list of companies," *Yale School of Management*, 20 May 2022 at https://www.yalerussianbusinessretreat.com/.
4. "Dodge v. Ford Motor Co.," *H2O*, 2 February 1919 at https://h2o.law.harvard.edu/cases/3965.
5. Adolf A. Berle and Gardiner Coit Means, *The Modern Corporation and Private Property*, Macmillan Co., 1933.
6. Milton Friedman, "A Friedman doctrine--The social responsibility of business is to increase its profits," *The New York Times Magazine*, 13 September 1970 at https://www.nytimes.com/1970/09/13/archives/a-friedman-doctrine-the-social-responsibility-of-business-is-to.html; Chris Myers, "The new 'rules of the game': Balancing profits and social responsibility in the 21st century," *Forbes Magazine*, 30 August 2016 at https://www.forbes.com/sites/chrismyers/2016/08/30/the-new-rules-of-the-game-balancing-profits-and-social-responsibility-in-the-21st-century/?sh=1c4bd8a33afb.
7. Larry Fink, "Larry Fink's 2018 letter to CEOS," *BlackRock*, 17 January 2018 at https://www.blackrock.com/corporate/investor-relations/2018-larry-fink-ceo-letter.
8. "Statement on the purpose of a corporation," *Business Roundtable*, 19 August 2019 at https://www.businessroundtable.org/archive/health-care-innovation.
9. Lucian Bebchuk and Roberto Tallarita, "Was the business roundtable statement on corporate purpose mostly for show?—(1) Evidence from lack of board approval," *The Harvard Law School Forum on Corporate Governance*, 12 August 2020 at https://corpgov.law.harvard.edu/2020/08/12/was-the-business-roundtable-statement-on-corporate-purpose-mostly-for-show-1-evidence-from-lack-of-board-approval/.
10. Jared Cohen and George Lee, "The new wealth of nations: How instrumental capital is reshaping the world," *Foreign Policy*, 3 December 2025 at https://foreignpolicy.com/2025/12/03/instrumental-capital-sovereign-wealth-funds-gulf/.
11. Brad Smith and Carol Ann Brown, "Digital diplomacy: The geopolitics of technology," in *Tools & Weapons: The Promise and the Peril of the Digital Age*, Penguin Press, 2019.
12. "Executive Order 13780: Protecting the nation from foreign terrorist entry into the United States initial Section 11 Report," *U.S. Homeland Security*, January 2018, https://www.dhs.gov/publication/executive-order-13780-protecting-nation-foreign-terrorist-entry-united-states-initial#:~:text=Trump%20issued%20Executive%20Order%2013780,of%20actions%20to%20enhance%20the.
13. Laura King, Barbara Demick, and Molly Hennessy-Fiske, "Trump's ban on some U.S. entries sparks confusion and protest worldwide, and legal rebukes at home," *Los Angeles Times*, 28 January 2017 at https://www.latimes.com/nation/la-na-pol-refugee-policy-20170128-story.html.
14. Sundar Pichai, Twitter post, 23 June 2020, 6:20 AM at https://publish.twitter.

com/?query=https%3A%2F%2Ftwitter.com%2Fsundarpichai%2Fstatus%2F1275192075214966784&widget=Tweet.

15. Darrell Etherington, "Lyft donates $1M to the ACLU, condemns Trump's immigration actions," *TechCrunch*, 29 January 2017 at https://techcrunch.com/2017/01/29/lyft-donates-1m-to-the-aclu-condemns-trumps-immigration-actions/.
16. Maya Kosoff, "How the #DeleteUber campaign revealed Uber's fatal flaw," *Vanity Fair*, 16 February 2017 at https://www.vanityfair.com/news/2017/02/how-the-deleteuber-campaign-revealed-ubers-fatal-flaw.
17. Aaron K. Chatterji and Michael W. Toffel, "The new CEO activists," *Harvard Business Review*, January–February 2018 at https://hbr.org/2018/01/the-new-ceo-activists.
18. Ibid., at 28.
19. Jeffrey N. Gordon and Curtis J. Milhaupt, "China as a "national strategic buyer": Toward a multilateral regime for cross-border M&A," *Columbia Business Law Review*, 1, 2019, p. 192.
20. Larry Fink, "Larry Fink's annual 2022 letter to CEOS," *BlackRock*, 18 January 2022 at https://www.blackrock.com/corporate/investor-relations/larry-fink-ceo-letter.
21. Edelman Trust Barometer, Edelman at https://www.edelman.com/trust/trust-barometer.
22. Ibid., at 34.
23. Alan Murray and David Meyer, "The trust equation is becoming more complex," *Fortune*, 15 March 2022 at https://fortune.com/2022/03/15/businesses-trust-societal-demands-ceo-daily/.

9. A PERILOUS MOMENT

1. In fact, "Asia dominated the world economy for most of human history until the 19th century." At the end of the seventeenth century, the region accounted for two-thirds of global wealth and three-quarters of its population. Valentina Romei and John Reed, "The Asian Century is set to begin," *Financial Times*, 26 March 2019.
2. The McKinsey Global Institute provides a convincing, if not mind-numbing, assessment in "The Future of Asia: Asian flows and networks are determining the next phase of globalization," 18 September 2019, at https://www.mckinsey.com/featured-insights/asia-pacific/the-future-of-asia-asian-flows-and-networks-are-defining-the-next-phase-of-globalization.
3. Robert Gilpin, *U.S. Power and the Multinational Corporation: The Political Economy of Foreign Direct Investment*, Basic Books, 1975, p. 42, 44.
4. Robert Blackwill and Thomas Wright, "The end of world order and American primacy," Council on Foreign Relations special report #86, May 2020, p. 12.
5. M. Taylor Fravel, Bonnie Glaser and Evan Madeiros, "Chinese diplomacy," in Asia Society Center on US–China Relations, UC San Diego School of Public Policy and Strategy 21st Century China Center, "China's new direction: Challenges and opportunities for US policy: Task force report," 2021, p. 21.

6. Cited by Marek Hrubec, "From China's reform to the world's reform," *International Critical Thought*, 10(2), 2020, p. 294.
7. Timofei Bordachev, "Illiberal world order," Valdai Discussion Club, 2 July 2020, at https://valdaiclub.com/a/highlights/illiberal-world-order/.
8. Embassy of the PRC in Norway, "National high-tech R&D program (863 Program)," at https://www.fmprc.gov.cn/ce/ceno/eng/kj/program/t715317.htm.
9. Barry Naughton "Testimony before the U.S.–China Economic and Security Review Commission," *The Chinese View of Strategic Competition with the United States*, 24 June 2020, pp. 2–3, at https://www.uscc.gov/sites/default/files/2020–06/Naughton_Testimony.pdf.
10. Ibid., p. 10.
11. Takita Yoichi, "China expands its economic 'gravitational force" amid US retreat," *Nikkei Asia*, 30 November 2020, at https://asia.nikkei.com/Spotlight/Comment/China-expands-its-economic-gravitational-force-amid-US-retreat.
12. Bloomberg News, "Secretive Chinese committee draws up list to replace US tech," 17 November 2021 at https://www.bloomberg.com/news/articles/2021–11–16/secretive-chinese-committee-draws-up-list-to-replace-u-s-tech?sref=6ZE6q2XR and Zijing Wu, "China adds domestic AI chips to official procurement list for first time," *Financial Times*, 9 December 2025.
13. Chinese Communist Party, "Resolution of the Central Committee of the Communist Party of China on the major achievements and historical experience of the party over the past century," *China Daily*, 17 November 2021, http://www.chinadaily.com.cn/a/202111/17/WS61944b49a310cdd39bc75c01_12.html.
14. Dan Rosen, "A post-engagement US-China relationship?" Rhodium Group, 19 January 2018, at https://rhg.com/research/post-engagement-us-china-relationship/.
15. Leo Lewis, "Companies prepare for a 'selective decoupling' with China," *Financial Times*, 10 October 2021, at https://www.ft.com/content/5ca525f7-cb40-468a-a294-5938d11af6a5.
16. Bain & Co. statistics in Cheng Ting-Fang and Lauly Li, "US-China tensions knock 96% off of bilateral tech investment," *Nikkei Asia*, 20 September 2021 at https://asia.nikkei.com/Economy/Trade-war/U.S.-China-tensions-knock-96-off-of-bilateral-tech-investment.
17. Una Galani, "US-China decoupling is crossing a Rubicon," Reuters, 16 April 2025 at https://www.reuters.com/breakingviews/us-china-decoupling-is-crossing-rubicon-2025-04-16/.
18. Nan Li and John D van Fleet, "Follow the money, not the rhetoric to make sense of China-U.S. ties," *Nikkei Asia*, 11 December 2021 at https://asia.nikkei.com/Opinion/Follow-the-money-not-the-rhetoric-to-make-sense-of-China-U.S.-ties.
19. US Chamber of Commerce, China Center, "Understanding US-China decoupling: Macro trends and industry impacts," 2021, p. 69.
20. International Monetary Fund, "Chapter 3: Geopolitics and financial fragmentation: Implications for macro-financial stability," Global Financial Stability Report, April 2023.

21. European Union Chamber of Commerce in China and the Mercator Institute for China Studies, "Decoupling: Severed ties and patchwork globalisation," 2021, p. 5.
22. "Transcript: NPR's full conversation with CIA Director Bill Burns," 22 July 2021, at https://www.npr.org/2021/07/22/1017900583/transcript-nprs-full-conversation-with-cia-director-william-burns
23. Ellen Nakashima, "US officials caution companies about risks of working with Chinese entities in AI and biotech," *Washington Post*, 22 October 2021, at https://www.washingtonpost.com/national-security/us-officials-caution-companies-about-risks-of-working-with-chinese-entities-in-ai-and-biotech/2021/10/21/d8e8e300-32c1-11ec-9241-aad8e48f01ff_story.html.
24. Jana Titievskaia, "US-China geoeconomic rivalry intensifies: A risk or an opportunity for European companies?" Finnish Institute for International Affairs, Briefing Paper 323, 18 November 2021 at https://www.fiia.fi/en/publication/us-china-geoeconomic-rivalry-intensifies.
25. Branko Milanovic, *Capitalism Alone: The Future of the System that Rules the World*, Harvard University Press, 2019, p. 5.
26. Aaron Friedberg, "The growing rivalry between the US and China and the future of globalization," *Texas National Security Review*, 5(1), Winter 2021–22, p. 117.
27. Edward Fishman and Siddharth Mohandas, "A council of democracies can save multilateralism," *Foreign Affairs*, 3 August 2020, at https://www.foreignaffairs.com/articles/asia/2020–08–03/council-democracies-can-save-multilateralism?utm_medium=social&utm_source=twitter_posts&utm_campaign=tw_daily_soc.
28. Matthew Goodman, "Toward a T12: Putting allied technology cooperation into practice," CSIS, 13 October 2021 at https://www.csis.org/analysis/toward-t12-putting-allied-technology-cooperation-practice.
29. Han Qiu, Dora Xia, and James Yetman, "Deconstructing global trade: The role of geopolitical alignment," *BIS Quarterly Review*, September 2024, pp. 35–50.
30. William Bratton, "The West will not contain Chinese innovation," *Nikkei Asia*, 24 March 2021, at https://asia.nikkei.com/Opinion/The-West-will-not-contain-Chinese-innovation.
31. Kai Fu Lee, *AI Superpowers: China, Silicon Valley and the New World Order*, Mariner Books, 2018 and "National Security Commission on Artificial Intelligence: Final Report," 19 March 2021.
32. The White House, National Security Strategy, October 2022, p. 14.
33. Department of Defense, Defense Science Board—Task Force on Cyber Deterrence, February 2017, https://dsb.cto.mil/reports/2010s/DSB-CyberDeterrenceReport 02–28–17 Final.pdf.
34. Kevin Wolf and Emily Weinstein, "COCOM's daughter," WorldECR, May 2022, pp. 24–8.
35. Eric Sayers and Brad Glosserman, "Collective resilience is the way to address the China challenge," *The Japan Times*, 14 August 2020 at https://www.japantimes.co.jp/opinion/2020/08/14/commentary/world-commentary/collective-resilience-way-address-china-challenge/.

36. Kathleen Claussen, "Trade security exceptionalism," *Stanford Law Review*, 72, May 2020, p. 1106.
37. Jimmy Soni, *The Founders: The Story of Paypal and the Entrepreneurs Who Shaped Silicon Valley*, Simon & Schuster, 2022, p. 380.
38. David Skidmore, "Rep. Bentley; None dare call it Toshiba," Associated Press, 2 July 1987, at https://apnews.com/article/c5f3c744b7b368b7a83b895f43872afe.
39. Richard Milne, "War with Russia? Finland has a plan for that," *Financial Times*, 8 March 2022, at https://www.ft.com/content/c5e376f9–7351–40d3-b058–1873b2ef1924.
40. National Defence University, "National Defence Courses," at https://maanpuolustuskorkeakoulu.fi/en/national-defence-courses.
41. Adam Satariano, Paul Mozur, and Aaron Krolik, "When Nokia pulled out of Russia, a vast surveillance system remained," *The New York Times*, 28 March 2022, at https://www.nytimes.com/2022/03/28/technology/nokia-russia-surveillance-system-sorm.html.
42. Jessica Tasman-Jones, "Reputational risks were already heading up the agenda… then Russia invaded Ukraine," *Financial Times*, nd, https://enterprise.ft.com/en-gb/blog/reputational-risks-were-already-heading-up-the-agenda-then-russia-invaded-ukraine/.
43. Despite the variation, trends all point in the same direction. See for example, "Average number of devices and connections per person worldwide in 2018 and 2023," 28 February 2022, at https://www.statista.com/statistics/1190270/number-of-devices-and-connections-per-person-worldwide/ and Smart Cities Association, "EE: Average UK smart home will have 50 connected devices by 2023," nd, at https://smartcitiesassociation.org/index.php/media-corner/news/169-ee-average-uk-smart-home-will-have-50-connected-devices-by-2023.
44. This is the name of a specific program by the IREX, the International Research and Exchanges Board. Information available at https://www.irex.org/project/learn-discern-l2d-media-literacy-training.
45. Elisabeth Braw and Peter Roberts, "Societal resilience as a deterrent," NATO Science and Technology Organization, STO-MP-SAS-141, 25 March 2019, pp. 11–1.
46. Steven Lee Myers, "A panel to combat disinformation becomes a victim of it," *The New York Times*, 22 May 2022.
47. Taylor Lorenz, "How the Biden administration let right-wing attacks derail its disinformation efforts," *The Washington Post*, 18 May 2022 at https://www.washingtonpost.com/technology/2022/05/18/disinformation-board-dhs-nina-jankowicz/.
48. Numbers vary among surveys, but they are all in this range. These figures are from an American Chamber of Commerce in South China poll. See, for example, "Xinhua Headlines: Foreign companies keen to expand business in China despite pandemic," Xinhuanet.com, 16 March 2022 at https://english.news.cn/20220316/9be0b567451740e188e2badb7c108b16/c.html or "Foreign companies show confidence in

Chinese market," Xinhuanet.com, 12 March 2022 at https://english.news.cn/20220312/a2043abec97644a0af3dd498d4078fd3/c.html.

49. Chris Nuttall, "Tech faces $3.5tn cold war costs," *Financial Times*, 16 July 2020, at https://www.ft.com/content/ac59c289–70ee-414d-ae1a-6864507c0b7f.
50. Randy Brown, "Don't underestimate the rise of US-China tech tensions," *Forbes*, 3 September 2020, at https://www.forbes.com/sites/randybrown/2020/09/03/dont-underestimate-the-rise-in-us-china-tech-tensions/#74b0b6ba3ebd.
51. Ana Swanson and David McCabe, "Trump effort to keep U.S. tech out of China alarms American firms," *The New York Times*, 16 February 2020 at https://www.nytimes.com/2020/02/16/business/economy/us-china-technology.html.
52. Ibid.
53. Antonio Vargas and Raj Varadarajan, "How restrictions to trade with China could end US leadership in semiconductors," Boston Consulting Group, March 2020, p. 1 at https://web-assets.bcg.com/img-src/BCG-How-Restricting-Trade-with-China-Could-End-US-Semiconductor-Mar-2020_tcm9–240526.pdf.
54. Robert Williams, "In the balance: The future of America's national security and innovation ecosystem," *Lawfare*, 30 November 2018 at https://www.lawfareblog.com/balance-future-americas-national-security-and-innovation-ecosystem.
55. Author conversation with Silicon Valley executive, 28 September 2021.
56. Kate Bartlett, "Grapes of wrath: Australian winemakers loss in China is South Africa's gain," *VOA*, 26 March 2022, at https://www.voanews.com/a/grapes-of-wrath-australian-winemakers-loss-in-china-is-south-africa-s-gain-/6502020.html.
57. Jack Thompson, "An ambidextrous EU approach to transatlantic relations," in Michiel Foulon and Jack Thompson, eds, *The Future of European Strategy in a Changing Geopolitical Environment: Challenges and Prospects*, The Hague Center for International Security, December 2021, pp. 27–9.
58. For Europe, see, for example, Noah Barkin, "Export controls and the US-China tech war," Merics, 18 March 2020 at https://merics.org/en/report/export-controls-and-us-china-tech-war. Brendan Thomas-Noone expressed Australian concerns in "Tech Wars: US-China technology competition and what it means for Australia," United States Studies Center, University of Sydney, 16 June 2020, at https://www.ussc.edu.au/analysis/us-china-technology-competition-and-what-it-means-for-australia.
59. Karen Sutter, "China's recent trade measures and countermeasures: Issues for Congress," Congressional Research Service, 21 December 2021, p. 6, 1, at https://sgp.fas.org/crs/row/R46915.pdf.
60. Darren J. Lim and Victor Ferguson, "Conscious decoupling: The technology security dilemma," in *China Dreams: The China Story 2019*, at https://www.thechinastory.org/yearbooks/yearbook-2019-china-dreams/chapter-4-conscious-decoupling-the-technology-security-dilemma/.
61. Ibid.
62. Mike Green and Scott Kennedy, "US business leaders not ready for the next US-China crisis," CSIS commentary, 16 May 2022, at https://www.csis.org/analysis/us-business-leaders-not-ready-next-us-china-crisis.

63. Li Zheng, "How to relieve the security dilemma of Sino-US technology competition," China-US Focus, 30 October 2019, at https://www.chinausfocus.com/peace-security/how-to-relieve-the-security-dilemma-of-sino-us-technology-competition.
64. Xiangning Wu, "Mutual distrust: fostering cooperation in tech competition," Australia National University National Security College, Futures Hub, 2020, at https://futureshub.anu.edu.au/mutual-distrust-fostering-cooperation-in-tech-competition/.

INDEX

Note: Page numbers followed by '*n*' refer to notes.

Abe Shinzo, 133
Academy Award for Best Picture, 181
Accenture, 206
Africa, 128
Aguilar, Francis, 188
Aixtron, 67–8
Alexander, Keith, 30, 119
AliPay, 171
Allen, Craig, 80–1
Alphabet (Google), 176, 208
Amazon Web Services, 166
Amazon, 176
American Chamber of Commerce, 74
American Civil Liberties Union, 208
American credit bureau, 169
Anbang Insurance Group Company, 46–7
Anglo-American capitalism, 87
Annual Threat Assessment (2025), 30, 52
APCO Worldwide, 235
APEC summit (2025), 15, 16
Apple Card, 176
Apple, 176, 193, 203, 205, 208
Article 19 Company Law, 77
Article 9 Tokyo, 36, 37, 250*n*101
artificial intelligence (AI), 21, 22, 25, 59, 95, 100, 103, 167, 189–90
  related technologies, 195–6
ASEAN, 111
Asia Africa Growth Corridor, 133
Asia, 128, 133, 172, 217–18
Asian Development Bank, 128
ASML, 96
Astvansh, Vivek, 199
ATop Tech, 69, 84
Australia Group, 93
Australia National University, 32
Australia, 60, 106, 111, 114, 122, 133
Australian government report, 56
Australian Strategic Policy Institute (ASPI), 28–9, 106
"Availability of Goods Act", 70
Avatar Integrated Systems, 69

Banco Delta Asia, 62
Bank of America, 23–4

Barroso, José Manuel, 155–6
Bauer, Daniel, 68
Bay Area, 211
Bebchuk, Lucian, 206
BeiDou satellite system, 36, 131
Beijing, 7, 32–3, 177, 182, 223, 227, 238
  supply chain resilience, 110–15
  See also China
Belt and Road Initiative (BRI), 7, 128–32, 139, 141
Belt and Road Space Information Corridor, 131
Bennis, Warren, 184
Berkeley Roundtable, 72
Berle, Adolf, 204
Berlin government, 104
Berlin Wall, 187
Bertelsmann Stiftung, 84
Bessent, Scott, 16, 18
Better Utilization of Investments Leading to Development (BUILD) Act, 133
Biden, Joe, 32–3, 55, 62, 87, 95, 111, 113, 117, 140, 142, 145, 146, 209, 225
  administration, 145–7
  AI-related technologies, 195–6
  American-made goods, 174
  on tariffs, 124
  trade and investment tools, 123–7
  See also FDI (foreign direct investment); Trump, Donald; US (United States)
big data, 59
biotechnology, 5, 21–2, 25, 28, 36, 41, 97, 99–100, 103, 146
BIS (Bureau of Industry and Security) Entity List, 94–5, 98
Black Lives Matter, 202
Black Sea, 203
BlackRock, 173, 205, 210
Blanchette, Jude, 78, 79
Bloomberg terminal, 190
Blue Dot Network (BDN), 133–4, 227
Bordachev, Timofei, 219
Boston Consulting Group, 115, 236
Bowen, Mike, 173–4
Braw, Elisabeth, 47
Brazil, 125
Bremmer, Ian, 164
Brexit, 167
Britain, 104
British East India Company, 48, 50, 192
British government, 108
Brown, Kerry, 129
Brown, Michael, 83
Brussels, 127
BTS, 181
Build Back Better World (B3W), 135, 227
Burns, Bill, 222
Busan, 15
Bush, George W., 31
Bush, Vannevar, 6
business purpose, 204–7
  politics of, 207–10
Business Roundtable, 198, 205–6
businesses, 40–1, 44, 49–50, 55, 67–8, 94, 95, 98–9, 112, 157–8, 213, 227, 237
  business purpose, 204–10
  Chinese businesses, 126, 138
  domestic businesses, 91–2
  European businesses, 139–40, 148
  execution risk, 179–80
  foreign businesses, 75, 220, 235
  foundational principles, 186–98
  Japanese businesses, 37–8

new national security economy, 161–79
See also capitalism; FDI (foreign direct investment)
Bütikofer, Reinhard, 135
ByteDance, 191

Cabolis, Christos, 174
Caine, Patrice, 40
Cambridge Analytica, 167
Campbell, Kurt, 146
Canada, 7, 41, 49, 101, 104, 106, 108, 122, 125
Canadian government, 107
Canberra, 135
Cannes Film Festival, 181
capitalism, 1, 164, 176, 177, 188, 222, 223
CCP Inc, rise of, 73–81
China's investment, 82–6
Chinese behavior, 86–90
cybersecurity strategies, 92–3
Caribbean, 128
Caroit, Eléonore, 118
Carter Doctrine, 6
Carter, Nick, 64
CCP Inc, 73–81
Center for Security and Emerging Technology, 90
Central Committee for Financial and Economic Affairs, 220
Central Economic Work Conference (2020), 114
CFIUS (Committee on Foreign Investment in the United States), 69, 84, 102–3, 199
Chatterji, Aaron, 209
Chemical Weapons Convention, 93
Cheng Hui, 18
Chey Tae-won, 193
Chicago School of Economics, 204
China Daily (newspaper), 151
China General Nuclear Power Corporation (CGN), 180
China Institutes of Contemporary International Relations, 238
China Standards 2035 Plan, 28
China, 2, 7, 10, 37–8, 41, 47, 125, 139, 181, 217–18, 220–1, 225, 235, 238–9
Belt and Road Initiative (BRI), 128–32
CCP Inc, rise of, 73–81
China's investment, 82–6
Chinese behavior, 86–90
Chinese concerns, 182–3
competition, 27–33
cybersecurity strategies, 92–3
digital technologies, 17–27
disinformation, 115–18
foundational principles, 186–98
globalization, 159–61
industrial policy, 143–5
intelligence agencies, 48–51
national security, epistemology of, 221–3
new national security economy, 161–79
rare earth elements (REE), 57–9
strategic trade controls, 93–9
supply chain resilience, 110–15
supply chains, 55–7
tariffs, 15–17
United States, important role of, 101–6
university research, 106–10
See also COVID-19 pandemic; FDI (foreign direct investment); Made in China (2025) (MIC2025)
China's State Council, 138
Chinese Academy of International

Trade and Economic Cooperation, 18
Chinese Action Plan for Standards Connectivity for the Joint Construction of the Belt and Road, 131
Chinese Antitrust Exceptionalism (Zhang), 180
Chinese Communist Party (CCP), 34, 35, 76–8, 114, 116, 128, 222
See also China
Chinese TV, 60
Chipman, John, 164–5, 192
Chips and Science Act (CHIPS Act), 146
CIA (Central Intelligence Agency), 49, 50, 119–20, 222
Citizens United vs. the Federal Election Commission (2010 US Supreme Court case), 192
Clapper, James, 45
Clausewitz, 63
Clean Network, 142
Clinton, Bill, 160
cloud computing, 59
Cohen, Jared, 207
Cold War law (1952), 70
Colonial Pipeline ransomware attack (2021), 121
Columbia Law School, 204, 210
Comey, James, 50
Command Vision (2018), 123
Communist Party, 77, 220
Comprehensive and Progressive Agreement on Trans-Pacific Partnership (CPTPP, the successor to the Trans-Pacific Partnership), 89
"comprehensive national power" (CNP), 30
Cook, Tim, 193, 205, 208
Coordinating Committee for Multilateral Export Controls (CoCom), 93
Cornwall Summit (June 2021), 107
Cornyn, John, 84
Council on Foreign Relations Task Force on Innovation and National Security, 145
COVID-19 pandemic, 6, 17–18, 99, 104, 143, 147, 148, 150, 155, 163
biotechnology, 25
case study, 39–42
cyberattacks, 170
framing uncertainty, 184–6
power and hope of purpose, 212–13
production processes, protecting, 172–5
supply chain resilience, 110–15
supply chains, 55–7
Crimea, 8, 165, 201–2
Crisis Response Network, 113
Critical Materials Strategy (2010), 113–14
Critical Raw Materials Alliance, 114
Cruz, Ted, 167
C-suite, 164
Cuba, 126
Cyber Deterrence Initiative, 122
Cyberspace Administration of China, 178
Cyentia Institute, 169
Czech Republic, 41

Data Breach Investigations Report (2020), 43
Davos, 155
de Margerie, Christophe, 201
Deese, Brian, 145

Delaware bankruptcy, 69
#DeleteUber campaign, 208
Deng Xiaoping, 60, 73
Deng, Wesley, 199
Denmark, 207
Didi Chuxing ("Didi"), 168, 177
Digital Markets Act, 126
Digital Services Act, 176
Digital Silk Road, 131, 139
Dimon, Jamie, 169, 205
Ding Xuexiang, 35
Disinformation Governance Board, 117, 234
Doctrine of Information Security (Moscow), 116
Dodd, Merrick, 204
Dodge v. Ford Motor Company (1919 Michigan Supreme Court case), 204
Donahue, John, 159
Doshi, Rush, 243–4*n*13
Dr. Strangelove, 120
Draghi, Mario, 149
Drezner, Daniel, 62–3
Dual Circulation Strategy (2020), 219
Dual Use Regulation 2021/821, 96
Dusseldorf, 41
Dutch government, 69

East Africa, 128
East Asian economies, 187
East Germany, 51
Eastern Europe, 165, 191
Economic Sciences, 204
Economic Security Promotion Law, 150
economic security, 3–4, 37–8, 52, 221–2, 225, 228–9
Economist, The (journal), 165–6, 189
Edelman, Richard, 212–13
Électricité de France (EDF), 180
End of History and the Last Man, The (Fukuyama), 158
Equifax, 169
Ericsson, 137
EU goods, 7
EU market, 84
Eurasia Group, 164
Europe, 50–1, 67, 84, 96–7, 111–12, 127, 128, 142, 145, 222, 233
  GDPR principles, 176
  geopolitical risk, 164–6
  industrial policy, 147–9
  production processes, protecting, 172–5
European Alliance for Industrial Data and Clouds, 148
European Battery Alliance, 148
European Central Bank, 149
European Chamber of Commerce, 57
European Clean Hydrogen Alliance, 148
European Commission (EC), 77, 103, 112, 127, 134, 139–40, 180
  industrial policy, 143–5
European Council, 3
European Economic and Security Strategy (2023), 3
European External Action Service, 117
European governments, 69
European Parliament, 1, 96, 135
European Raw Materials Alliance, 148
European Union (EU), 3, 19, 96, 103–4, 112, 117, 125, 148, 167
  competition, 27–33
  cybersecurity strategies, 92–3

Europe, industrial policy, 147–9
intelligence agencies, 48–51
supply chains, 55–7
See also NATO (North Atlantic Treaty Organization)
European Union Chamber of Commerce, 74, 172
Evanina, William, 30
Executive Order 13769, 207
Exotix, 158
Export Control Reform Act (ECRA), 27–8, 94

Facebook, 47, 167, 171, 176, 197
Farrell, Henry, 18, 21, 62
FBI, 2, 43, 44, 47
FDI (foreign direct investment), 72, 100, 103–4
Chinese behavior, 86–90
cybersecurity strategies, 92–3
strategic trade controls, 93–9
Federal Strategy to Ensure Secure and Reliable Supplies of Critical Minerals, 114
Federation of German Industries, 89
Ferguson, Victor, 238
15th Five-Year Plan, 34
Fiji, 113
Financial Times (newspaper), 32, 78, 155
Fink, Larry, 173, 205, 210–11
Finland, 231
Five Eyes, 107, 122, 123
5G (fifth-generation) communications technology, 24, 53–4, 100, 140, 157, 233
Ford Motor Company, 175
Ford, Christopher, 31–2
Foreign Direct Product Rule, 95
Foreign Exchange and Foreign Trade Act, 104, 251*n*109
Foreign Investment Risk Review Modernization Act (FIRRMA), 102–3, 105
Foreign Policy (magazine), 207
Foreign Subsidies Regulation (FSR), 127
Fortune 1000, 169
14th Five-Year Plan (2021–2025), 34
"the Fourth Industrial Revolution", 20
fourth-generation communications networks (4G), 26, 128
France, 41, 49, 122, 125, 194–5
"A Friedman Doctrine: The Social Responsibility of Business Is to Increase Its Profits" (article), 204
Friedman, Milton, 204–5, 207
Friedman, Thomas, 159
Fujian Grand Chip, 67, 68
Fukushima, 163
Fukuyama, Francis, 158

G10, 223
G20, 80, 128
G7 nations, 107, 117–18
G7 summit, 114, 134–5
GAE, 68–9
Gartner, 169
Gates, Robert, 49–50
GE, 209
Georgetown University, 90, 168
Gerasimov, Valery, 63
Germany, 17, 20, 41, 49, 67–8, 104, 106, 118, 122, 127, 174
Gewirtz, Julian Baird, 33–4
GIC, 191
Global Financial Crisis (2008), 163, 194
Global Financial Crisis (2008–9), 147

Global Gateway Initiative, 134–5
Global Risks Report (2021), 63
Global Times (newspaper), 98
Global Trends report (2021), 19
"Golden Shield Project", 116
Goldman Sachs, 205, 207
Goodrich, Jimmy, 151
Google Cloud, 166
Google, 21, 40, 72, 146, 190
Gordon, Jeffrey, 210
Gordon, Susan, 2
Governance in a Globalizing World (Nye and Donahue), 159
GPNs (global production networks), 220
Grab, 171
GRAPL, 187, 198
"gray zone" tactics, 8
Great Britain, 49, 71
Great East Japan Earthquake (2011), 173
Great Firewall, 116
Greeks, 48
green energy, 59
green technology, 80, 150
Greer, Jamieson, 16, 18
Grimm, Ryan, 60
Grossman, Derek, 135
Group of Seven, 3

Habib, Adnan, 199
Hagel, Chuck, 120
Hague, 70
Hallyu, 181–2
Hambantota port, 132
Hamilton, Alexander, 71
Hannas, William, 90
Hao Peng, 78
Harris, Shane, 45–6
Harvard Business Review, 162
Harvard Business School, 162, 188
Harvard Law Review, 204
Harvard Law School, 204, 206
Hayden, Michael, 45, 119–20
Heilmann, Sebastian, 68
Helberg, Jacob, 33
Hellstrom, Jerker, 69
High End Foreign Expert Recruitment Plan, 85
Holding Foreign Companies Accountable Act, 178–9
Holslag, Jonathan, 135
Hong Kong Stock Exchange, 168
Hong Kong, 69, 168
HSBC, 156–7
Huawei, 26, 36, 62, 141–2, 157
Huey-meei Chang, 90
Hufbauer, Gary, 90
Hultquist, John, 50
"hybrid capitalism", 7

IBM (2019), 54
IMD World Competitiveness Center, 174
Immelt, Jeff, 209
Important Project of Common European Interest (IPCEI) framework, 148–9
India, 41, 49, 111, 113, 133, 143
See also "Made in India" strategy
Indo-Pacific Economic Framework for Prosperity, 113
Industrial Revolution, 71
Industrie 4.0, 20
Innovation Driven Development Strategy (China), 28
Instagram, 171
Institute for Deep Learning, 85
Integrated Circuit Fund (IC Fund), 68–9
Integrated Innovation Strategy (2020), 151
Intel, 191
Intellectual property rights (IPR), 50, 125, 138

"The Interim Provisions on Guiding the Orientation of Foreign Investment", 105
International Economy, 17, 72
International Electrotechnical Commission (IEC), 131, 139
International Emergency Economic Powers Act (IEEPA), 125
International Institute for Strategic Studies, 192
International Monetary Fund (IMF), 80, 143, 217, 221
International Organization for Standardization (ISO), 131, 138–9
International Telecommunication Union, 139
Internet of Things (IoT), 24, 53–4, 138, 140, 233
IPOs, 167, 178, 208
Iran, 18, 41, 53, 62, 126, 207
Iraq, 207
Ireland, 75
Islamic State, 63
Israel, 18, 49, 60
Italy, 127, 149

James Bond films, 48–9
Jankowicz, Nina, 234
Japan, 17, 20, 36–8, 49, 56, 72, 75, 95, 99, 101, 104–5, 111, 113, 114, 133, 139, 143, 145, 163, 223–4, 251*n*109
  application, 171
  Chinese behavior, 86–90
  economic policies, 149–51
  REE shipments, 58
  See also China
Japanese Coast Guard, 37
Japanese government, 107
Japanese law enforcement authorities, 108
Japanese plant, 110
Japanese universities, 107
Jefferson, Thomas, 71–2
Jenson Huang, 196
Ji Li, 177
Jiangxi, 115
Johns Hopkins University, 158
Johnson, Boris, 47
JP Morgan, 169, 205

Kakao, 171
Kania, Elsa, 109
Kantor, Mickey, 48
Kaplan, Robert, 162–3
Kaspersky, 54
Kaye, Charles "Chip", 179
Kennedy, Scott, 84
Kessler, Jeffrey, 95
KGB, 44, 48–9
Kim, YJ, 198
Knight, Alissa, 54
Knudsen, Edward, 62
KPMG, 164
KPop Demon Hunters, 181
Kremlin, 165
Kubrick, Stanley, 120

Labor Rights Advisory Board, 113
Langham, Tony, 232
Lanson, 232
Latin America, 128
Lee, George, 207
Lee, Jay Y., 192–3
Lenin, Vladimir, 5
Leonard, Herman, 163
Lewis, James, 142
Lexus and the Olive Tree: Understanding Globalization, The (Friedman), 159
Li Keqiang, 114
Li Zheng, 152
Libra, 176

Libya, 207
Lim, Darren, 32, 238
Line, 171
LinkedIn, 47
Liu, Lizhi, 168
London, 211
Los Angeles, 208
Lotte Group, 181, 184, 187, 192
Lotte Mart, 181
Lowell, Francis Cabot, 50
Lu Wei, 116
Luttwak, Edward, 4
Lyft, 208

M&A, 210
Macedonia, 50
Made in China (2025) (MIC2025), 28, 30, 57, 68, 79, 84, 90, 94, 102, 109, 115, 173
  industrial policy, 143–5
"Made in India" strategy, 143
Magnachip Semiconductor Corporation, 198–9
magnetic resonance imaging (MRI), 146
Malaysia, 50
Manhattan Project, 28
Manning, Robert, 27
Mao Zedong, 73
Maoism, 34
Marriott, 46
Marxism, 34
Mazzucato, Mariana, 150–1
McAfee-CSIS study, 42
McArver, Trey, 78–9
McDonald's, 187–8
McGregor, Jim, 235
McKinsey & Co, 21, 22, 205
Means, Gardiner, 204
Measures of Security Review of Foreign Investment, 105
Medium- to Long-Term Plan for the Development of Science and Technology (2006), 28
Meng Wanzhou, 156–7
Mercator Institute of Chinese Studies, 68
Merics, 57
Meta, 167, 171, 176, 197
METI, 113, 150
#MeToo movement, 202
Metzl, Jamie, 22
Mexico, 7, 49
Microsoft Azure, 166
Microsoft operating system, 137
Microsoft, 8–9, 41, 208
Middle East, 60
Mikes, Anette, 162–3
Milanovic, Branko, 223
Milhaupt, Curtis, 210
Miliband, David, 147–8
military–civil fusion (MCF), 31–2
Ministry of Commerce (China), 82
Ministry of Science and Technology (China), 108
Missile Technology Control Regime, 93
Model S (Tesla), 185
Modern Corporation and Private Property, The (Berle and Means), 204
Moonshot Research and Development Program, 151
Morrison, Scott, 47
Moscow, 9, 10, 60, 96, 116, 165, 173, 187, 219, 238
  See also Russia
Mueller, Robert, 2, 42
"Muslim travel ban", 208
Myanmar, 212

Nadella, Satya, 208
nanotechnology, 22–3

Nanus, Burt, 184
National Bureau of Economic Research, 75
National Counter-intelligence and Security Center, 50
National Counterintelligence Strategy, 52
National Cyber Strategy (2018), 117
National Defense Authorization Act (FY 2012), 120
National Defense Authorization Act (FY2019), 94, 102
National Defense Courses (Finland), 231
National Economic Council, 145
National Five-Year Plan for Rare Earth Industry (1986–90), 114–15
National Industrial Structure Vision, 20
National Institutes of Health, 146
National Intelligence Community, 30
National Mineral Resource Plan for 2016–20, 115
National Pension Service (South Korea), 191
National People's Congress, 98
National Rugby League, 135
National Science Foundation, 6, 145–6
National Security Agency (NSA), 44, 45, 119–20, 121
National Security and Investment Act, 104
National Security Strategy Interim Guidance, 117
National Security Strategy, 37, 225
National Standardization Development Strategy Research, 138
National Standards Strategy for Critical and Emerging Technology, 140
"national strategic buyers" (NSBs), 88
NATO (North Atlantic Treaty Organization), 10, 117, 122
Naughton, Barry, 219–20
NavTech, 68
Netherlands government, 70
Netherlands, 71, 75, 82, 83–4, 95, 96, 106
Neuffer, John, 235–6
New Generation of Artificial Technology Development Plan (China), 28
New Growth Strategy, 150
New Maritime Silk Road, 128
new national security economy, 161–79
  execution risk, 179–80
  foundational principles, 186–98
  geopolitical risk, 164–6
  law and regulations, challenge of, 175–9
  production processes, protecting, 172–5
  protecting customers, 169–72
  reconceptualizing assets and their protection, 166–8
New York City, 87, 208, 211
New York Stock Exchange (NYSE), 168, 199
New York Times (newspaper), 159, 167, 175, 231–2
New York Times Magazine, The (magazine), 204
Newman, Abraham, 18, 62
Nexperia, 70
19th CPC National Congress, 34
Nippon Steel, 61, 191
Nobel Memorial Prize, 204

Nobel Peace Prize (2010), 60
Nokia Threat Intelligence Report (2020), 54
Nokia, 142, 231–2
Norges Bank Investment Management, 191
North America, 233
North Korea, 18, 41, 62, 126, 170
North Korean ballistic missiles, 182
North Korean government, 62
Norway, 191
NotPetya, 55
Nuclear Suppliers Group, 93
NVIDIA, 196
Nye, Joseph, 159

Obama, Barack, 31, 45, 121
Odesa, 203
OECD (Organisation for Economic Co-operation and Development), 27, 92, 99–100, 101, 220
Office of Scientific Research and Development, 6
Office of the Director of National Intelligence, 44–5
Office of the US Trade Representative, 125
"One Belt, One Road" initiative, 128
OPEC (Organization of the Petroleum Exporting Countries), 18, 60
Open Radio Access Network (O-RAN), 142
OpenNet Initiative, 116
Oppenheimer & Co, 199
Orwellian techniques, 9
overseas foreign direct investment (OFDI), 82, 127

Pacific, 168, 173
Palme d'Or, 181
Papua New Guinea, 135
Paquet, Jean-Eric, 108
Parasite (film), 181
Paris, 134
Park Geun-hye, 193
Partnership for Global Infrastructure and Investment (PGII), 135, 227
Partnership for Quality Infrastructure, 133
Party Central Committee, 151
PDF Expert, 203
Pebble Beach golf club, 87
Pentagon, 121
People's Bank of China, 10
People's Congress, 78
People's Liberation Army (PLA), 106, 121
People's Political Consultative Conference, 78
People's Republic of China (PRC). *See* China
personally identifiable information (PII), 172
PEST framework, 188
Peterson Institute for International Economics, 90
Pichai, Sundar, 208
PIPE (people, identity, process, and engagement), 189, 198
Plastic Alliance, 148
Poland, 101
Polanyi, Karl, 71
Portugal, 101
Pottinger, Matthew, 134
Prestige Ameritech, 173
Project 111 (Chinese government), 85
Protecting the Nation from Foreign Terrorist Entry Into the United States, 207

Public Investment Fund, 191
Putin, Vladimir, 8, 21, 121, 218–19, 232
  See also NATO (North Atlantic Treaty Organization); Russia; Ukraine; US (United States)
Pyongyang government, 170
Pyongyang, 53

Qing China, 48
Quadrilateral Security Dialogue, 105, 111
Qualcomm, 137, 142
quantum computing, 22, 29, 103, 146, 151
Quinn, Noel, 156–7

Rachman, Gideon, 155, 156
Raimondo, Gina, 96
Rakuten, 105
RAND, 151
rare earth elements (REE), 57–9, 114
Raw Materials Initiative (European Union), 113
Readdle, 203
Red Cross, 203
Research Collaboration Advice Team, 107
Responsible Innovation Labs ("RIL"), 197
Rice, Condoleezza, 169
Rid, Thomas, 44
robotics, 20, 23–4, 28, 82, 83, 84, 100, 146
Rockefeller Center, 87
Ross, Wilbur, 133–4
Ruitong Semiconductor Technologies, 68
Rumsfeld, Donald, 185–6
Russia, 8–9, 41, 44, 52–3, 63, 87, 96, 115, 126, 155, 161, 165, 201–2, 212
  business in, 10–11
  disinformation, 115–18
  intelligence agencies, 48–51
  See also NATO (North Atlantic Treaty Organization); Ukraine; US (United States)

S&P 500, 167
Salt Typhoon, 45
Samsung, 113, 142, 192–3
San'an Optoelectronics, 67–8
Saudi Arabia, 87, 191
Scanner Pro, 203
Schmidt, Eric, 21, 25–6, 29, 72
Schneier, Bruce, 44, 53–4
Schwab, Klaus, 20
Semiconductor Industry Association, 112, 235
"A Sense of Purpose", 205
Shanghai, 74, 178, 182
Shenzhen, 47, 178, 223
Shin Kyuk-ho, 181
Sia Partners, 194–5
Sibos, 156
Siemens, Werner von, 136
Silex Microsystems, 68–9
Silicon Valley Leadership Group, 110
Silicon Valley, 72, 85, 110, 175, 236
Silk Road Economic Belt, 128
Singapore, 75, 88, 106, 113
Singh, Daleep, 9
Singh, Pavneet, 83
SK Group, 193
Smith, Brad, 8–9
Snowden, Edward, 45
Society 5.0, 20
Softbank Vision Fund, 191
Softbank, 191
SolarWinds, 44–5
Somalia, 207

SonicWall Cyber Threat Report (2024), 54
Sonnenfeld, Jeffrey, 11, 201–2, 212
South Korea, 15, 16, 41, 49, 60, 75, 99, 113, 142, 170, 181, 197, 236
  application, 171
  Chinese concerns, 182–3
  See also BTS
South Korean government, 60
South Korean law, 101
South Korean pop music (K-Pop), 181–2
Southeast Asia, 37, 111, 171
sovereign wealth funds (SWFs), 191
Soviet Union, 5, 158–9, 184, 238–9
Spain, 41, 122, 127
Spark, 203
Special Competitive Studies Project, 29
Sri Lanka, 132
stakeholder capitalism, 188, 206–7, 209, 211, 213
Stanford Law School, 210
Starwood Preferred Guest (SPG), 46
State Administration of Foreign Exchange, 82
State Department Global Engagement Center, 117
State-owned Assets Supervision and Administration Commission (SASAC), 76
state-owned enterprises (SOEs), 76–7, 80, 89
STEM (science, technology, engineering, and math), 86, 110, 226
Stoltenberg, Jens, 10
Strategic Emerging Industries Plan (China), 28
Sudan, 208
Suez Canal, 110
Suga Yoshihide, 142, 150
Sullivan, Jake, 145
Sun Tzu, 63
Sunway TaihuLight, 36
Supply Chain Agreement, 113
Supply Chain Council, 113
Supply Chain Disruptions Task Force, 111
Supply Chain Resilience Initiative, 111
Sutter, Karen, 238
SVR, 44
Sweden, 69, 106
Swedish Defense Research Agency, 69
Sweet, Julie, 206
SWIFT, 156
Syria, 208
Syrian government, 42

T12, 223
Taiwan Semiconductor Manufacturing Company (TSMC), 113
Taiwan, 49, 110
Takaichi Sanae, 151
Tallarita, Roberto, 206
TechCode, 85
"TechPlomacy" initiative, 207
Temasek, 88, 191
Tencent, 105, 170–1
Terminal High Altitude Area Defense (THAAD), 182–3
Tesla, 185
Texas, 110
Thailand, 41
Thatcher, Margaret, 158
"thick globalism", 159

"thin globalism", 223
Third Generation Partnership Project (3GPP), 141
Thousand Talents Program, 85
3D printing, 23, 97, 174
Tianhe-2, 36
TikTok, 33
TINA doctrine, 158
Toffel, Michael, 209
Tohoku earthquake (2011), 163
Tokyo, 36–8, 50, 149
Tooze, Adam, 35
Toshiba, 230
Toyota North American, 60
Trade Expansion Act (1962), 123–4
Trans-Pacific Partnership, 133
Trump National Security Strategy, 222
Trump, Donald, 7, 29, 31–2, 70, 95, 109, 117, 133, 167, 202, 209, 229
  administration, 145–7
  AI-related technologies, 195–6
  Belt and Road Initiative (BRI), 128–32
  disinformation, 115–18
  industrial policy, 143–5
  on national security, 228
  order and ban, 207–8
  on stakes, 191
  strategic trade controls, 93–9
  supply chain resilience, 110–15
  tariffs, 15–17
  trade and investment tools, 123–7
  United States, important role of, 101–6
  university research, 106–10
Trust Barometer, 213
Tufts University, 62
Tusk, Donald, 61
Twitter (X), 47, 176, 202, 208

UAE (United Arab Emirates), 87
Uber, 168, 191, 208
UK (United Kingdom), 41, 104, 106, 122, 167, 180
UK Centre for the Protection of National Infrastructure, 106
Ukraine, 8–11, 53, 60, 87, 96, 155, 165, 211, 220
  Ukrainian people, aid to, 203
  See also NATO (North Atlantic Treaty Organization); US (United States)
UNCTAD, 101
United Nations, 117, 217
Universities UK, 107
University of California San Diego, 219–20
University of Southern California, 180
US (United States), 3, 7, 9, 41, 49, 75, 220–1
  administration, 145–7
  agenda for success, 224–9
  Belt and Road Initiative (BRI), 128–32
  CCP Inc, rise of, 73–81
  China's investment, 82–6
  competition, 27–33
  digital technologies, 17–27
  disinformation, 115–18
  framing uncertainty, 184–6
  globalization, 159–61
  industrial policy, 143–5
  intelligence agencies, 48–51
  new national security economy, 161–79
  strategic trade controls, 93–9
  supply chain resilience, 110–15
  supply chains, 55–7
  trade and investment tools, 123–7

Trump's tariffs, 15–17
United States, important role of, 101–6
university research, 106–10
See also COVID-19 pandemic; FDI (foreign direct investment)
US Army War College, 184
US Congress, 57, 230
US Congressional Budget Office, 144
US Council on Foreign Relations Task Force on Innovation and National Security, 27
US Cyber Command, 123
US Department of Homeland Security, 69, 234
US Director of National Intelligence, 63
US Federal Trade Commission, 169
US Innovation and Competition Act, 225
US National Commission on Innovation and Competitiveness Frontiers, 19, 28
US National Counterintelligence Executive, 49
US National Cyber Strategy, 122
US National Intelligence Council, 19
US National Security Agency, 30
US Office of Personal Management (OPM), 45
US Steel, 191
US Supply Chain Review, 113
US trade law, 123–4
US Trade Representative, 16, 74, 78, 87
US Treasury official, 47
US–China Business Council, 80
US–China Clean Energy Research Center, 85
US–China Economic and Security Review Commission, 57, 140
US–China joint studies (2023), 29
US–EU Trade and Technology Council, 105

Valdai Club, 219
Venezuela, 87, 126
Venice, 71
Verizon, 43–4
Vietnam, 125
View to a Kill, A (film), 48
von der Leyen, Ursula, 56, 97, 134–5
VUCA (volatility, uncertainty, complexity, and ambiguity), 184–5, 198

Wales, Duncan, 158
WAPI, 137
Wassenaar Arrangement, 93
"weaponized interdependence", 18
WeChat, 170–1
Weinberger, Caspar, 5
Weiss, Linda, 27
Wessel, Michael, 57
Western Europe, 165, 191
Western liberal democracy, 160
WhatsApp, 171
Wilson, Jeffrey, 58–9, 259–60*n*112
Wingtech, 69–70
Wise Road Capital, 198
Wolf, Kevin, 94
World Bank, 73, 217
World Economic Forum (May 2022), 155
World Economic Forum, 20, 166
World Health Organization (WHO), 40
World Intellectual Property Organization, 55

World Is Flat: A Brief History of the Twenty-first Century, The (Friedman), 159
World Trade Organization, 89
World War I, 48
World War II, 5, 37, 48, 61, 71, 145, 204, 217
Worldwide Threat Assessment, 63
Wray, Christopher, 47
Wu, Mark, 87–8
Wuttke, Jörg, 172

X. *See* Twitter (X)
Xi Jinping, 15, 16, 18–19, 31, 34, 56, 70, 76–9, 114, 121, 128–9, 138, 177, 218, 220
  on production chains, 89
  See also China
Xiamen, 68
Xiao Yaqing, 76, 78
Xinhua news agency, 10

Yale Business School, 11
Yale School of Management, 201
Yang Jiechi, 218
Yang, Martin, 199
Ye Yujiang, 108
Yellen, Janet, 3
Yemen, 208
Yom Kippur War (1973), 18, 60
Yoon Suk-yeol, 193
Yu Yongding, 10

Zangger Committee, 93
Zero-Sum World (Rachman), 155
Zhadanov, Igor, 203
Zhang Xuezheng, 70
Zhang Yesui, 129
Zhang, Angela, 180
Zhao Lijian, 10
Zhenhua Data, 47
Zieger, Jon, 197
ZTE, 62
Zuckerberg, Mark, 208